Next-Generation Solar Cells

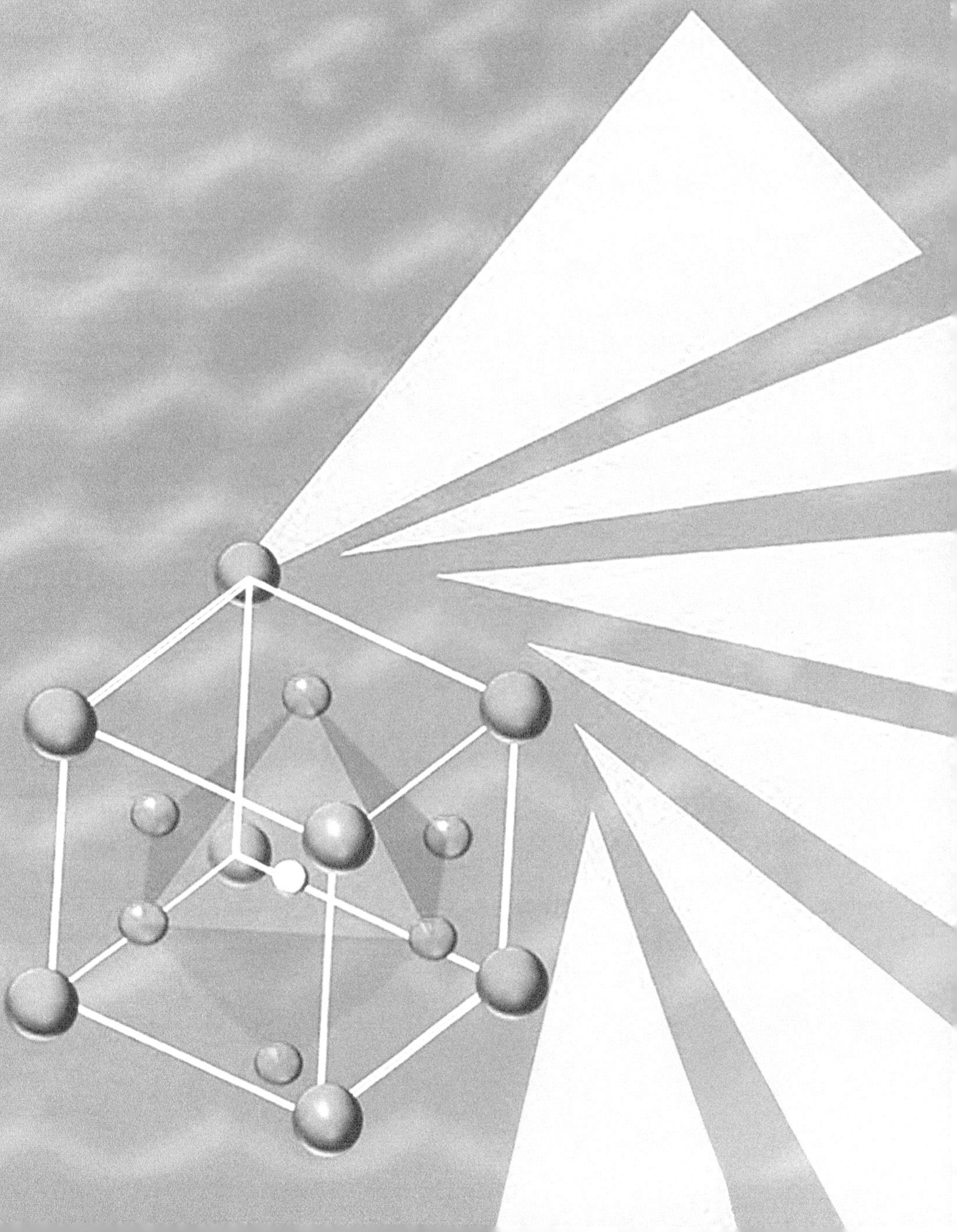

Next-Generation Solar Cells

Principles and Materials

Yoon-Bong Hahn
Tahmineh Mahmoudi
Yousheng Wang

JENNY STANFORD
PUBLISHING

Published by

Jenny Stanford Publishing Pte. Ltd.
101 Thomson Road
#06-01, United Square
Singapore 307591

Email: editorial@jennystanford.com
Web: www.jennystanford.com

British Library Cataloguing-in-Publication Data
A catalogue record for this book is available from the British Library.

Next-Generation Solar Cells: Principles and Materials

ISBN 978-981-4968-66-9 (Hardcover)
ISBN 978-1-003-37238-7 (eBook)

Contents

Preface

Building a sustainable energy system is one of the great challenges of our time that has prompted both academia and industry to seek alternative energy and renewable energy solutions. Recently, advanced materials and technologies for the next-generation solar cells have been exploited to develop economically viable, high-performance solar cells. This book addresses the principles and materials for the development of the next-generation solar cells for a sustainable global society. It will appeal to advanced undergraduate- and graduate-level students as well as researchers interested in next-generation solar cells.

The first two chapters review electromagnetic radiation (EMR) and the physics and properties of semiconductors. Understanding the fundamental properties of EMR is essential for the design and fabrication of photovoltaic energy devices. Semiconductor materials are the most important parts of a solar cell because the absorption of photons, generation of charge carriers, and separation of photo-generated charge carriers take place in semiconductor layers of photovoltaic cells. Thus, understanding the physics and properties of semiconductors is very important for designing the materials and photovoltaic cells.

Chapter 3 deals with the structures of solar cells as well as their working principles and limitations. It describes in detail the methods to improve power-conversion efficiency. Chapters 4 to 7 introduce the various generations of solar cells. These photovoltaic devices can be classified into three generations based on the key materials used and the level of commercial maturity of the PV technology. Third-generation solar cells, such as organic solar cells, quantum dots solar cells, and organic–inorganic hybrid solar cells, are especially described in detail in Chapters 5, 6, and 7, respectively.

Chapters 8 to 12 focus on the emerging perovskite solar cells (PSCs). The advent of PSCs has paved the way for significant PV developments due to impressive power-conversion efficiency. The extraordinary high efficiency of PSCs is attributed to several predominant breakthroughs, such as optimization of perovskite films with additives and innovative design of device configurations with ideal interfacial materials. Chapter 8 introduces properties of perovskite materials as well as structures and basic working principles of PSCs. Chapter 9 reviews cell configurations, transport materials, and fabrication processes of PSCs in detail.

Chapter 10 discusses defects and ions' migration because the performance of PSCs is greatly affected by the quality of interfaces, crystallographic defects, and ions' migration. Ions' migration in perovskite layer is especially a major cause of anomalous device behaviors, including light-soaking effect, current-voltage hysteresis, and slow decay of open-circuit voltage in active layer.

Chapter 11 focuses on quantum dots, tandem, and lead-free PSCs. PSCs suffer from efficiency degradation and device instability due to deterioration of perovskite molecules when operating under harsh conditions for a long time. To solve such problems, quantum dot photovoltaics and tandem PSCs have been developed. Another critical problem with PSCs is lead (Pb) toxicity, which prevents commercial use of PSCs. To replace lead with non-toxic materials, the Pb-free absorbers should have low toxicity and proper stability.

Chapter 12 covers composite-based, efficient, and stable perovskite solar cells with interface engineering. Perovskite-based composites combined with metals, metal oxides, and graphene are effective in solving the performance degradation and instability problems in PSCs. The chapter also introduces interface engineering as an approach to resolve interfacial instability, resulting from the interaction between the perovskites and charge transport materials.

Chapter 13 introduces characterization techniques of solar cell materials and devices. To develop reliable and efficient cells and devices, the materials should be examined in term of chemical, physical, and structural analysis. In addition, photovoltaic parameters should be defined to evaluate device performance.

Last but not least, we would like to acknowledge the publishers and authors who kindly permitted reprint of their work, Jenny Stanford Publishing for carefully editing of this book, and all of our families. Especially, I, Yoon-Bong Hahn, would like to pay special thanks to my co-authors for their constant efforts and enthusiasm.

Yoon-Bong Hahn
Tahmineh Mahmoudi
Yousheng Wang
March 2023

Chapter 1

Electromagnetic Radiation

Understanding the fundamental properties of electromagnetic radiation (EMR) is essential for the design and fabrication of photovoltaic energy devices. EMR is the emission and transmission of energy in the form of electromagnetic waves, i.e., electric field and magnetic field components. EMR is made when atoms absorb energy. The absorbed energy causes one or more electrons to change their position within an atom. When electrons return to their original position, electromagnetic waves are generated. Solar radiation, or sunlight, is the EMR emitted by the sun that can be captured and turned into various types of energy (e.g., heat and electricity) using energy devices. This chapter introduces the characteristics of EMR and the quantities that can be used to evaluate radiation and photon energy.

1.1 Light and Photon

EMR travels through space at the speed of light in packets of radiant energy called photons, which contain neither mass nor charge. EMR can vary in strength from low energy to high energy, including radio waves, microwaves, infrared light, visible light, ultraviolet light, X-rays, and gamma rays. EMR is classified by wavelength (λ)

Next-Generation Solar Cells: Principles and Materials
Yoon-Bong Hahn, Tahmineh Mahmoudi, and Yousheng Wang

ISBN 978-981-4968-66-9 (Hardcover), 978-1-003-37238-7 (eBook)
www.jennystanford.com

or frequency (ν). Figure 1.1 illustrates the spectrum of EMR with visible light highlighted.

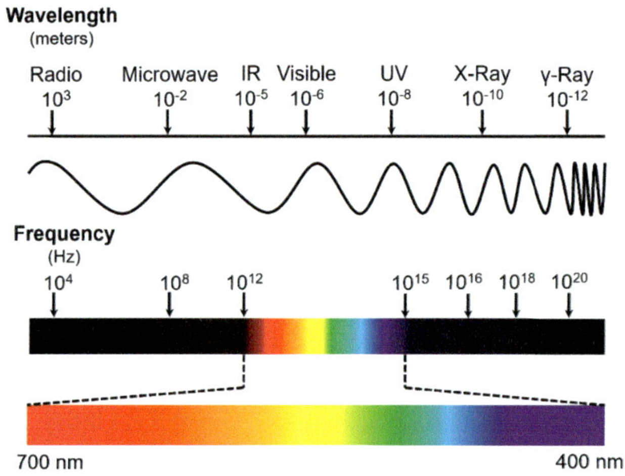

Figure 1.1 Spectrum of the electromagnetic radiation with visible light highlighted.

The optical region, which is the wavelength region responsible for light and solar energy conversion, consists of the infrared region (IR), the ultraviolet region (UV), and the visible-light region ranging in wavelengths from 400–700 nm. Wavelength and frequency hold the relationship [1]:

$$\lambda = \frac{c}{\nu} \tag{1.1}$$

where c represents the speed of light (2.998 × 10^8 m/s), which is a function of the refractive index of the medium (n_r) through which the radiation travels.

Photon energy is calculated by:

$$E = h\nu = \frac{hc}{\lambda} \tag{1.2}$$

where h is Planck constant (h = 6.62607015 × 10^{-34} J·s).

The dimension of photon energy (E) is electronvolt (eV). It is worth noting that the bandgap of a solid (i.e., minimum energy required to promote an electron from the valence to the conduction band) and the wavelength of light that is reflected by the solid are related to one another by Eq. (1.2).

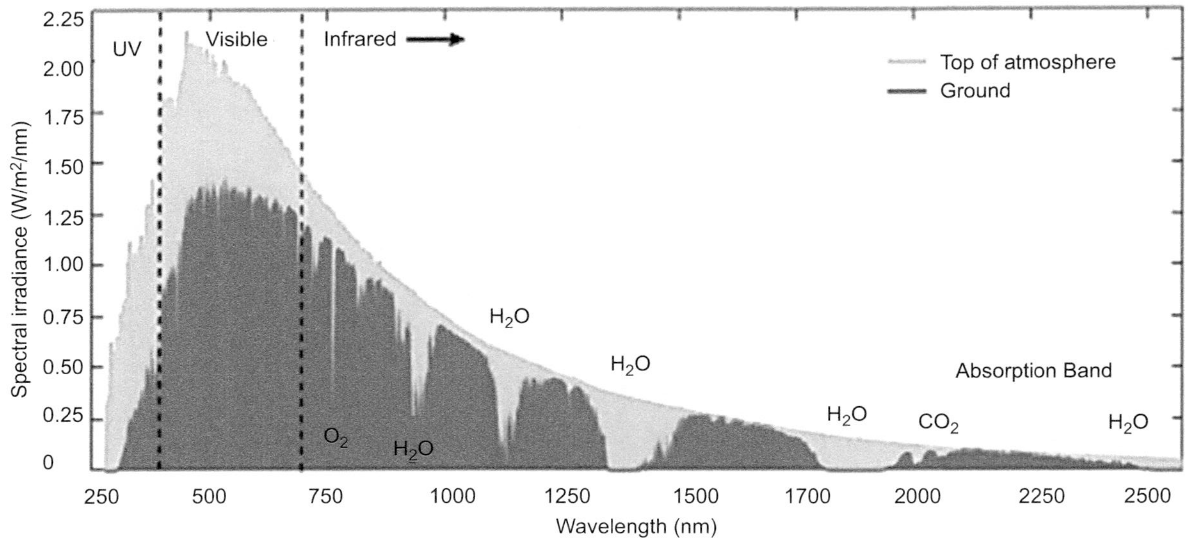

Figure 1.2 Solar spectral photon flux densities at the top of Earth's atmosphere, at the Earth surfaces with estimated absorption of water, in vivo BChl pigment, carotenoids. BChla and BChlb and green sulfur bacteria. Reprinted with permission from Ref. [4], Copyright 2021, Elsevier.

The wavelength of photon energy emitting from a solid can be calculated using the following formula with the bandgap (E_g) of solid:

$$\lambda\,(\text{nm}) = \frac{1240}{E_g\,(\text{eV})} \qquad (1.3)$$

Through the atmosphere, light is influenced by gases, aerosols, and dust as they absorb photons with specific absorption bands. As sunlight reaches the atmosphere, most UVs (below 300 nm) get absorbed by ozone (O_3) and most IRs (above 2 μm) are absorbed by atmospheric gases such as water vapor (H_2O) and carbon dioxide (CO_2) [2, 3]. Figure 1.2 shows the solar spectrum with several photons arriving at the atmosphere and Earth surfaces as a function of wavelength.

1.2 Photometry

Photometry is the science of measurement of light in a way that takes the sensitivity of human eyes. Contrary to radiometry, which measures light in wider wavelengths including UV and IR regions, photometry only measures in the visible spectral region (360 nm ~ 830 nm), where human eyes are sensitive. Human eyes are sensitive to three regions of frequencies, known as blue (~419 nm), green (~531 nm), and red (~558 nm). Observed colors are composed of a broad range of these three regions. The most dominant wavelength in a color is the color's *hue*. The saturation of the color is the amount of distribution in wavelengths in the color. For example, a highly saturated color contains a narrow set of wavelengths [5].

The concept of color by the human eyes is measurable by tristimulus values, which is the foundation of color language, and refers to the Commission Internationale de l'Éclairage (CIE) color system. Tristimulus values measure light intensity based on defined primary RGB (red-green-blue) color values, presented as X, Y, and Z coordinates, and can define precise color values around the world (Fig.1.3a).

A chromaticity diagram is a full plot of all visible colors sensible by average human eyes (Fig. 1.3b). The *xy* values indicate the contribution of the three primaries (RGB), and the boundaries indicate the highest saturation of spectral colors. In the chromaticity diagram, hot objects emit electromagnetic radiation at different

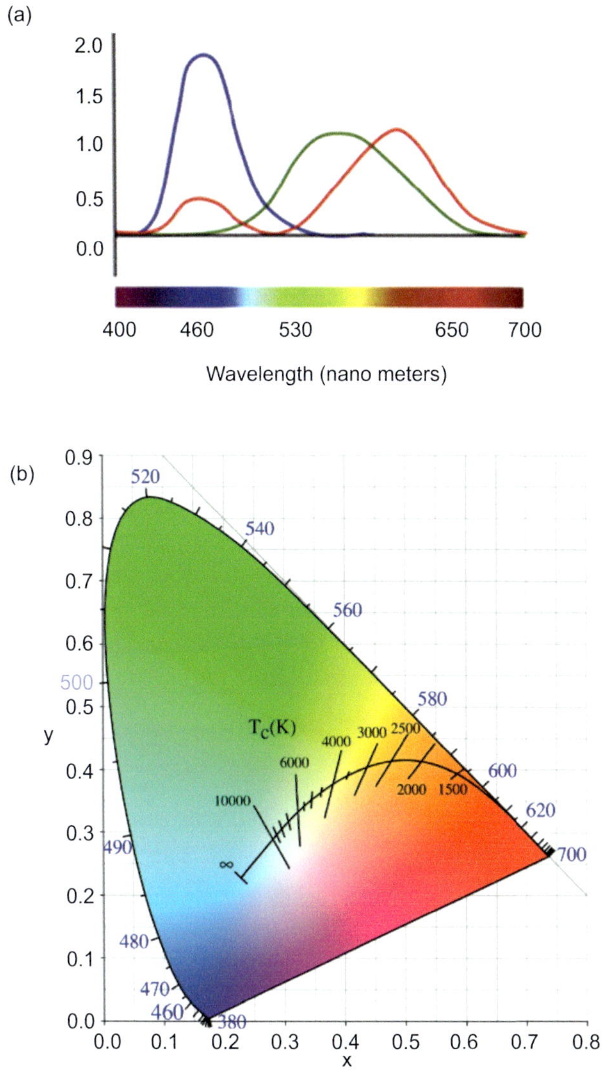

Figure 1.3 (a) The CIE standard observer color matching functions to map blackbody into XYZ coordinates. Reprinted with permission from [6], Copyright 2021, Elsevier. (b) The CIE 1931 color space chromaticity diagram. The Planckian locus is the path that a blackbody color emission through the diagram as the blackbody temperature rises.

ranges of frequency and wavelength. Hot body emission is called *incandescence*, which comes from the Latin verb *incandescere*, meaning 'to glow white.' It specifically refers to the visible region, unlike thermal radiation which refers to IR or other electromagnetic radiation. For example, the white light of the tungsten filament bulb is equal to blackbody radiation at around 2900 K. Irradiating blackbodies in the chromaticity diagram form the Planckian locus. From low to very high temperatures, the emission of irradiating blackbodies changes from deep red to orange, yellowish-white, white, and bluish-white at the end [7–9].

1.3 Blackbody Radiation

All objects at a temperature above absolute zero (0 K, –273.15°C) emit energy in the form of EMR. A blackbody is a hypothetical object that completely absorbs all radiation energy, without reflecting, scattering, or transmitting it and reaches an equilibrium temperature, emitting all that energy quickly. Thus, a blackbody is a theoretical object which is a perfect absorber, and the emission of thermal energy radiated from the blackbody depends only on its temperature. The electromagnetic energy radiated from the surface of a hot object is called thermal radiation [10].

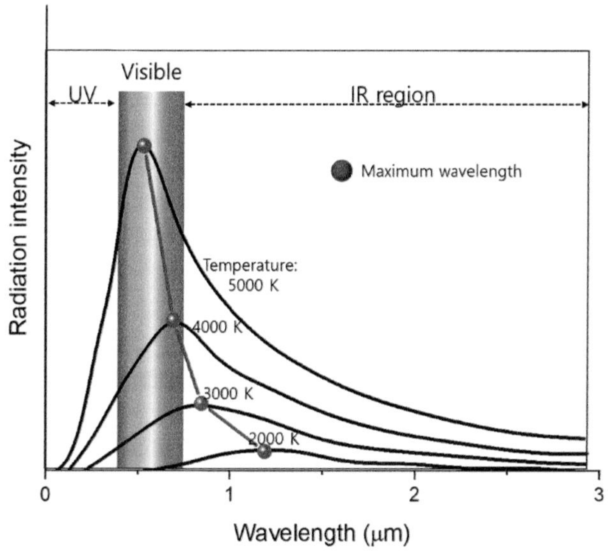

Figure 1.4 Blackbody radiation intensity curves as a function of wavelength and temperature.

The blackbody radiation at different temperatures is shown in Fig. 1.4. The features of blackbody can be explained by several laws:

- Planck's radiation law: Spectral-energy density of the emission *(E_λ)* is a function of wavelength (λ) and absolute temperature (T).

$$E_\lambda = \frac{8\pi hc}{\lambda^5\left(e^{(hc/\lambda\kappa T)} - 1\right)} \tag{1.4}$$

  where h is Planck's constant (h = 6.62607015×10^{-34} J·s), c is the speed of light, and k is Boltzmann constant (k = 1.38064852×10^{-23} m^2 kg s^{-2} K^{-1}).
- Wien's displacement law: The blackbody radiation plot for different temperatures shows peaks at different wavelengths that are inversely proportional to the temperature.
- Wien's law: As the temperature of a blackbody increases, the peak wavelength decreases. The peak of the emission at all wavelengths increases as the temperature of the blackbody increases.
- Stefan–Boltzmann law: The total energy being radiated (the area under the curve) increases rapidly as the temperature increases ($E \sim T^4$). At any temperature above absolute zero energy is theoretically emitted at all wavelengths, but the blackbody radiation intensity curves never reach zero.

According to Wien's displacement law, the spectral radiance of blackbody radiation per unit wavelength (i.e., the wavelength of emission peak) decreases as the temperature of the radiation object increases, given by:

$$\lambda = \frac{a}{T} \tag{1.5}$$

where T is the absolute temperature and a is Wien's displacement constant, equal to 2.9 × 10^{-3} m·K. Wien's law states that the peak wavelength is inversely proportional to temperature but the peak frequency is directly proportional to temperature [11, 12]. When the temperature of an object rises, total radiation increases, and the emission peak shifts into the visible radiation range, causing the object starts to emit dark red and orange-red colors, and at very high temperatures the emission becomes white, as shown in the chromaticity diagram, Fig. 1.3.

1.4 Planck's Radiation Law

Planck's radiation law, formulated by Max Planck in 1900, is a mathematical equation to explain the spectral-energy distribution of emission by a blackbody. Planck assumed that the energy of oscillating atoms is the source of radiation. He further assumed that when an oscillating atom changes a state of energy E_1 to a lower state of energy E_2, a discrete amount of energy between two states (i.e., $E_1 - E_2$) is equal to the product of frequency (ν) of radiation and a constant h (Planck's constant). Then the blackbody radiation was determined by $E_1 - E_2 = h\nu$ [9, 13–18, 20]. Planck's radiation law for the energy radiated from a blackbody per unit volume in the wavelength interval $\Delta\lambda$ is given by Eq. (1.5).

The blackbody spectrum describes thermal radiation which is an essential concept for device engineering in physics and chemistry. For solar energy harvesting devices, in order to design photon energy-to-electricity devices, it is important to know the spectral distribution of solar irradiation, its intensity, and the amount of energy. In semiconductors, thermalized radiation is a useful tool to obtain transition rates for light absorption and emission.

1.5 Solar Spectrum

The solar spectrum covers wavelengths ranging from gamma rays to radio waves, as shown in Fig. 1.1. Solar radiation has irradiation energy as much as a blackbody radiator at around 5800 K.

As the solar radiation travels through the atmosphere, it is attenuated by absorption and scattering by active components in the atmosphere such as H_2O, N_2, O_2, and CO_2 (see Fig. 1.5). The absorption and scattering by active components in the atmosphere weakens the intensity of short-wavelength light and when sunlight reaches the Earth's surface, the spectrum is severely narrowed between the far-IR and near-ultraviolet [21–26].

The optical path length through the atmosphere is determined by the air mass (AM) coefficient, which is defined as a ratio relative to the path length vertically upwards. The solar spectrum outside the atmosphere, approximated by the 5,800 K black body, is named

'AM0,' meaning air mass at 'zero atmosphere.' Solar cells used for space power applications, like those on communications satellites, are generally characterized using AM0.

The spectrum after passing through the atmosphere to sea level with the sun directly overhead is referred to 'AM1,' meaning air mass at 'one atmosphere.' Solar cell panels do not generally operate under exactly one atmosphere's thickness: if the sun is at an angle to the Earth's surface the effective thickness will be greater. Many of the world's major population, solar installations, and industries lie in this area. Therefore, an AM number representing the spectrum at mid-latitudes is much more common, i.e., AM1.5 is used for mid-latitudes, meaning 1.5 atmosphere thickness. AM1.5 corresponds to a solar zenith angle of 48.2°. The solar industry has been using AM1.5 for all standardized testing or rating of terrestrial solar cells or modules, including those used in concentrating systems.

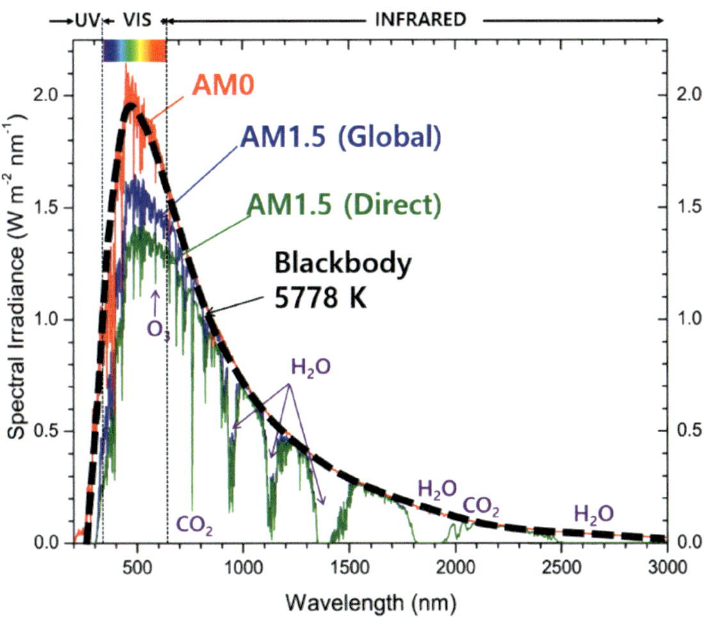

Figure 1.5 Standard solar irradiance spectrum in zero atmosphere AM0 (reference spectrum) and AM1.5. The dashed line refers to the blackbody at 5778 K.

The reference spectrum of AM0 signifies the spectral of solar irradiance above the atmosphere. In 2000, the American Society for Testing and Materials established the reference spectrum of E490-00a (2019) as a standard solar constant for AM0 (zero air mass spectral irradiance). Furthermore, two reference spectra are included for the irradiance under absolute air mass of AM1.5G or Global spectrum (ASTM G173). AM1.5G is the spectrum that includes the blue sky and the surrounding ground. The integrated power of AM1.5G is 1000 $W\cdot m^{-2}$ with a photon flux of 4.31E21 $S^{-1}m^{-2}$. The standard AM1.5D or Direct spectrum—unlike AM1.5G—includes radiation coming only from small surroundings of the sun and projected orthogonally onto the cell. The difference between the AM1.5D and AM1.5G spectra is mainly the blue sky and light scattering in the atmosphere. The integrated power density of AM1.5 D is 900 $W\cdot m^{-2}$ [27–30].

References

1. Castrejón-García, R., Castrejón-Pita, J. R., Castrejón-Pita, A. A. (2010) *Rev. Sci. Instrum.*, **81**, pp. 055106.
2. Chen, Z., Zhu, L., Raman, A., and Fan, S. (2016) *Nat. Commun.*, **7**, pp. 13729.
3. Doscher, H., Young, J. L., Geisz, J. F., Turner, J. A., and Deutsch, T. G. (2016) *Energy Environ. Sci.*, **9**, pp. 74–80.
4. Cleveland, C. J. and Morris, C. G. (2013) *Handbook of Energy: Chronologies, Top Ten Lists, and Word Clouds*, Vol. 1, Elsevier.
5. Dyachenko, P. N., Molesky, S., Petrov, A. Y., Störmer, M., et al. (2016) *Nature Communications*, **7**, pp. 11809.
6. Mark, N. and Aguado, A. (2019) *Feature Extraction and Image Processing for Computer Vision*. Academic Press.
7. Fowles, G. R. (1989) *Introduction to Modern Optics*. Dover: New York.
8. Howell, J. R., Siegel, R., and Menguc, M. P. (2010) *Thermal Radiation Heat Transfer*. CRC Press: New York.
9. Kangro, H. (1976) *Early History of Planck's Radiation Law*. Taylor & Francis: London.
10. Kennard, E. H. (1918) *Phys. Rev.*, **11**, pp. 29–38.
11. Kiang, N. Y., Siefert, J., and Govindjee, B. (2007) *Astrobiology*, **7**, pp. 222–251.

12. Landsberg, P. T. (1978) *Thermodynamics and Statistical Mechanics.* Dover: New York.
13. MacIsaac, D., Kanner, G., and Anderson, G. (1999) *Phys. Teach.*, **37**, pp. 520–523.
14. Markvart, T. (2007) *Appl. Phys. Lett.*, **91**, pp. 064102.
15. Markvart, T. (2008) *J. Opt. A*, **10**, pp. 015008.
16. Mooney, J. and Kambhampati, P. (2013) *J. Phys. Chem. Lett.*, **4**, pp. 3316–3318.
17. Murphy, A. B., Barnes, P. R. F., Randeniya, L. K., Plumb, I. C., et al. (2006) *Int. J. Hydrog. Energy*, **31**, pp. 1999–2017.
18. Overduin, J. M. (2003) *Am. J. Phys.*, **71**, pp. 219–219.
19. Petty, G. W. (2006) *Atmospheric Radiation*, 2nd Ed. Sundog Publisher: Madison.
20. Planck, M. (1914) *The Theory of Heat Radiation.* P. Blakiston's Son & Co: Philadelphia, PA.
21. Ries, H. and McEvoy, A. J. (1991) *J. Photochem. Photobiol. A*, **59**, pp. 11–18.
22. Robitaille, P.-M. (2008) *Prog. Phys.*, **3**, pp. 36–55.
23. Ross, R. T. (1967) *J. Chem. Phys.*, **46**, pp. 4590–4593.
24. Shirasaki, Y., Supran, G. J., Bawendi, M. G., and Bulovic, V. (2012) *Nat. Photonics*, **7**, pp. 13–23.
25. Sizmann, R., Köpke, P., Busen, R. (1991) Solar radiation conversion: In: Winter, C.-J., Sizmann, R. L., Vant-Hull, L. L. (eds.), *Solar Power Plants. Fundamentals, Technology, Systems, Economics.* Springer: Berlin.
26. Smestad, G. and Ries, H. (1992) *Sol. Energy Mater. Sol. Cells*, **25**, pp. 51–71.
27. Waldman, G. (1983) *Introduction to Light.* Dover Publications: New York.
28. Würfel, P. (2009) *Physics of Solar Cells. From Principles to New Concepts*, 2nd Ed. Wiley: Weinheim.
29. Yablonovitch, E. (1982) *J. Opt. Soc. Am.* **72**, pp. 899–907.
30. Zanetti, V. (1984) *Am. J. Phys.*, **52**, pp. 1127–1130.

Problems

1.1 Explain the relationship between wavelength, frequency, and photon energy.

1.2 The difference in energy between the valence band and the conduction band of solid material is called the 'bandgap.' What is the minimum wavelength in nm required to promote an electron from the valence to the conduction band of a semiconductor with a bandgap of 1.5 eV?

1.3 Explain Planck's radiation law.

1.4 The solar spectrum intensity is defined by the AM parameter, briefly explain the AM parameter and the difference between AM1.5, AM1, and AM0.

1.5 If the absolute temperature of a black body is increased from T to 1.414T, by what factor is the total emitted power per unit area (RT) increased? (Use Stefan's law.)

1.6 Light with wavelength of 525 nm is green. Calculate the energy in joules for green light photon.

1.7 What is the wavelength of light with frequency of 3.42×10^{11} Hz?

Chapter 2

Physics and Properties of Semiconductors

Harnessing the massive amounts of energy of sunlight that reaches the Earth's surface is challenging, but the future energy needs of our planet must be ensured. Semiconductor material combines the properties of metals and insulators with materials that uniquely convert sunlight into electricity. Semiconductor materials are the most important parts of a solar cell. In most photovoltaic (PV) cells the absorption of photons, generation of charge carriers, and separation of photo-generated charge carriers take place in semiconductor layers. The first efficient solar cell was made of silicon (Si), which is the most commonly used material in the PV industry. In order to understand the working principles of semiconductor-based solar cells, semiconductor physics and characteristics that are important for fabrication and operation of solar cells operation have to be specified.

To determine the configuration and performance of a solar cell, several semiconductor parameters need to be taken into account:

- Bandgap energy (E_g) and absorption coefficient (α): They determine the ability of a semiconductor to absorb radiation spectra.

Next-Generation Solar Cells: Principles and Materials
Yoon-Bong Hahn, Tahmineh Mahmoudi, and Yousheng Wang

ISBN 978-981-4968-66-9 (Hardcover), 978-1-003-37238-7 (eBook)
www.jennystanford.com

- Doping: Semiconductor material can be doped by electron donor or electron-acceptor atoms. Dopant and its concentration determine the width of a depletion zone of p-n junction. This helps to engineer one semiconductor material to better match the absorber material.
- Mobility (μ) and diffusion coefficient (D): They determine how much charge carriers can transport and diffuse through different semiconductors. To design solar cell layers with optimum charge transport and less accumulation of carriers, mobility and diffusion coefficient must match for each semiconductor layer.
- Lifetime (τ) and diffusion length (L) of the excess carriers: They determine carriers' recombination and solar cell stability.

This chapter provides a summary of semiconductor materials properties using Si as an example, addressing carrier concentration, doping, transport properties, recombination, optical properties of light absorber materials, and the quantum efficiency that control the conversion between light and electricity [1].

2.1 Atomic Structure of Semiconductor

Si atom contains 14 electrons. In steady state, one Si atom has four valence electrons that it can share with other Si atoms and form covalent bonds. In crystalline Si, one atom is bonded with four neighboring Si atoms, forming four bonds of the same length and angles. Four Si atoms form a unit cell, and the unit cells stack next to each other to form a crystalline structure.

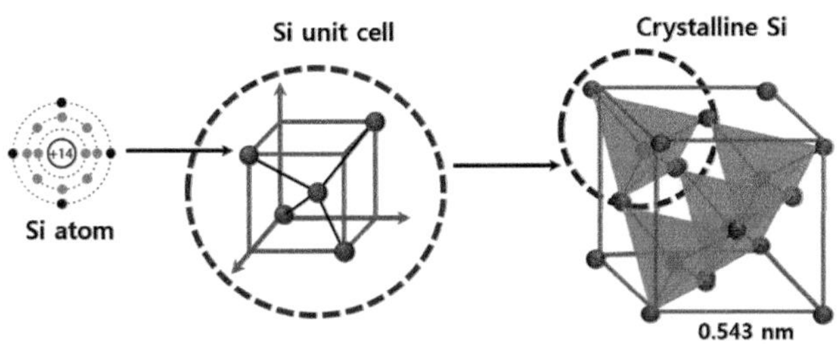

Figure 2.1 The atomic structure of Si, a unit cell of single-crystal Si, and crystalline Si structure.

Figure 2.1 shows the arrangement of the atomic structure and a unit cell of single-crystalline silicon. There are four Si atoms in the volume of the unit cell. The lattice constant of c-Si is 5.4 Å, so it is easy to calculate that Si atoms per unit volume in cm^3 are approximately 5×10^{22} [2].

2.2 Carrier Concentration

Figure 2.2 shows the bonding model of crystalline Si (c-Si). In this model, Si atoms are shown surrounded by four valence electrons connecting with four neighboring Si atoms. At 0 K all the valence electrons of Si are bonded in covalently with other atoms. This state is called an ideal crystal. At temperatures above 0 K, the bond begins to absorb thermal energy and breaks, freeing electrons from the bond. Free electrons move in the crystalline structure. The position of a missing electron from the bond is referred to as a hole with a positive charge.

When a bond breaks and a hole is generated, a valence electron of the neighboring bond can jump up and fill the empty position to restore the broken bond. Such action occurs through neighboring atoms and resulting jumps of valence electrons can be understood as the movement of the empty position (i.e., hole) in the opposite direction to the movement of the electron through the atomic bond. Figure 2.2 shows the movement behavior of carriers at $T > 0$ K when the covalent bond breaks and forms electron–hole pairs [3].

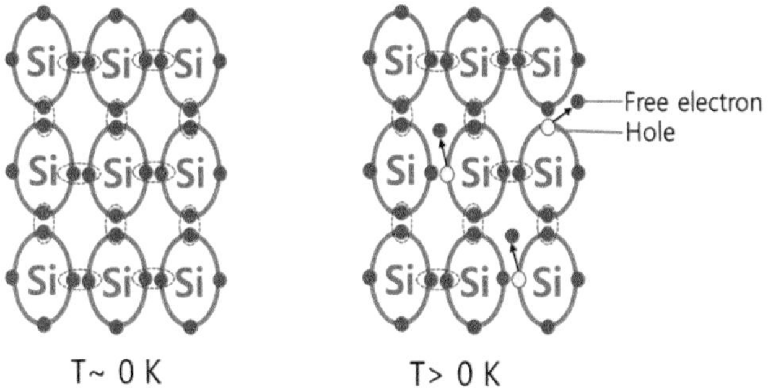

Figure 2.2 The bonding model for c-Si at $T = 0$ K and $T > 0$ K. At a temperature higher than 0 K a Si–Si bond breaks, forming a mobile electron and a hole.

Breaking of a Si–Si bond leads to the formation of an electron–hole pair. In pure c-Si the concentration of electrons and holes are equal. At 300 K in pure c-Si, there are approximately 1.5×10^{10} broken bonds per unit volume of cm^{-3}. This number defines the concentration of electrons and holes at 300 K, i.e., 1.5×10^{10} cm^{-3}, which is referred to as carrier concentration (n_i).

2.3 Doping

In semiconductors, doping is defined as replacing c-Si atoms with other elements to manipulate electron and hole concentrations. The replacement can be done by atoms in groups (III) or (V) of the periodic table with three and five valence electrons, respectively, such as boron (B) and phosphorous (P) [4]. The doping process is demonstrated using the bonding model shown in Fig. 2.3.

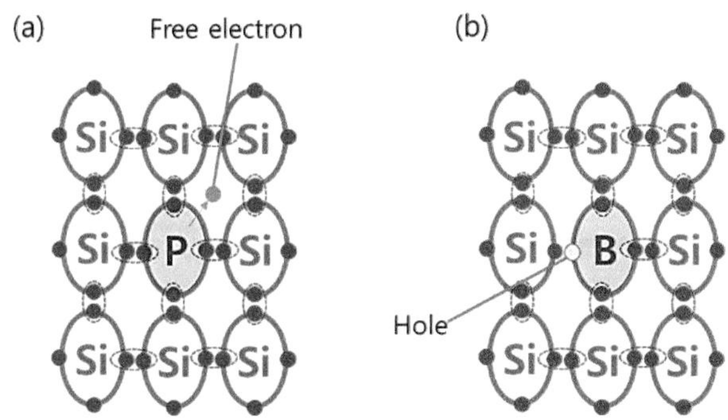

Figure 2.3 The doping process (a) P atom replaces a Si atom in the c-Si lattice resulting in a free electron, (b) B atom substitutes a Si atom resulting in a hole in the c-Si lattice.

By replacing a Si atom in c-Si with a P atom, four of five valence electrons of P participate in covalent bonding with neighboring Si atoms, while the remaining electron fails to bind and weakly attaches to the P atom. The unbonded electron can be easily freed from the P by thermal energy at room temperature. The free electron can move through the lattice. In this case, the P atom which replaces Si atom in c-Si, donates an electron to the c-Si lattice. The P atom is considered an impurity that has increased the electron concentration in the c-Si lattice and is referred to "donor."

On the other hand, an atom with three valence electrons, such as boron (B) atom, can replace Si in c-Si lattice. The B atom with three valence electrons cannot bond with four Si adjacent atoms. However, at a temperature ($T > 0$ K) B can accept an electron from neighboring atoms and complete the bonding with four Si atoms. In this case, a hole is created and can move in the c-Si lattice. The impurity atom that increases the concentration of the hole in the c-Si is referred to "acceptor."

It is worth noting that the concentration of free electrons or holes is proportional to the amount of impurity atoms, and it affects the conductivity and mobility of a semiconductor. When electrons or holes are dominant mobile charge carriers, the semiconductors are called n-type or p-type, respectively.

2.4 Drift and Mobility

Electrons in the conduction band and holes in the valence band are considered "free" carriers because they can move randomly throughout the semiconductor lattice. In the absence of an electric field, carriers move a certain distance at a constant velocity in a random direction. However, in the presence of an electric field or a carrier-concentration gradient, carriers move in a net direction. The movement of charge carriers in a semiconductor under an applied electric field is called "drift" [5]. As illustrated in Fig. 2.4, the positively charged carriers (i.e., holes) accelerate in the direction of the electric field and the negatively charged carriers (i.e., electrons) move in the direction opposite to the electric field. The net velocity at which the charged particles drift is known as 'drift velocity' (V_d) and it is proportional to an applied electric field (ξ).

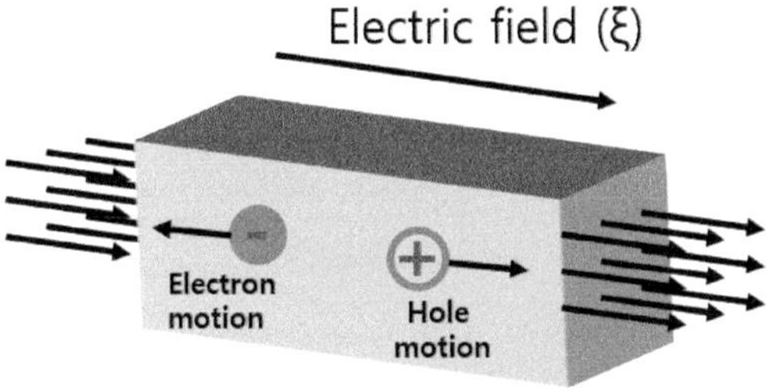

Figure 2.4 Net motion of carriers in the presence of an electric field.

$$V_d = \mu\xi \tag{2.1}$$

In Eq. (2.1), the proportionality factor (μ) is called the "mobility" of a carrier. Thus, the carrier *mobility* is equal to the drift velocity of a carrier for a unit electric field.

The flow of charge carriers due to the applied voltage or electric field is called 'drift current.' The drift current density of electrons and holes is obtained by Eq. (2.2), and the total drift current density by Eq. (2.3).

$$J_N = qn\mu_n\xi \text{ and } J_P = qp\mu_p\xi \tag{2.2}$$

$$J_{drift} = J_n + J_p = q(n\mu_n + p\mu_p)\xi \tag{2.3}$$

where n and p are number of free electrons and holes per unit volume in cm^3, μ is mobility (cm^2/V.s), ξ is applied electric field intensity (V/cm), and q is charge of an electron (1.6×10^{-19} coulomb). Subscripts n and p stand for electrons and holes, respectively [6].

In semiconductors, mobility implies how easily a charge can flow through the material. Mobility is highly dependent on temperature and dopant concentration. In presence of an electric field or thermal radiation, the mobility can be affected by collisions between flowing charges and vibrating impurity atoms in the lattice, which means increasing the doping concertation can increase collision possibility with dopant atoms, resulting in decreased mobility.

2.5 Diffusion

Diffusion is the carrier's movement from a higher concentration region to a lower concentration region to attain an equilibrium. In a doped semiconductor, the unbalanced distribution of the charge carriers is the driving force for diffusion. Diffusion does not need an external field and it occurs due to the repulsive forces among carriers with the same charge that is highly concentrated in an area. Diffusion current is in the same direction as the movement of holes and opposite to that of electrons. Diffusion current can be expressed by Eq. (2.4), and total current by drift and diffusion can be obtained by Eq. (2.5).

$$J_{diff} = qD_n\nabla n \tag{2.4}$$

$$J_n = J_{n\,\text{drift}} + J_{n\,\text{diff}} = qp\mu_n\xi + qD_n\nabla n \tag{2.5}$$

where q is the charge, μ_n is the mobility, n is the charge concentration, D_n is the diffusion coefficient and ∇n represents the concentration gradient.

The diffusion process and the resulting particle fluxes current are shown in Fig. 2.5.

The diffusion current moves toward the direction where there is initially a higher concentration of electrons or a lower concentration of holes. More simply, the current movement is opposite to that of electrons and in the same direction as that of holes [7].

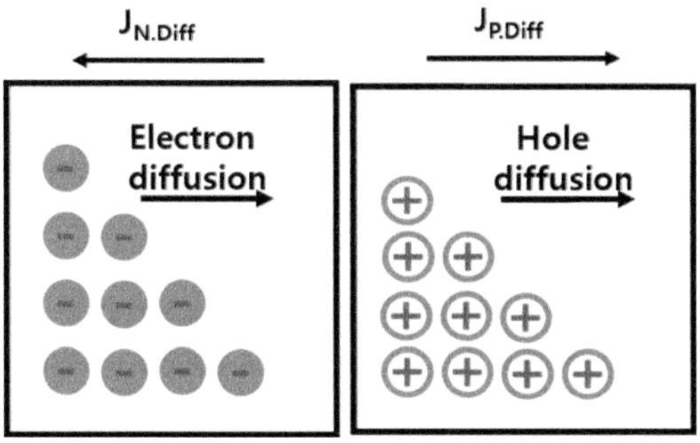

Figure 2.5 Carrier diffusion process and resulting charge-carrier fluxes.

2.6 Recombination

Electrons in the conduction band are in a metastable state and will eventually stabilize to a lower-energy position in the valence band. When this occurs, an electron has to move into an empty valence-band state. Therefore, when the electron stabilizes back down into the valence band, a hole is also effectively removed. This process is called recombination, i.e., the process of annihilation or destruction of electrons and holes. Contrary to recombination, 'generation' is a process where electrons and holes are created. In the case of radiative recombination, an electron in the conduction band recombines with a hole in the valence band and the excess energy is emitted in the form of a photon. In the case of non-radiative recombination, when an electron and a hole combine the excess energy is released in the form of a phonon (i.e., a mode of vibration in lattice structure) or thermal heat in the crystal lattice.

There are three types of recombination mechanisms; band-to-band, trap (or defect), and Auger.

- Band-to-band recombination (left, Fig. 2.6). This process occurs when an electron from the conduction band directly annihilates with the valence-band hole. In direct-bandgap semiconductors, this direct recombination is typically a radiation process.
- Trap-assisted recombination (middle, Fig. 2.6). This recombination occurs in presence of a defect or impurity atoms. As the name suggests, this trap-assisted recombination happens when an electron falls into a trap, which has an energy state within the bandgap. The trapped electron moves to an empty valence band and completes the recombination process. This process is a two-step transition and both electron and hole carriers recombine in the trapped state. Trap-assisted recombination is referred to Shockley–Read–Hall (SRH) recombination.
- Auger recombination (right, Fig. 2.6). In this process, electron and hole have a direct band-to-band annihilation, but the excess energy from the electron–hole recombination is transferred to electrons or holes that are subsequently excited to higher-energy states within the same band instead of giving off photons. The energy is absorbed by the second electron.

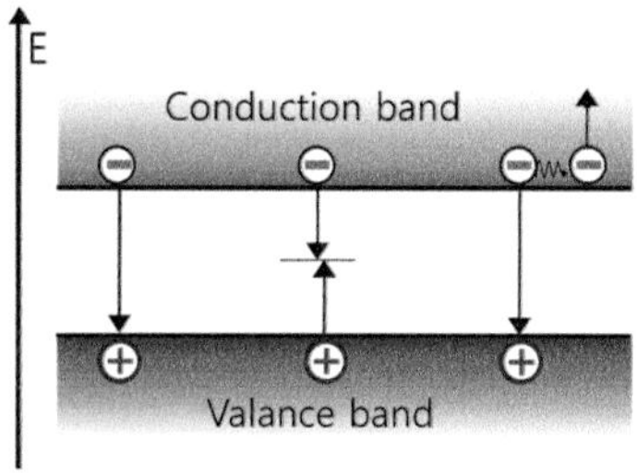

Figure 2.6 Carrier recombination mechanisms in semiconductors.

Band-to-band recombination is a light-emitting process, but trap-assisted and Auger recombination are non-radiative transitions that reduce solar cell efficiency below the ideal radiative limit [8, 9].

Since both carriers must be available to complete the recombination process, the rate of band-to-band radiative recombination is proportional to the concentrations of accessible electrons (n) and holes (p). In the thermal equilibrium state, the recombination and generation rate are equal, and $n \times p$ is equal to n_i^2. The band-to-band recombination rate (R_{b-b}) can be calculated by:

$$R_{\text{b-b}} = \beta\left(np - n_i^2\right) \tag{2.6}$$

where β is the recombination constant and n_i is electron density.

The rate of trap-assisted recombination or SRH recombination is defined as:

$$R_{\text{SRH}} = \frac{pn - n_i^2}{\tau_n\left(p + n_i\right) + \tau_p\left(n + n_i\right)} \tag{2.7}$$

where τ_p and τ_n are the hole and electron lifetimes.

Auger recombination involves three carriers: an electron and a hole that is annihilated in the band-to-band process and another electron or hole to which the excess energy is transferred. The manifestation for the Auger recombination rate is close to that of band-to-band radiative recombination including the density of the electrons and holes, which absorb the excess energy from the electron–hole recombination:

$$R_{\text{Auger}} = C_n n\left(pn - n_i^2\right) + C_p p\left(pn - n_i^2\right) \tag{2.8}$$

where C_p and C_n are the Auger recombination coefficients.

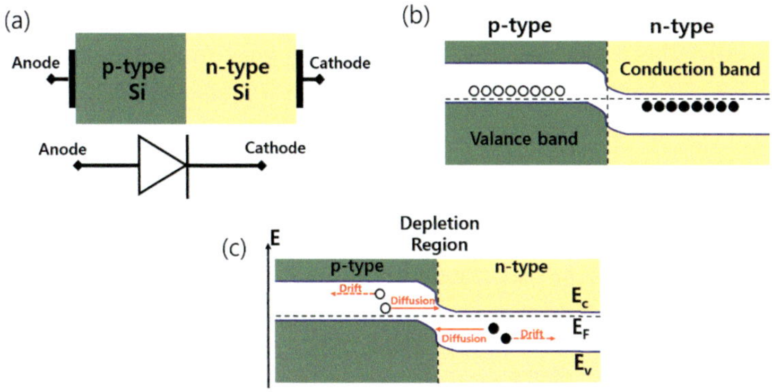

Figure 2.7 p-n junction: (a) The junction and its equivalent circuit symbol (triangle presents to the p-side), (b) Energy band diagram of p-n junction, and (c) Schematic band diagram illustrating diffusion and drift processes in the p-n junction.

2.7 p-n Junction

p-n junctions are formed when p-type and n-type semiconductors are placed in contact with each other (a, Fig. 2.7). The junction behaves differently from either of the semiconductors. Electrons flow from a high electron concentration region (i.e., n-type) to a high hole concentration region (i.e., p-type). Such non-reversing behavior rises from the natural charge-transport process between the two types of materials. p-n junction and its energy bands are visualized in Fig. 2.7.

For a p-n junction at equilibrium, the Fermi levels coincide on the two sides of the junction (b, Fig. 2.7). Electrons and holes reach equilibrium at the junction, forming a depletion region that is the layer where the flow of charges decreases. The depletion region acts as the barrier that opposes the flow of electrons from the n-side to the p-side of the p-n junction diode.

In the band diagram, the upward direction represents increasing electron energy, which means that energy supply is required to move an electron up on the diagram and to move a hole down (c, Fig. 2.7). The diffusion process continues till the concentrations of electrons and holes on both sides are the same. In the junction, when the electron and hole carry more to the other side, they leave behind empty sites on the dopant atom in the crystal lattice, which become charged and fixed in the lattice. Therefore, because of the concentration gradient holes diffuse from the p-side to the n-side across the junction, while electrons diffuse from the n-side to the p-side. The holes and the electrons diffusing toward each other combine near the junction, resulting in an electric field at the junction and forming a "depletion region" [10].

Depletion region forms in asymmetrical p-n junction. Symmetrical junctions have equal and opposite doping on two sides, while an asymmetrical p-n junction is highly asymmetrical in doping concentration in n and p sides. Asymmetrical junction is called a one-sided junction, i.e., n+p junction or p+n junction, where n+ and p+ signify the heavily doped sides. The depletion region forms in the lighter doping side and acts as insulating region, which inhibits further electron transfer unless it is helped by putting a forward bias on the junction.

2.8 Optical Properties

2.8.1 Absorption: Direct-Bandgap and Indirect-Bandgap Transitions

The bandgap of a semiconductor is the least energy required for an electron to move from the valence band (E_{VB}) to the conduction band (E_{CB}). "Band diagram" is a diagram plotting the energy levels of the electrons in a semiconductor device. The bandgap (E_g) is the gap in energy between the valence and conduction bands. Thus, the bandgap is the lowest amount of energy required to form free electrons in a semiconductor. Semiconductors have two bandgaps: a *direct bandgap* and an *indirect bandgap*. The lowest-energy state in the conduction band and the highest-energy state in the valence band are each characterized by a particular crystal momentum (k-vector). If the crystal momentums of the electron in the conduction band and the hole in the valence band are same, the material has a "direct bandgap" and the electron can directly quench and emit a photon (left, Fig. 2.8). On the other hand, if the momentums are different, the material has an "indirect bandgap," and the electron must go through a transitional state and change its momentum (right, Fig. 2.8). For example, amorphous Si is a direct bandgap with a bandgap of about 1.75 eV, but crystalline Si is an indirect bandgap that does not absorb light very well.

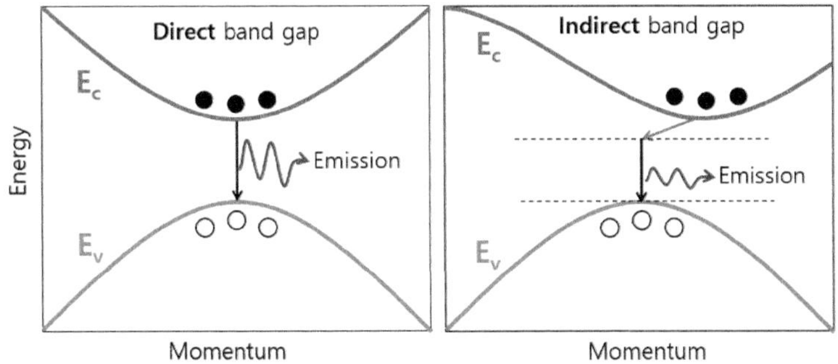

Figure 2.8 Energy vs. crystal momentum for a semiconductor with a direct (left) and indirect (right) bandgap.

Direct-bandgap semiconductors can absorb light efficiently because electrons can transfer directly from the valence band to the conduction band without changing momentum, which demands interactions that can drain energy. The most important direct-bandgap semiconductors are compounds of elements from groups III and V in the periodic table, such as GaAs and InP [11].

2.8.2 Luminescence Emission

Luminescence is an optical phenomenon in which materials emit light when they have absorbed energy from a source such as electrical energy, ultraviolet radiation, electron beams, electrochemical reactions, and so on. Unlike incandescence, which is radiation emission from hot materials, luminescence is radiation from a cold body. Luminescence requires the transition of the atoms of the material into an excited state. Since the excited states are unstable, the material undergoes another transition, back to its stable ground state, and then the absorbed energy is emitted in the form of either light or heat. Thus, luminescence efficiency depends on the degree of transformation of excitation energy into light.

Photoluminescence (PL) is a typical example of the transformation of excitation energy into light, an optical phenomenon in which semiconductors give light emissions by absorbing energy higher than the energy bandgap of the semiconductor. Light emission by the photons generated when excess electron–hole pairs are created by an externally applied bias is called electroluminescence (EL). There are two types of photoluminescence (PL): fluorescence and phosphorescence. Both emissions depend on the material's ability to absorb light and emit longer wavelengths of light, but the main difference is the time it takes to do so. Fluorescence is an immediate emission that occurs immediately after the photoexcitation of a material, while phosphorescence is a process that takes time to emit light, which can take several seconds to hours longer after the photoexcitation [3, 4].

Fluorescence emission occurs at a longer wavelength than the incident light, which means that absorption and emission do not have the same energy. Vibrational or thermal relaxation loss is a major cause of the shift to lower energy (or longer wavelength). This universal principle is named Stokes' law (Fig. 2.9).

Two forms of photoluminescence, fluorescence and phosphorescence, are illustrated in Fig. 2.10.

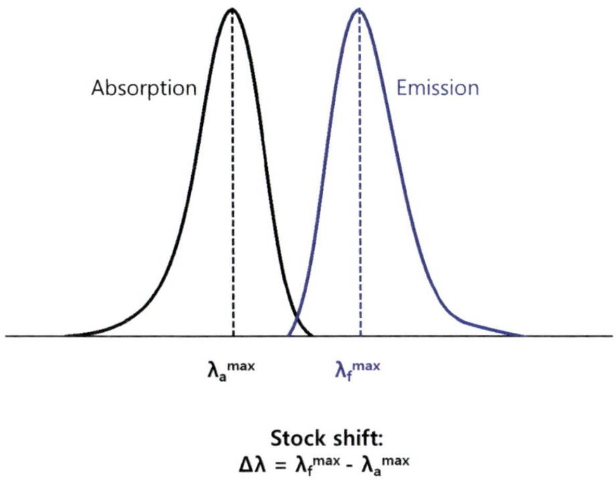

Figure 2.9 Description of the Stokes' shift for fluorescence.

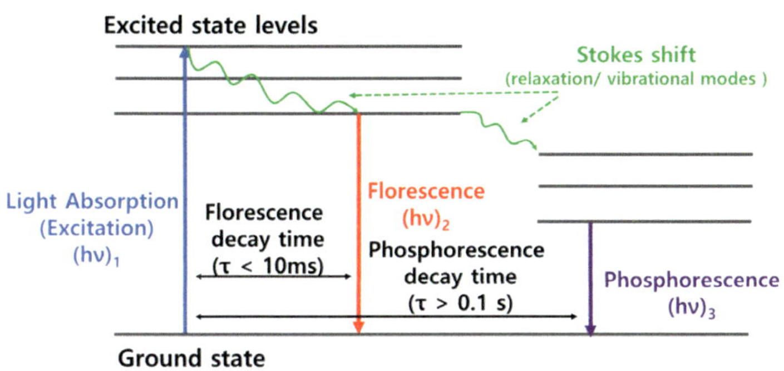

Figure 2.10 Photoluminescence: fluorescence and phosphorescence emission after photoexcitation.

An inorganic solid that exhibits luminescence when exposed to radiant energy is called 'phosphor.' Phosphor can be used for both fluorescent and phosphorescence materials when exposed to ultraviolet or visible light. The decay from an excited state does not necessarily lead to the emission of a photon because of non-radiative recombination. In such a case, the excitation energy is wasted in the lattice as heat. Non-radiative recombination is the main reason why only small amounts of atoms emit light from a luminescent material. In inorganic phosphors, emission centers can be created usually by adding small amounts of dopants or impurities to the host material.

The activator is an element added as a dopant to the semiconductor to create the desired type of emission center.

An activator forms an emission center which creates visible radiation by absorbing excitation energy [12]. When an activator with the preferred emission cannot have the required absorption from the available excitation energy, a sensitizer can be used, which absorbs the excitation energy and transfers the energy to the activator. Then the activator emits its characteristic luminescence. The luminescence process in a solid-state material with activator and sensitizer impurities is shown in Fig. 2.11.

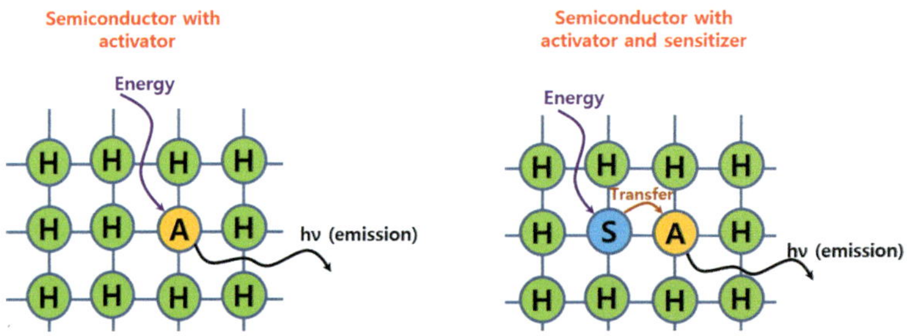

Figure 2.11 Schematic demonstration of the luminescence process in a solid-state material with activator and sensitizer impurities. "H" is a host lattice element, "A" is an activator ion, and "S" is a sensitizer ion.

2.8.3 Quantum Efficiency

Quantum efficiency (QE) is the measure of the effectiveness of a photosensitive device to convert incident photons into electrons. The QE of a PL material is the number of photons emitted per incident photon flux. The QE is defined for monochromatic radiation with frequency (ν). The QE of a solar cell indicates the respective percentage of current that a cell produces when irradiated by incident light. Integrated QE over the whole solar electromagnetic spectrum shows the amount of current that the cell will produce when exposed to sunlight. The light-harvesting efficiency (LHE) is the number of photons that are absorbed by the converter, proportional to absorption coefficient, thickness, and optical features such as reflection, texturing, scattering layers, etc.

The external QE (EQE) of a device is the ratio of the number of charge carriers collected by the solar cell to the number of photons incident on the cell.

$$\mathrm{EQE} = \frac{\text{charge carriers / s}}{\text{received photons / s}} \tag{2.9}$$

Internal QE (IQE) is the ratio of the number of charge carriers collected by the solar cell to the number of photons absorbed by the cell. Therefore, the EQE is always smaller than IQE.

$$\mathrm{IQE} = \frac{\text{charge carriers / s}}{\text{absorbed photons / s}} \tag{2.10}$$

In solar cells, low IQE implies poor carrier collection in the active layer. Charges generated after absorption of incident light must be separated from the junction. In most solar cells, the QE decreases due to the recombination process. Similar mechanisms that influence the charge collection also affect QE.

References

1. Chapin, D., Fuller, M. C., and Pearson, G. (1954) *J. Appl. Phys.*, **25**(5), pp. 676–677.
2. Lammert, M. D. and Schwartz, R. J. (1977) *IEEE Trans. Electron. Devices*, **24**(4), pp. 337–342, 1977.
3. Green, M. A. (1982) *Solar Cells: Operating Principles, Technology, and System Applications*. Prentice-Hall, Inc.: Englewood Cliffs, NJ, p. 288.
4. Aberle, A. G. (2000) *Prog. Photovolt.: Res. Appl.*, **8**(5), pp. 473–487.
5. Sze, S. M. (2002) *Semiconductor Devices: Physics and technology*, 2nd Ed. John Wiley Sons Inc: New York, NY.
6. Richards, B. S. (2004) *Prog. Photovolt.: Res. Appl.*, **12**(4), pp. 253–281.
7. Green, M. A. (2009) *Prog. Photovolt.: Res. Appl.*, **17**(3), pp. 183–189.
8. Franklin, K., Fong, K., McIntosh, A., Fell, A., et al. (2016) *Prog. Photovolt.: Res. Appl.*, **24**(4), pp. 411–427.
9. Delamarre, A., Paire, M., Guillemoles, J. F., and Lombez, L. (2014) *Prog. Photovolt.: Res. Appl.*, **23**(10), pp. 1305–1312.
10. Hecht, J. (2018) *Understanding Lasers: An Entry-Level Guide*. John Wiley & Sons.
11. Sizmann, R., Köpke, P., and Busen, R. (1991) Solar radiation conversion. In: Winter, C.-J., Sizmann, R. L., Vant-Hull, L. L. (eds.), *Solar Power Plants. Fundamentals, Technology, Systems, Economics*. Springer: Berlin.

Problems

2.1 To design a solar cell, several semiconductor parameters need to be taken into account. Explain briefly.

2.2 Explain carrier concentration and mobility concepts in a semiconductor.

2.3 Explain and draw the photoluminescence, fluorescence, and phosphorescence emission after photoexcitation.

2.4 What is QE in a photovoltaic device?

2.5 What is a bandgap, and what is the difference between direct and indirect bandgaps?

2.6 What is the role of doping in semiconductors?

2.7 Explain the recombination in semiconductors and what are the types of recombination mechanisms?

2.8 What is the role of doping in the electrical and optical properties of semiconductors?

2.9 In Group IV semiconductors what are the dopant elements?

2.10 How is the drift and diffusion currents produced in semiconductor sample?

2.11 Explain the difference between radiative and non-radiative recombination.

2.12 How p-n junction comes in equilibrium statement?

2.13 Explain the symmetrical and asymmetrical p-n junction and formation of depletion region.

2.14 Explain the difference between EQE and IQE. In solar cells, what does a low IQE imply?

Chapter 3

Working Principles and Limitations of Solar Cells

To design and fabricate an efficient and stable solar cell, understanding the basic structure, working principles and limitations of photovoltaic cells is essential. Structure and working principles are rather simple compared to complicated semiconductor electronic chips. However, the device performance in terms of efficiency and long-term stability is closely related to device structures and their working principles. In this chapter, structures, working principles, and limitations of solar cells and methods for improving the power-conversion efficiency are described in detail.

3.1 Basic Structure and Working Principle of Solar Cells

3.1.1 Basic Structure of Solar Cells

Solar cells are made from a variety of semiconductor materials, such as n-type, p-type, and intrinsic semiconductors. Basically, there are two main configurations of solar cells: p-n junction and non-p-n junction. The p-n junction photovoltaics can be divided into the Si colar cells that include monocrystalline, polycrystalline and

Next-Generation Solar Cells: Principles and Materials
Yoon-Bong Hahn, Tahmineh Mahmoudi, and Yousheng Wang

ISBN 978-981-4968-66-9 (Hardcover), 978-1-003-37238-7 (eBook)
www.jennystanford.com

amorphous solar cells, and the thin-film solar cells, which include gallium arsenide (GaAs), cadmium telluride (CdTe), copper-indium-gallium diselenide (CIGS) and amorphous Si (a-Si:H) thin-film solar cells. A good example of p-n junction solar cells can be found in commercially available solar panels made of crystalline silicon. The p-type semiconductor which has a 300 μm thick absorber directly contacts the highly doped n-type semiconductor which has an electron layer less than 1 μm thick. The non-p-n junction solar cells belong to emerging PVs, including dye-sensitized solar cells (DSSCs), organic solar cells (OSCs), inorganic solar cells (ISCs), quantum dot solar cells (QDSCs), and perovskite solar cells (PSCs).

Figure 3.1 shows a basic structure of the solar cells consisting of an absorber material that absorbs sunlight and generates electron-hole pairs, n- and p-type semiconductors for electrons and holes transport, and electrodes of anode and cathode.

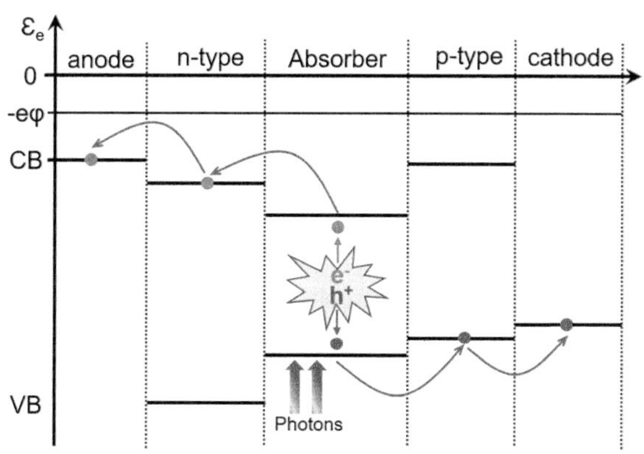

Figure 3.1 Basic structure of solar cells including electrodes (anode and cathode), n- and p-type semiconductors, and absorber.

3.1.2 Solar Cell Working Principles

The solar cell is an electric device that can absorb photons of the sunlight to generate free electron–hole (e-h) pairs (i.e., excitons). These e-h pairs can be separated and transported to circulate through an external load that supports a positive voltage. This process of conversion of light to electricity is called the photovoltaic effect.

The basic working principle of the p-n junction is of stacking p- and n-type semiconductors to form the p-n junction, as shown in Fig. 3.2. The free electrons from p-type semiconductor move to n-type semiconductor and create a depletion region. When the light photons break the depletion region, the electron and holes separately move toward p- and n-type semiconductors, respectively, and generate current when providing an external circuit.

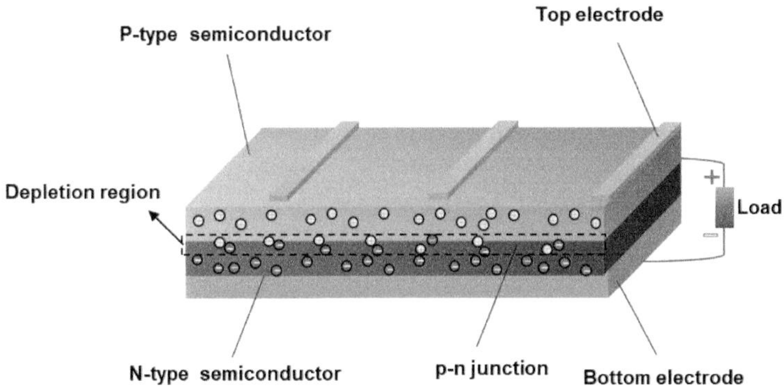

Figure 3.2 Configuration of p-n junction of semiconductors in solar cells.

In general, an absorber semiconductor is excited by photons more energetic than its bandgap and yield excitons (i.e, electron–hole pairs). Once the electrons and holes are photo-generated in the absorber, they have to move to the n- and p-type semiconductors, respectively. Finally, the current can be produced when giving an external circuit by connecting anode and cathode electrodes. The n-/p-type semiconductors with reduced transport resistances and suitable bandgaps can accelerate electrons/holes separation and transport efficiencies, respectively. Except for the basic solar cell structure, interfacial layers and carrier managements of the active layer are needed for improving device interface contacts and light absorption ability toward theoretical power-conversion efficiency (PCE) of solar cells. Note that n- and p-type contacts in a solar cell need to well match the conduction band (CB) and valence band (VB) for suppressing electron/hole recombination. For an optimal solar cell, the designed solar cell should satisfy the requirements:

(i) The absorber semiconductor should have an appropriate bandgap with better light absorption.
(ii) The electron contact layer has a large electron conductivity and small hole conductivity.
(iii) The hole contact layer has a large hole conductivity and small electron conductivity.
(iv) The anode/cathode has a matchable work function.

3.2 Limitations and Improvements of Energy Conversion in Solar Cells

3.2.1 Efficiency Limitation Factors in Solar Cells

3.2.1.1 Energy bandgap

In 1961, Shockley and Quessier (S-Q) first calculated the theoretical maximum efficiency for a silicon solar cell (E_g = 1.1 eV) to be 30% under the AM1.5G with 1000 W/m^2 for an ideal case in which the only recombination of hole-electron pairs is radiative as required by the principle of detailed balance [1]. The solar cell efficiency is limited by the following factors [2]:

(i) not-absorbed solar energy (19%),
(ii) thermalization loss due to high-energy photons (33%), and
(iii) extraction loss because of charge carrier recombination (15%).

The rest 33% of solar energy is the available free energy that can be converted to electrical energy.

It is well known that the bandgap (E_g) of a semiconductor determines how much solar energy can be converted to electrical power. Photons with energy below the E_g cannot be absorbed; photons with energy above the E_g can be absorbed, but cannot be fully converted to electrical energy because of thermalization loss. Therefore, the E_g of an absorber is of paramount importance in a solar cell, which determines the values of short-circuit current (J_{sc}) and open-circuit voltage (V_{oc}). However, not all incident light can be absorbed by the active layer, meanwhile, photo-induced generated carriers also cannot be fully collected. Hence, the practical J_{sc} is always lower than the theoretical value that can be calculated for a given E_g. Furthermore, the practical V_{oc} is below the E_g, attributed to non-radiative recombination, Auger recombination, and trap-assisted recombination [3–6]. In addition, interface resistance and contact losses can reduce the fill factor (*FF*), which further lower PCE. Therefore, the above factors yield practical PCEs that are always lower than the S-Q limit for a given E_g.

3.2.1.2 Interface defects

Interfaces in stacked configurations are considered to be fundamentally important in designing highly efficient and operationally stable solar cells. As shown in Fig. 3.3 (left), imperfect interface contacts due to energy-level mismatch lead to unnecessary traps, energy barriers, and poor ohmic contacts which hinder the extraction and transportation of charge carriers. The photo-induced holes and electrons have to transport across the interfaces in a device, resulting in occurrences of charge loss, charge recombination, and interfacial degradation because of possible interfacial defects (Fig. 3.3, right).

Hence, it is believed that interface engineering/passivation is crucial to ameliorate the interface contacts and passivation, in other words, it can effectively reduce electron or hole-transport barriers, non-radiative recombination, interfacial defects, and ion migration in various solar cells.

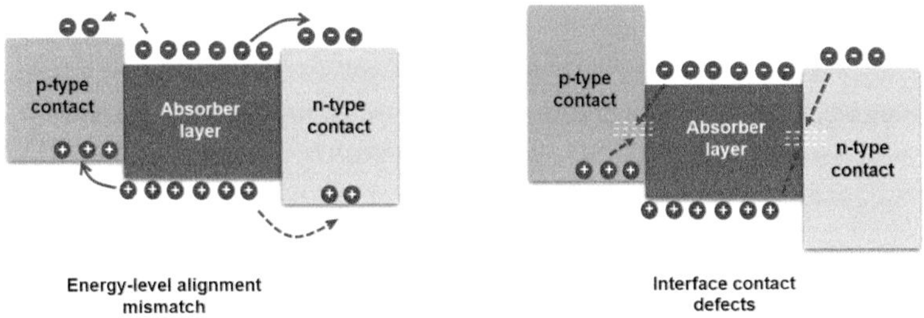

Figure 3.3 Interface defects in a solar cell.

3.3 Maximum Efficiency of Solar Cells

The efficiency of solar cells is mainly based on three parameters V_{oc}, J_{sc}, and *FF* (Fig. 3.4a). *FF* can be determined at a maximum power point (MPP):

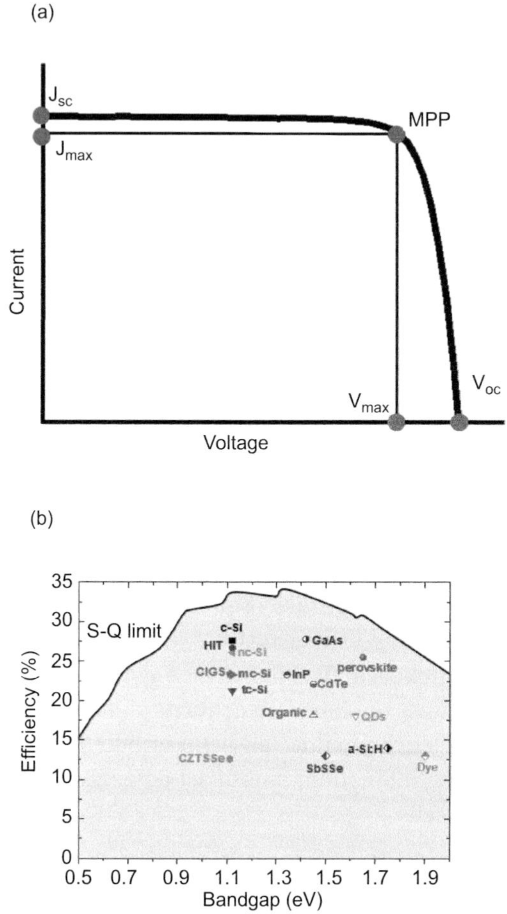

Figure 3.4 (a) General *J–V* curve of solar cells, (b) Record efficiencies of solar cells for various absorber materials as a function of the bandgap, in comparison to theoretical Shockley–Queisser detailed balance efficiency limit (S-Q limit) (black curve) (The record efficiencies are accessed in July 2021).

$$FF = \frac{J_{max} V_{max}}{J_{sc} V_{oc}} \tag{3.1}$$

For an ideal FF $(0.8 \le FF < 0.9)$, the maximum V_{oc} and J_{sc} determine the maximum efficiency of solar cells. The maximum efficiency (η_{max}) can be achieved by maximizing the values of V_{oc}, J_{sc}, and FF, and it is calculated by

$$\eta_{max} = V_{oc} \times J_{sc} \times FF_{max} \; (0.8 \le FF_{max} < 0.9) \tag{3.2}$$

Therefore, it is significant to balance the bandgap to achieve the maximum V_{oc} and J_{sc}. Figure 3.4b shows the current record efficiencies of solar cells for various absorber materials as a function of E_g, in comparison to the theoretical S-Q limit (black curve). Note that the maximum theoretical PCE that can be achieved for an absorber with an E_g of 1.34 eV will be 33.7%.

3.4 Improvement of the Efficiency of Solar Cells

3.4.1 Tandem Solar Cells

To further enhance the PCE, the thermalization and recombination losses should be reduced, but the absorption efficiency has to be improved. To overcome efficiency-limiting factors, the design of efficient tandem PVs is the most effective strategy to break the efficiency limit [Fig. 3.5(a,b)]. As shown in Fig. 3.5c, the larger E_g absorber-based solar cell at the top of the device can collect most of the higher energetic photons, while the smaller E_g absorber-based solar cell at the bottom of the device will absorb the remaining lower-energy photons, leading to the maximum utilization of the solar spectrum and minimum thermalization loss. Until now, multi-junction tandem PVs have been widely studied and obtained spectacular record efficiencies [7]. A detailed discussion of emerging tandem solar cells will be presented in Chapter 11.

3.4.2 Concentrator Solar Cells

The basic concept of concentrator solar cells is to use optics to concentrate sunlight on a smaller area of cells. The unscattered solar light on the smaller area of solar cells can significantly improve photocurrent intensity, leading to a higher PCE of solar cells. The unscattered solar light can cause higher temperatures for small area cells, so the absorbers must be able to absorb a sufficiently large part of the solar spectrum. Besides, the concentrator cells merely require fewer materials, which enables more expensive materials-based multi-junction cells economically viable. Thus, GaAs and Si are well suited for the development of concentrator PVs.

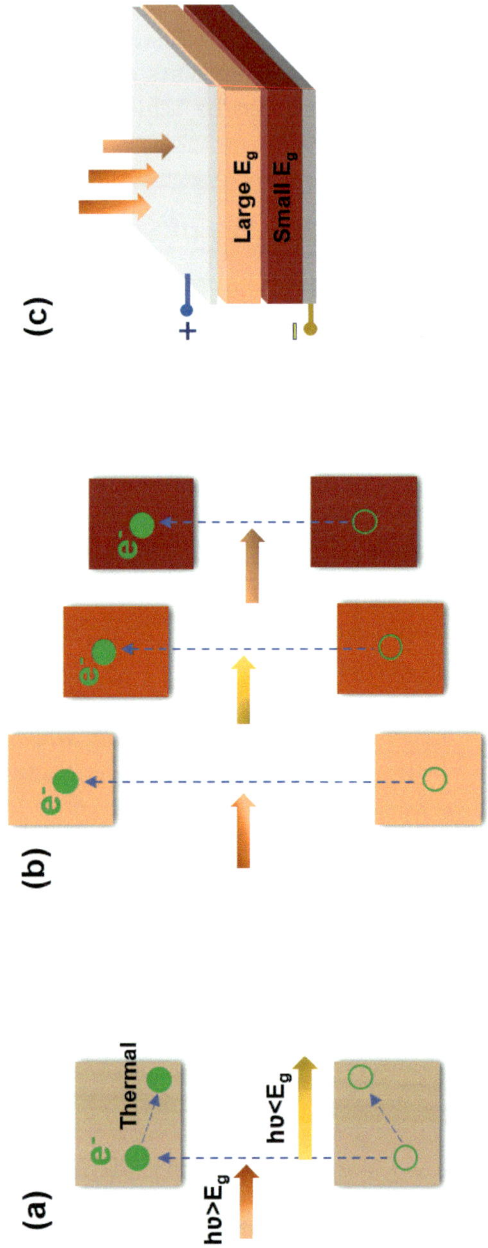

Figure 3.5 Light absorption for single (a) and multi-junction (b) solar cells, (c) basic structure of tandem solar cells.

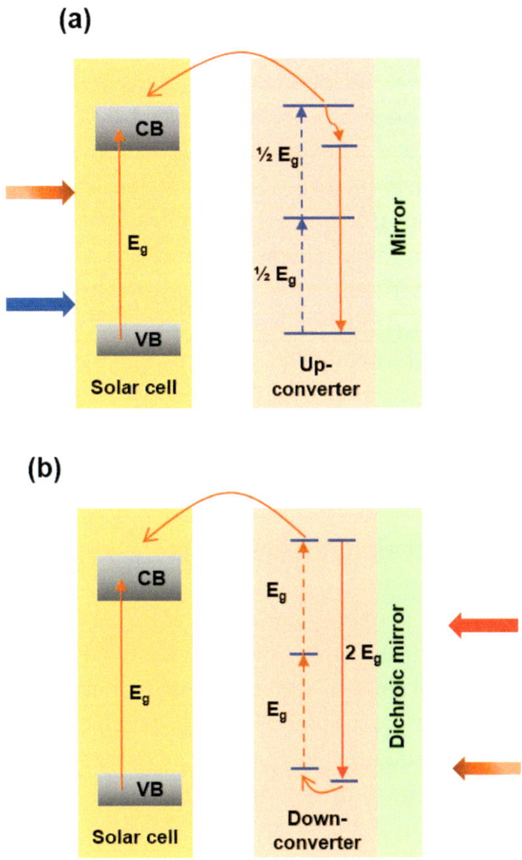

Figure 3.6 Basic structures and mechanisms (a) up-conversion and (b) down-conversion solar cells.

3.4.3 Up- and Down-Conversion of Photons

As we already discussed in Section 3.2.1, a large part of efficiency loss in a solar cell is attributed to photons with energy less than E_g, which cannot be absorbed. Only the absorption of energy above the E_g can yield electron–hole pairs contributing to electric current. In this case, two or more of these unabsorbed photons could be converted into one photon with energy no less than E_g, which could then be absorbed by the solar cells (Fig. 3.6a). This process is called up-conversion photons, which can extend the spectral absorption from the visible to near-infrared (NIR) range and minimize photon transmission loss [8]. The up-conversion of photons process can significantly improve the photocurrent density of solar cells. To ensure all emitted photons can be absorbed by the solar cells, an up-converter is needed. The up-converter usually is organic dyes or luminescent materials [9, 10], which can significantly convert two sub-bandgap photons into one above-bandgap photon.

As shown in Fig. 3.6b, down-conversion is a process where a high-energy photon is split into several lower-energy photons with energies above the E_g. Down-conversion materials are mostly located at the front side of the solar cells, which can prevent thermalization energy loss. Most luminescent down-conversion materials can absorb high-energy ultraviolet (UV) photons and then convert them to low-energy photons.

3.5 Photovoltaic Generations

In PV technology, a photovoltaic cell produces electricity from sunlight radiation. Semiconductor materials have a key role in PV technology and cell design. Among semiconductors, silicon is the most reliable material for solar PV fields, which takes over 80% of commercialized PV devices. In general, the current PV technologies can be categorized into three generations based on their technical characteristics. Figure 3.7 demonstrates the three generations of PV cells.

First generation These are based on c-Si wafers. Si-based devices are the most efficient solar cells and last longer than non-silicon-based cells. Besides silicon-based PV cells, various groundbreaking kinds of PV cells have been developed and improved over time to improve PCE and lower manufacturing and maintenance costs.

First-generation cells include three types of Si-based PV: mono-crystalline Si (m-Si), polycrystalline Si (poly-Si), and amorphous Si (a-Si). The PCEs of Si-based PV cells are different from each other. The monocrystalline PV cells show the highest conversion efficiency than polycrystalline and amorphous cells.

Second generation These are called thin-film solar cells. Compared to c-Si-based cells these devices are made from thin layers of semiconductor materials with few micrometers thickness. In contrast to the first-generation cells, their fabrication costs in terms of semiconductor material are significantly cheaper. Second-generation PV cells are made of thin layers of semiconductor materials. The production of these cells is simpler and semiconductor materials cost less than the first-generation cells. However, their efficiencies are lower than those of Si-based first-generation cells. Three types of devices are considered in this category: cadmium telluride (CdTe), copper-zinc-tin-sulfur (CZTS), and copper indium gallium diselenide (CIGS). Thin-film solar cells conquer about 20% of the total PV market. CdTe PV cells benefit from having the lowest manufacturing cost among all the thin-film technologies.

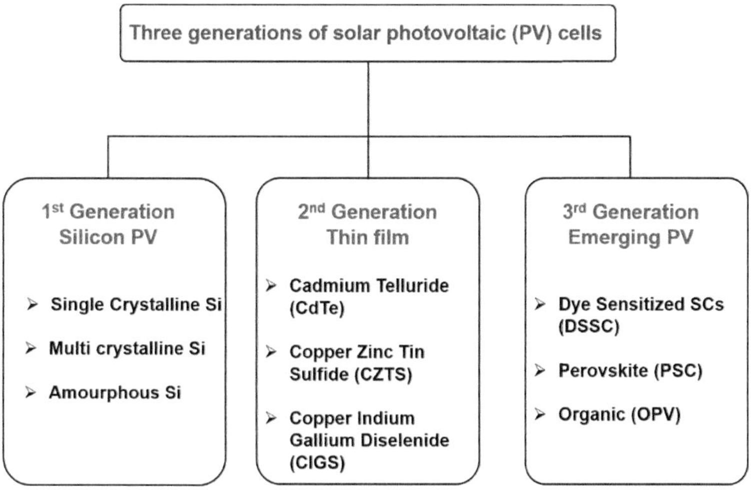

Figure 3.7 The generations of photovoltaic cells.

Third generation This category includes solar cells made from a variety of new materials including nanotubes, solar inks, organic dyes, and conductive inorganic using conventional coating technologies. The main goal of third-generation PV devices is to fabricate more efficient devices that still use thin-film technology. Most of the research and development on third-generation cells are now done in laboratories, therefore, it is not commercially unavailable and the majority of PV markets are belonging to the first- and second-generation PV cells. In Chapter 4, three generations of solar cell will be discussed in detail.

References

1. Shockley, W. and Queisser, H. J. (1961) *J. Appl. Phys.*, **32**, pp. 510–519.
2. Semonin, Q. E., Luther, J. M., and Beard, M. C. (2012) *Mater. Today*, **15**, pp. 508–515.
3. Polman, A., Knight, M., Garnett, E. C., Ehrler, B., and Sinke, W. C. (2016) *Science*, **352**, pp. aad4424.
4. Rau, U., Paetzold, U. W., and Kirchartz, T. (2014) *Phys. Rev. B*, **90**, pp. 035211.
5. Tiedje, T. (1982) *Appl. Phys. Lett.*, **40**, pp. 627.
6. Rau, U. and Werner, J. H. (2004) *Appl. Phys. Lett.*, **84**, pp. 3735.
7. Green, M., Dunlop, E., Hohl-Ebinger, J., Yoshita, M., et al. (2021) *Prog. Photovolt.: Res. Appl.*, **29**, pp. 3–15.
8. Auzel, F. (2004) *Chem. Rev.*, **104**, pp. 139–174.
9. Wild, J., Meijerink, A., Rath, J. K., van Sark, W., and Schropp, R. E. I. (2011) *Energy Environ. Sci.*, **4**, pp. 4835–4848.
10. Huang, X., Han, S., Huang, W., and Liu, X. (2013) *Chem. Soc. Rev.*, **42**, pp. 173–201.

Problems

3.1 What is a p-n junction in semiconductors?

3.2 What is the main difference between p-n junction and non-p-n junction solar cells?

3.3 Write the main reasons for solar cell efficiency, its limitation, and the solutions?

3.4 Calculate the fill factor for the solar cell with following parameters: I_{mpp} = 285 mA, I_{sc} = 315 mA, V_{mpp} = 1.65 V, and V_{oc} = 2 V.

Chapter 4

Generations of Solar Cells

Photovoltaic (PV) devices can be classified into three generations, i.e., first-, second-, and third-generation cells, depending on the key materials used and the level of commercial maturity of the PV technology. The first-generation cell is the commercially predominant PV technology, based on crystalline silicon (c-Si) wafers, that includes materials such as polysilicon and monocrystalline silicon. The second-generation cells are thin-film solar cells, that include amorphous silicon (a-Si) based thin-film PVs, cadmium telluride/cadmium sulfide (CdTe/CdS) solar cells, and copper indium gallium selenide (CIGS) solar cells. The third-generation solar cells, defined as emerging PV technology that is still in the research or development phase, include several thin-film technologies based on organic materials, organometallic compounds, and inorganic substances. Notwithstanding relatively low efficiency and poor stability for commercial applications, a lot of research investments have been made in the development of these solar cells because of the great potential to achieve the manufacturing goals of low-cost and high-efficiency solar cells.

This chapter focuses on the understanding of three generations of SC technology and discusses their operational principles. It allows us to compare and evaluate the different types of SCs.

Next-Generation Solar Cells: Principles and Materials
Yoon-Bong Hahn, Tahmineh Mahmoudi, and Yousheng Wang

ISBN 978-981-4968-66-9 (Hardcover), 978-1-003-37238-7 (eBook)
www.jennystanford.com

4.1 First Generation: Solar Cells Based on Silicon Wafers

4.1.1 Basic Material

Next to oxygen, silicon is the most abundant element in nature (27.8% of the Earth's crust). Silicon can be found in soil, rocks, and sand, combined with oxygen as silicon dioxide (SiO_2), or with oxygen and other elements as silicates. Silicon is the most common material in PV technology and is mainly produced from quartzite, using carbon in an arc furnace process. The process takes place according to the following reaction:

$$SiO_2 + 2C \xrightarrow{\text{Arc}-1800°C} Si + 2CO$$

Since Si production requires a large amount of energy (14 kWh/kg), production takes place in parts of the world where there is excess hydroelectric power (e.g., Noway, Canada).

Polycrystalline silicon is fabricated from high-quality Si according to the following process:

$$4SiHCl_3 + 2H_2 \xrightarrow{1350°C} 3Si + SiCl_4 + 8HCl$$

Crystalline silicon (c-Si) is a main semiconductor material in the PV industry for the production of solar cells. The c-Si-based cells are combined into panels and PV systems to produce solar power from sunlight sources.

c-Si Solar cells, as first-generation solar cells, were developed in the 1950s and still are the most common type of solar cells. These solar cells are produced from ~170 μm thick Si wafers, thus they are often called wafer-based solar cells. Solar cells based on c-Si are single-junction devices and are more efficient than their competing technologies, i.e., the second-generation thin-film solar cells.

Monocrystalline silicon (m-Si) is typically produced through several steps, including melting high-purity semiconductor-grade silicon and then using a seed to start a single-crystal growth. To avoid impurities and maintain crystal uniformity, the entire process is carried out in an inert atmosphere in an inert crucible (i.e., quartz). Among the various techniques, the Czochralski method is the most common method, in which seed crystals mounted on a directional rod are immersed in molten Si, and then slowly pulled upward while rotating at the same time. The pulled material is solidified into a monocrystalline cylindrical ingot. For further uniformity of the crystallization, a magnetic field is used to control and minimize turbulent flow. In the end, the solidified ingots are sliced into thin wafers (0.75 mm thick). This process is very slow and expensive, but there is a high demand for m-Si due to the superior electronic properties and enhanced performance of integrated circuits and PV devices. For semiconductor device fabrication, the wafers are polished for a flat surface, on which micro-electronic devices can be fabricated through microfabrication processes. The overall process steps of production of c-Si by the Czochralski method are visualized in Fig. 4.1.

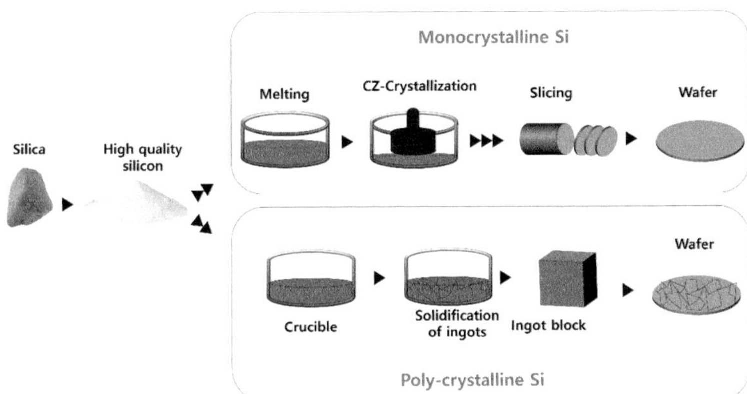

Figure 4.1 Overall process steps of production of c-Si by Czochralski method.

The quality of crystalline Si affects the performance and reliability of a device, and the presence of grain boundaries, crystallographic defects, and/or impurities interferes with the electronic properties of the device, therefore continuous single-crystalline Si is crucial for electronic devices.

4.1.2 Crystalline Silicon Solar Cell

A Si solar cell consists of two layers of p-type silicon and n-type silicon, placed next to each other (Fig. 4.2). The n-Si layer has an excess of electrons, while the p-Si layer has an excess of positively charged holes due to the lack of valence electrons. In a p-n junction, electrons in the n-type layer move to the holes in the p-type layer. The area around the junction is called the depletion region. In the depletion zone, the p-type side of the depletion zone receives electrons, and the n-type side gets more holes without electrons. This situation creates an internal electric field with repulsion that prevents charge flow at the junction. Connecting these two layers with a metallic wire (in a circuit) causes electrons to flow from the n-type Si to the p-type Si, crossing the depletion zone, and then return to n-type Si through the wire, generating a current of electricity.

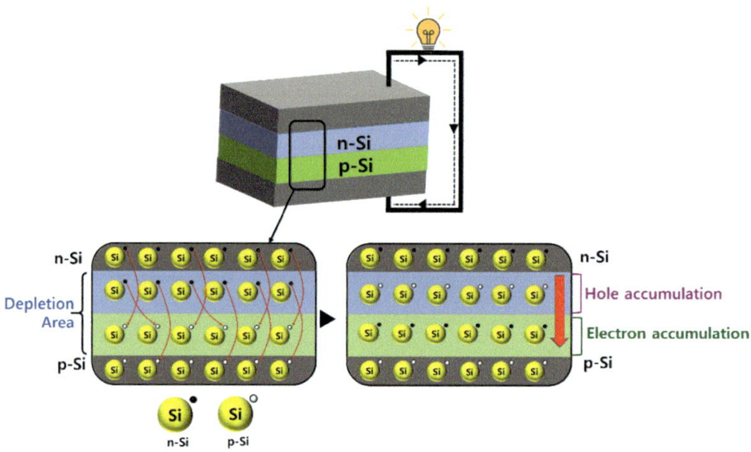

Figure 4.2 Schematic structure of a Si solar cell, with n-type and p-type layers, and a close-up presentation of the depletion area at the junction of the n-Si and p-Si layers.

Crystalline silicon (c-Si) solar cells are the most popular solar cells and deliver the utmost energy conversion efficiencies of all commercial PV cells. The configuration of typical c-Si PV cells is presented in Fig. 4.3. Standard cells are produced using one of two different boron-doped p-type monocrystalline silicon (m-Si) and polycrystalline silicon (p-Si) substrates. A typical c-Si cell consists of:

- Anti-reflection coating (ARC) on a textured surface: The top surface of the PV cell is textured in the shape of micron-sized pyramid structures to minimize reflection loss of incident light. An anti-reflection layer made of silicon nitride or titanium oxide is formed on the textured surface, thereby further reducing reflection loss.

- p-n junction: p-n junction is formed with the phosphorous-doped Si (i.e., n-type, electron-donor) layer deposited on the surface of boron-doped Si (i.e., p-type, electron-acceptor) layer.
- Back-surface field (BSF): the back surface of the silicon substrate is required to suppress the recombination of photo-generated carriers.

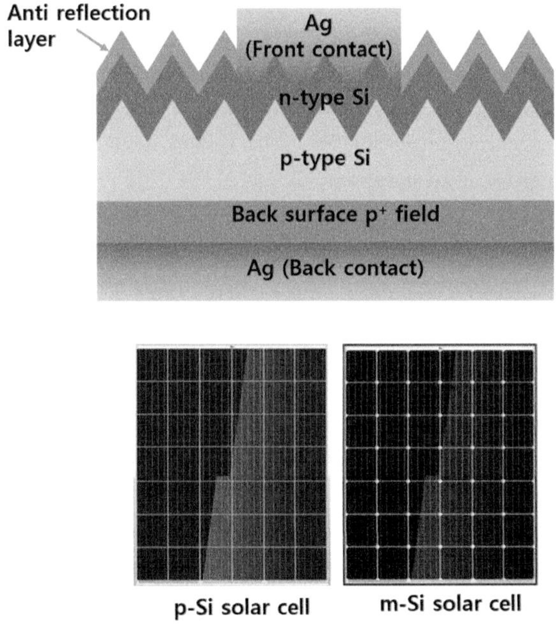

Figure 4.3 (Top) Configuration of a typical c-Si solar cell. (Bottom) Modules of polycrystalline Si (p-Si) cell and monocrystalline Si (m-Si) cell.

4.1.2.1 Monocrystalline silicon solar cell

Many technologies that deliver high-performance cells with m-Si solar cells have one of the three main features: the passivated emitter rear localized (PERL) cell, the heterojunction with intrinsic thin layer (HIT) cell, and the back contact, back junction (BC-BJ) cell.

- PERL cell: To increase the current flow in this cell, the front and back surfaces are covered with insulating layers, which are named the passivating layers. Passivating layers suppress defects and help the cell to produce more electricity from the solar cell. This cell approaches the boundary of the current technologies for absorption of solar photons and collection of charge carriers generated from the cell emitter and base. The PERL cell is the most efficient type of m-Si solar cell and has been a common laboratory structure of high-efficiency c-Si PV cells. However, the full PERL design is not cost-effective for industrial production because of the inevitable multiple photolithography stages.
- HIT cell: This cell has an exclusive heterojunction structure consisting of very thin, amorphous p- and n-doped layers and amorphous silicon (a-Si) layers on the front and back surfaces of an n-type m-Si substrate. The HIT solar cell has a low-cost and low-temperature process of amorphous silicon (a-Si) combined with highly efficient and stable c-Si. In this cell, transparent conductive oxide (TCO) is used as a window layer, which with high transmittance and low resistivity and work function, improves the performance of the HIT solar cell. Compared to other high-efficiency PV cells, the HIT cells show lower photocurrent density due to the suppression of photocurrent collection by the front a-Si layers and/or the lower internal quantum efficiency of the a-Si layer.
- BC-BJ cell: The BC-BJ cell has front and back-surface passivation layers, a random-textured light-trapping surface, n- and p-doped regions on the rear surface, n- and p-contacts on n- and p-doped regions, an anti-reflection layer, and an n-type single-crystalline silicon substrate with a minimum minority carrier lifetime >1 ms. Among all the c-Si solar cells on the market, only BC-BJ cells provide the possibility of unit efficiency of 20% and more. BC-BJ cells have several advantages compared to the conventional structure: no sub-electrode or main electrode shading, good passivation on the front surface due to the absence of front contact, flexible design of back electrodes, and better look with no front electrodes.

There is a large difference in power-conversion efficiency between industrial cells and the best research c-Si cells. Standard industrial m-Si PV cells have an efficiency of 16–18%, much lower than the best research cells. Industrial cells have limitations due to simple high-speed automatic production using inexpensive materials. To improve cell efficiency, many solar cell industries are striving to adapt attempts to apply high-efficiency structures such as improved gridlines, selective emitter materials, or more shallowly doped n^+ regions.

4.1.2.2 Polycrystalline silicon solar cell

Polycrystalline silicon (p-Si) wafers have been used to reduce costs for silicon production, and have been explored since the mid-1970s. Currently, p-Si cells are the most extensively produced cells and count for 48% of global solar cell production in 2008.

Due to the difference in quality between m-Si and p-Si, the efficiency of p-Si cells is lower than that of m-Si cells. Higher-efficiency p-Si solar cells require functions such as minimizing photon loss with less carrier and electrical loss. To achieve these features, several improvements have been proposed, including:

- To reduce photon loss, texture the front surface (or honeycomb texturing) by etching technique (i.e., acid etching or reactive ion etching), i.e., chemical etching of the back surface and deposition of a reflective metal layer on the etched back surface.
- To reduce carrier loss, use a surface passivating layer with shallowly doped n+ regions with silicon nitride, local BSF and laser fired contacts, or back-surface passivation with silicon oxide and/or SiN layers.
- To minimize electrical loss, make a gridline for front contact printing methods, use a selective emitter underneath front contact, or deposition of seed layer films and copper coating for metallization to replace screen printing silver.

4.1.3 Solar Cell Performance

The commercial goal of PV power generation is to generate electricity at a cost equivalent to industrial grid electricity. The energy conversion efficiency of PV cells is an important issue because it greatly affects the production cost of a PV system, from materials fabrication to system installation.

The solar cell performance is limited by the loss mechanisms:

- Photon losses, caused by surface reflection, transmission through Si bulk, and back contact absorption;
- Charge-carrier loss, due to recombination in the bulk and at the surfaces;
- Heating joule loss is caused by series resistance in the gridlines, electrode, and silicon interfaces, and in the Si bulk and diffusion region.

In the designing of solar cell structure and fabrication processes, these losses should be minimized without reducing the productivity of cells.

As mentioned in Chapter 3, the overall performance of the solar cell is determined by PV parameters: open-circuit voltage (V_{oc}), short-circuit current (J_{sc}), fill factor (FF), the voltage at the maximum power point (V_{max}), current at the maximum power point (J_{max}), maximum power (P_{max}), and energy conversion efficiency (η). These parameters were obtained under standard test conditions for terrestrial solar cells (AM1.5). The solar cell FF and efficiency are calculated by the following equations.

$$FF = \frac{P_{max}[\mathrm{W}]}{J_{sc}[\mathrm{A}] \times V_{oc}[\mathrm{V}]} \tag{4.1}$$

where $P_{max}[\mathrm{W}] = J_{max}[\mathrm{A}] \times V_{max}[\mathrm{V}]$, and

$$\eta = \frac{P_{max}[\mathrm{W}] \times 100}{1000\left[\mathrm{Wm}^{-2}\right] \times \mathrm{Cell\ area}[\mathrm{m}^2]} \tag{4.2}$$

4.1.4 Cell Fabrication Technology

There are eight stages of manufacturing Si solar cells from wafers including, wafer check, texturing, diffusion, edge isolation, anti-reflective coating, contact printing, sintering and testing, and cell sorting. Figure 4.4 shows the schematic steps of solar cell manufacturing, from silicon wafers to the ready solar cell [1, 2].

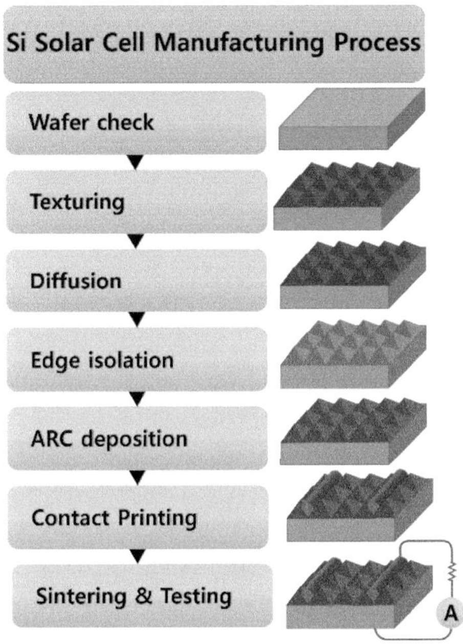

Figure 4.4 Schematic steps of solar cell manufacturing; from silicon wafers to the ready solar cell.

4.1.4.1 Wafer check

The quality of the Si wafer directly affects the efficiency of the solar cell, so it is essential to test the starting Si wafer. In this process, the technical parameters of wafers, such as surface roughness, p/n type, resistivity, and micro-cracks will be measured.

4.1.4.2 Texturing

Texturing is the first step in manufacturing procedures and is critical for all types of Si solar cells, which helps to reduce surface reflection. Surface texturing is roughening of the surface to reduce reflection and enhance the possibility of reflected light bouncing back onto the surface.

Texturing reduces the optical loss in Si solar cells as illustrated in Fig. 4.5. The reduction mechanism of optical loss is:

(i) Texturing of the front surface increases the surface area which helps absorb short-wavelength light, resulting in more generation of electron–hole pairs and easier charge collection near the junction.

(ii) Inclined surfaces cause external reflectance of light, increasing the likelihood that light will be reflected by the device.

(iii) Texturing can promote the 1st and 2nd order of diffraction into Si, while the diffraction order of incident light is zero on a flat surface.

(iv) For long wavelength (twice the Si thickness), the length of the optical path increases, and the probability of absorption increases.

(v) Internal reflection of long wavelengths increases the amount of light that can be converted.

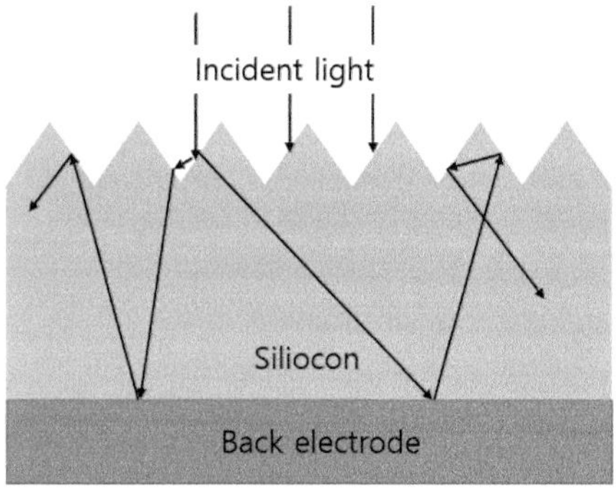

Figure 4.5 Mechanism of the reduction of optical loss in a Si solar cell by surface texturing to reduce surface reflectance and obtain a high current density.

The shape, size, and uniformity with the appropriate height/ width ratio of texturing are important to reduce surface reflectance and obtain a high current density. On the other hand, higher roughness with a large aspect ratio can lead to a large current saturation and poor performance.

The most common and low-cost texturing method in the industry is wet chemical etching and mechanical grooving. Chemical texturing relies on acidic or alkaline treatment of the Si surface, which damages the Si wafer surface to initiate texturing. Alkaline texturing is developed to form Si solar cells, while acidic texturing, was proven effective for p-Si solar cells. Acidic and alkaline treatments of Si wafer can reduce the reflection to around 20%, but it is ineffective for infrared and blue rays.

A medium-sized pyramid pattern with ~5 μm size produced by alkali texturing is optimal for solar cells. Submicron patterns cannot be obtained by chemical etching. Smaller textures (~1 micron) have shown higher light reflectance and more shunt path due to their delicacy. Small textures cause many crystal defects on the surface with deeper emitters.

The optimal texture size can lower the reflectance with proper texturing density, uniformity, optimal coverage, and homogeneity. The type of structure is also prominent in device performance. For example, Si cells can generate higher current density with regular inverted pyramid texturing compared to a regular, random upward pyramid. It has been shown that the texture of a regular inverted pyramid has lower reflectance compared to random upward pyramids. Nanostructure texturing with upward pyramids improves internal quantum efficiency. Figure 4.6 shows several types of texturing features.

Figure 4.6 Texturing features on Si surface: (a) random conical, (b) nanosize pyramids, (c) inverted pyramid, (d) porous upward pyramids, (e) random upward pyramids, (f) random inverted pyramids, (g) honeycomb, and (h) grooves. Reprinted with permission from [3], Copyright 2021, Elsevier.

4.1.4.3 Diffusion

Diffusion is a solid-state process to introduce impurity dopant into semiconductors crystal lattice to alter the electrical properties. Dopant atoms are introduced into the crystal lattice of the host in either substitutional or interstitial sites (see Fig. 4.7). Dopant atoms such as boron (B) and phosphorus (P) occupy substitutional or interstitial positions to donate free electrons or holes to the Si lattice.

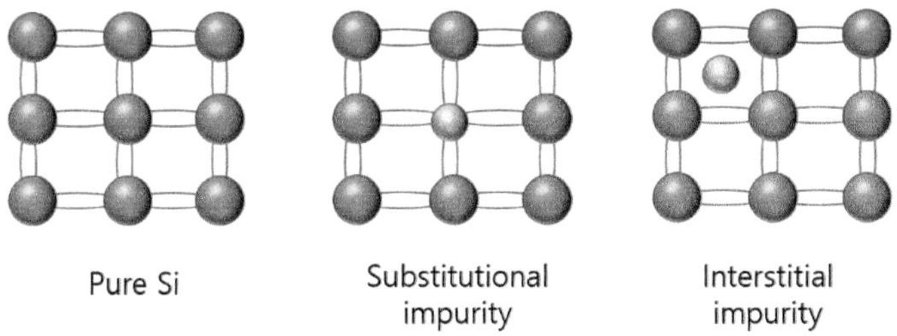

Figure 4.7 Substitutional or interstitial impurities in Si crystal lattice.

Diffusion and ion implantation are the two common processes to introduce an accurate amount of dopant into a semiconductor. In the diffusion, the dopant impurity atoms enter the host lattice from a gas phase and then diffuse into the Si bulk. The doping concentration decreases monotonically, and the in-depth concentration of the dopant highly depends on the diffusion process time and temperature. Boron is a common p-type impurity in Si and P ans As are n-type dopants. These elements are dissolved in silicon with a high speed of 50×10^{20} atoms/cm^3 in the temperature range of 800–1200°C. The main sources for these dopants are solid sources of BN, P_2O_5, and As_2O_3, liquid sources of BBr_3, $POCl_3$, and $AsCl_3$, and gas sources of B_2H_6, PH_2, and AsH_3.

4.1.4.4 Edge isolation

The edge isolation process helps remove the impurity atoms from the edge of the cell so that the front emitter and back contact are electrically isolated. A common method for edge isolation is plasma etching by CF_4 and O_2.

In plasma etching, samples are stacked on each other in a way that only the edges are exposed. Afterward, the uncovered surface of the wafer is exposed to plasma, and the SiO_2 layer containing impurity elements is removed. Plasma etching is time-dependent, a short process results in insufficient etching and causes shunt resistance. On the other hand, a long process can result in extreme damage to the edges, which can cause a high recombination rate in this area, disturbing the ideality factor of the diode cell.

In another process, wet chemical etching is performed to remove doped Si from the wafer edge using HF and HNO_3. This process is highly efficient and there is no risk of extensive damage like plasma etching. For this reason, wet chemical etching is the most common process used for edge isolation.

4.1.4.5 Anti-reflection coating

Optical loss is one of the major parameters that affect the short-circuit current in a solar cell. Optical loss can be caused by shading the front contact of the Si surface, or by the light reflected from the wafer surface and the light reflected out of the cell after exiting the rear surface.

The reflectance of the Si surface without anti-reflection coating (ARC) film is 35%. To reduce the reflection, a layer of ARC, which is a thin layer of insulating dielectric material, must be coated. Texturing reduces the reflection by ~15%, and an ARC can further improve and minimize the reflection from the surface. The thickness of ARC, d_1, is calculated via the refractive index, n_1, of the ARC and wavelength λ_0 by the equation:

$$d_1 = \frac{\lambda_0}{4n_1} \tag{4.3}$$

n_1 can be calculated from:

$$n_1 = \sqrt{n_0 n_2} \tag{4.4}$$

where n_0 and n_2 are refractive indexes of the substrate and the dielectric layer, respectively.

Typically, the ARC is optimized to reduce the reflection at λ = 600 nm, which is the peak of AM 1.5 light irradiance. Silicon nitride (SiN_x) is a commonly used material for ARC. SiN_x layer can reduce the carrier recombination at the interface, and release H_2 from the layer during co-firing steps, which helps to passivate defects such as bulk defects, laser damage, or boron–oxygen defects. In industrial production, the plasma-enhanced chemical vapor deposition method is a common method to deposit ARC. In the ARC process, the energy source is low-temperature plasma under low-pressure samples that are heated to a determined temperature and exposed to SiH_4 and NH_3 gases. After a series of chemical reactions, a solid thin film of silicon nitride is formed on the Si surface with a thickness of ~70 nm. The SiN_x thin film is optically functional and can significantly reduce light reflection, but the short-circuit current and the efficiency can be improved [4].

4.1.4.6 Contact printing

The front metal contact electrode helps the current flow of the generated carriers. The front electrode is made of fingers and busbars (see Fig. 4.8a). It is important to optimize the design of the front contact based on the *shading loss* and *the contact resistance loss.* The front contact loss can affect either J_{sc} or R_S. The front contact (silver) transports the charges from fingers to the busbars. Afterward, cells are interconnected to form modules (see Fig. 4.8b). Finally, the interconnect cell ribbon is soldered to the busbars and electrodes to the back contacts of the adjacent cell in the cell line.

Front contact: This electrode is printed on an anti-reflection coating (SiN_x). The front electrode is required to infiltrate into the ARC film to make a contact with the silicon. The electrical contact is deposited in an inline firing furnace at a temperature range of 750 to 870°C for a few seconds. In this process, the paste carves and penetrates through the ARC coating layer to make ohmic contact with the silicon. It is essential to optimize the deposition time and temperature. At higher temperatures or longer process times, the front electrode can infiltrate deeper and come into contact near the junction. In this case, the contact resistance increases (so higher R_S) because the electrode is making contact with a more resistive area in the wafer.

Back contact: For the rear electrode, the back surface of the cell is printed with aluminum paste. The thickness of the rear contact is about 30 μm and its optimization helps improve the cell performance. Currently, most industrial Si cells have all Al back electrodes (called aluminum back-surface field (Al-BSF) cells). During the deposition process at a temperature exceeding 570°C an aluminum-silicon eutectic is formed, and after cooling Si recrystallization and an Al-doped Si layer are formed. In this case, the Al concentration depends on process temperature. This process thus results in *p*-type doping at the back of the solar cell which helps the collection of holes and also reduces back-surface recombination [5].

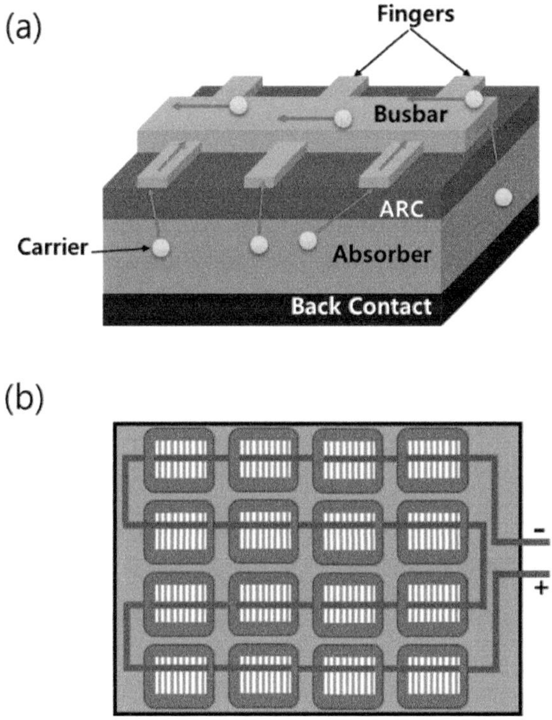

Figure 4.8 (a) Charge collection by the screen-printed contact, busbar, and finger. (b) A module contains 16 interconnected cells in series.

4.1.4.7 Testing and sorting

In the end, cells ready to be assembled are tested under a sunlight simulator, classified, and sorted based on their efficiencies. This part is controlled by a cell testing device that systematically tests and arranges the cells. Depending on the production process and quality of the Si wafer, the final solar cell is graded into various solar cell quality grades.

4.2 Second Generation: Thin-Film Solar Cells

4.2.1 Materials

The most commercialized thin-film solar cells are based on a-Si, CdTe, and CIGS. The most common feature of these materials is the direct bandgap. The a-Si is the oldest thin-film technology, fabricated by the chemical vapor deposition (CVD) of an a-Si thin film onto a substrate. The a-Si is a non-toxic, abundant material, with a high absorption coefficient with excellent performance in lower light intensity. The a-Si thin films are mostly used in low-watt, portable solar products. The highest efficiency of a-Si solar cell is ~13.6%.

The next common thin-film solar cell is based on rare metal compounds of cadmium and tellurium. The cadmium telluride (CdTe) thin-film solar cells are less expensive than a-Si ones and have a world-record efficiency of >22% and average module efficiency of 17%.

Copper indium selenide (CIS) and copper indium gallium diselenide (CIGS) are other types of semiconductor materials used in thin-film technology, fabricated by co-deposition of Cu, In, Se, and (sometimes) Ga onto substrates. CIGS achieved efficiency of more than 23% and the average module efficiency was 16.5%. CIGS cell manufacturing is more expensive than other types of cells, and for this reason, CIGS is not widely used [6]. Gallium arsenide (GaAs) thin-film solar cells have a high efficiency of ~30% in laboratory production but are becoming a big challenge in a limited market due to high manufacturing costs. The main uses of GaAs solar cells are satellites and spacecraft.

4.2.2 Si-Based Thin-Film Solar Cells

Si atoms tend to bond tetrahedrally to four neighboring Si atoms. In c-Si, Si atoms form a continuous well-ordered tetrahedral structure over a large range. Unlikely, in a-Si, this order is not present and atoms have a random disordered network. In disordered a-Si, the atoms still have a tetragonal structure but the bonds and their angle are distorted compared to c-Si. Due to the disordered network, Si atoms have a dangling bond, which represents defects and causes weak photoconductivity and electrical properties. In hydrogenated a-Si, the hydrogen atoms bond to the dangling bonds and reduce their density (Fig. 4.9) [7]. A hydrogenated alloy of a-Si:H has a very low amount of defects by several orders of magnitude and can be used effectively in PV devices. However, hydrogenation can cause light-induced degradation, known as the Staebler–Wronski effect.

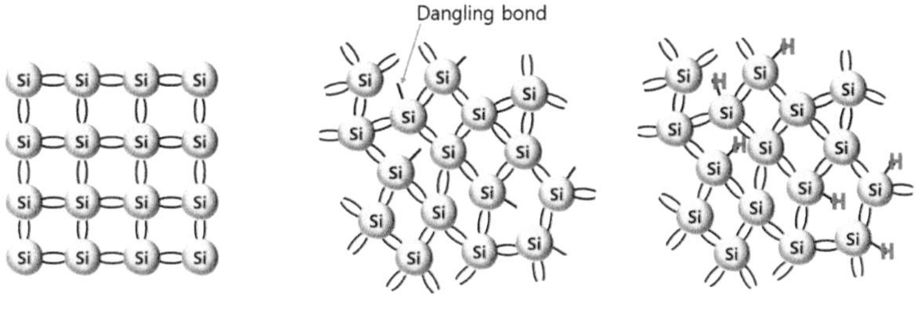

Figure 4.9 Schematic of the c-Si (left), a-Si (middle), and a-Si:H (right).

Compared to the bandgap of c-Si (≈1.12 eV), the bandgap of a-Si can be tuned by hydrogen alloying in the range of 1.6 to 1.8 eV. The higher bandgap of a-Si than that of c-Si is due to distortions in Si bond length and angle. Unlike c-Si with an indirect bandgap, a-Si is a direct-bandgap material, which leads to higher absorption and a much larger absorption coefficient than c-Si. Therefore, it is possible to fabricate a-Si films thinner than c-Si [8, 9].

As mentioned, a-Si has a disordered structure with dangling bonds, which act as defects and limit the lifetime of charge carriers. Such high defect density leads to a short diffusion length of photo-generated carriers (~200 nm). Unlike c-Si solar cells with a p-n junction structure, a-Si solar cells have a p-i-n or n-i-p structure, where "i" is an intrinsic (undoped) layer between thin p-doped and n-doped films (Fig. 4.10).

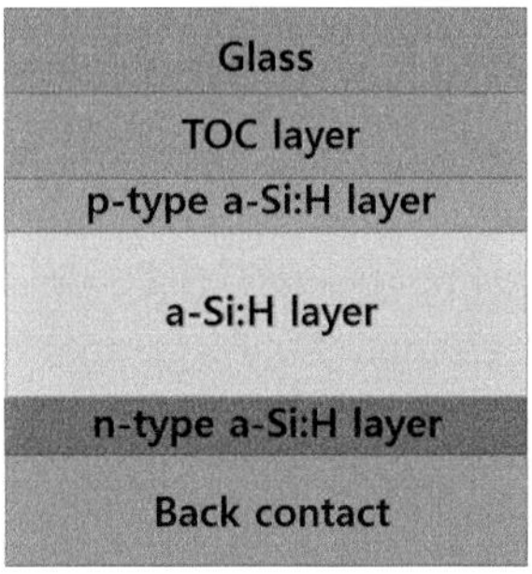

Figure 4.10 Schematic illustration of a-Si thin-film solar cell with p-i-n structure.

It is worth noting for a-Si thin-film solar cells [10, 11]:

- Thin-film silicon solar cells have nano-textured interfaces.
- Textured interfaces scatter the light and prolong the light path length, resulting in more light absorption and photocurrent.
- The p layer is a material with a higher bandgap to reduce parasitic absorption of the blue part of the spectrum using boron as a dopant.
- The n layer is often silicon oxides or a-Si:H with phosphorus as a dopant.
- Transparent n-SiO_x:H is used as a back reflector. Its thickness affects destructive interference at the i-n interface and reduces the electric field strength and parasitic absorption in the n layer.
- Al or Ag are mainly used for metallic back contact with a higher reflectivity effect.
- Besides a-Si:H, nc-Si:H films are also used for the intrinsic absorber layers that absorb wavelengths of 700 to 950 nm.
- Neither a-Si:H nor nc-Si:H has an optimum absorption. Hence, in thin-film technology, their junction is used, which is called a double-junction or micrograph cell including both a-Si:H and nc-Si:H. In this design, the solar cell with the highest bandgap is used as a top cell to convert the high-energy photons, while the low-bandgap cell is used at the bottom to convert the lower-energy photons.

4.2.3 Chalcopyrite-Based Solar Cells

The chalcogenide solar cell is another type of thin-film solar cell. In this type of thin-film device, the main focus is on CIGS and cadmium telluride (CdTe) solar cells. Chalcogenide is a chemical compound consisting of chalcogen anion, which is an element from the group VI in the periodic table: O, S, Se, and Te. This class of solar cells is based on chalcopyrite materials. Chalcopyrites are ternary semiconductors with elements from groups I, III, and VI. Generally, all these combinations can be used:

$$\begin{bmatrix} \mathrm{Cu} \\ \mathrm{Ag} \\ \mathrm{Au} \end{bmatrix} \begin{bmatrix} \mathrm{Al} \\ \mathrm{Ga} \\ \mathrm{In} \end{bmatrix} \begin{bmatrix} \mathrm{S} \\ \mathrm{Se} \\ \mathrm{Te} \end{bmatrix}_2$$

The CIGS chalcopyrite is a mixture of CIS (copper indium diselenide, $CuInSe_2$) and CGS (copper gallium diselenide, $CuGaSe_2$). CIGS is formulated as $Cu(In_xGa_{1-x})Se_2$, where x is between 0 and 1. There is another chalcopyrite that contains sulfur, i.e., CIGSS (copper-indium-gallium diselenide/disulfide, $Cu(In_xGa_{1-x})$ $(Se_yS_{1-y})_2$), where y is between 0 and 1 [12].

In these compounds, the bandgap changes from 1.0 eV to 1.7 eV by tuning the x in In:Ga and y in Se:S. For example, the bandgap of $CuInSe_2$ is 1.0 eV, but when Se is replaced with S (i.e., $CuInS_2$) the bandgap increases to 1.5 eV. On the other hand, when Ge is replaced with In the bandgap increases to 1.7 eV. CIGS(S) is a p-type semiconductor with a direct bandgap, thus a large absorption coefficient and electron diffusion length of several micrometers. The p-type property is the result of intrinsic defects related to Cu deficiencies.

Figure 4.11 shows a typical CIGS thin-film solar cell composed of a glass substrate, a molybdenum layer (Mo) as the electric back contact, a p-type CIGS absorber layer, and a thin n-CIGS layer. The p-n junction is formed by stacking a thin cadmium sulfide (CdS) buffer layer, TCO layer to extend the n-type region, intrinsic zinc oxide (ZnO) layer, and Al-doped ZnO layer as n-type TCO as the transparent front contact for the solar cell.

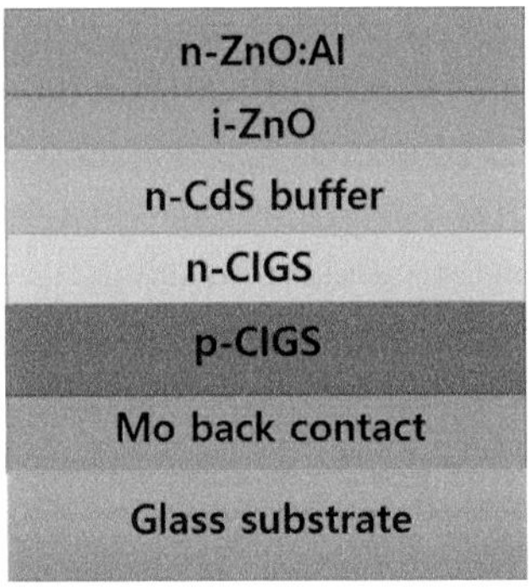

Figure 4.11 Configuration of a CIGS solar cell.

The light enters the cell from the top ZnO layer. The p-type CIGS absorber has a bandgap of 1.1–1.2 eV, using $Cu(In_xGa_{1-x})Se_2$ with $x{\sim}0.3$. The n-type CdS buffer layer has a bandgap of 2.5 eV. Since, n- and p-type materials have different bandgaps, CIGS devices can be considered heterojunctions. The high bandgap of ZnO (3.2 eV) reduces the parasitic absorption loss in the cell. In p-CIGS (Cu-deficient) layer, the Shockley–Read–Hall recombination (SRH) is dominant in bulk while in Cu-rich CIGS films the SRH recombination is dominant at interfaces of CIGS/CdS. The low recombination results in a higher bandgap and higher open-circuit voltage. A low amount of contamination (i.e., Na, K) helps to reduce the recombination in p-type CIGS materials [13].

CIGS films can be fabricated with different technologies. One method is co-evaporation of Cu, Ga, In, and Se precursors onto a heated substrate under vacuum conditions. Another approach is sputtering Cu, Ge, and In at ambient temperature on the substrate, followed by thermal annealing under the presence of selenium or sulfur vapor, to form the CIGS structure. The last step is called the selenization/sulfurization process. In the sputtering method, because of a variety of complex reactions taking place in the selenization step, the properties of the final CIGS are hard to control. An approach is the wafer bonding technique where two different films are coated onto a substrate and a superstrate, and under high pressure, two films are pressed together, while annealing the superstrate releases the film, and a CIGS forms on the substrate.

4.2.4 Cadmium Telluride (CdTe) Solar Cells

Cadmium telluride (CdTe) is a II-VI semiconductor, formed in a zinc blende crystal lattice structure where each Cd atom is bonded to four Te atoms. CdTe has a direct bandgap of 1.44 eV, very close to the optimum semiconductor bandgap for single-junction cells, which is beneficial for a high absorption rate. Cd atoms in the CdTe crystal network are II valence and can be replaced with III, and VI valence Te atom can be replaced with VII valence atoms and form n-type CdTe. III valence atoms like Al, Ga, or In and VII valence atoms like F, Cl, Br, or I act as shallow donors.

Cd(II) and Te(VI) atoms also can be replaced with I-valence (Ag, Au) and V-valence (N, P, Ar) atoms, respectively to form p-type CdTe materials. The p-type CdTe is used for solar cells. CdTe solar cells contains four layers of 1) TCO, front contact (SnO_2, Cd_2SnO_4), on glass substrate, 2) n-type CdS (bandgap ~2.4 eV) layer, 3) p-type

CdTe (bandgap ~1.44 eV) layer, and 4) Cu or Sb based contact. Since the bandgap n-CdS and p-CdTe are different CdTe thin-film cell is a heterojunction, the same as CIGS.

One of the major concerns about CdTe thin-film solar cells is the toxicity of Cd. However, Cd atoms in the form of CdS or CdTe are less toxic but Cd contamination is one challenging factor in this class of solar cell.

4.3 Third Generation: Organic, Quantum Dot, Organometallic Solar Cells

The third-generation solar cells are based on solution process technology, which is most likely to promote solar electricity generation on a large scale. Emerging photovoltaics include organic solar cells, quantum dot solar cells, dye-sensitized solar cells, and perovskite solar cells. Research achievements in this class of solar cell, especially, perovskite solar cells have significantly improved cell efficiency by over 25%. Materials in these solar cells offer a wide spectrum with inexpensive production processes.

4.3.1 Organic Solar Cells

In organic solar cells, the absorber materials are either conductive organic polymers or molecules based on carbon. Conductive organics are large conjugated systems, in which the carbon atoms have an alternative single bond and double bonds with p orbitals containing delocalized electrons. In such carbon-based conjugated materials, the delocalized electrons can move through the atom in the chain which has a p orbital. Figure 4.12 shows some organic compounds for PV application.

Organic solar cell materials are large conjugated organics, in which the valence electron in the p orbital is shared over all alternating carbons in the chain with p orbitals. In the conjugated carbon materials, the electrons in the p orbital do not belong to a single atom but can be shared with a group of atoms. All these carbon atoms have three equivalent sp^2 hybrid orbitals with a bond angle of 120°. Later we will discuss carbon orbitals in detail [14, 15].

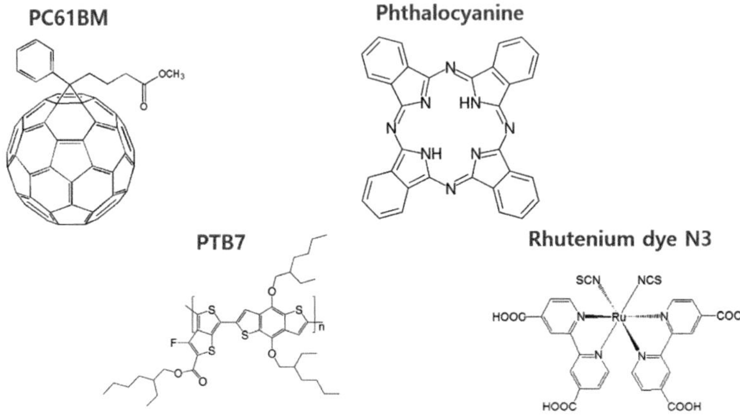

Figure 4.12 Examples of organic molecules for organic photovoltaics.

Figure 4.13 (a) shows the structure of an organic solar cell, consisting of organic acceptor material and organic donor materials. Same as semiconductors, donor and acceptor materials create a heterojunction. Typical diffusion lengths in organic materials are very short (~10 nm). Hence, device performance is strongly limited by the thickness of the cell. Due to this limiting factor, organic solar cells perform better with bulk heterojunction configuration (see Fig. 4.13 (b)), where both electron-donor and -acceptor are blended in one layer. Blending both electron-donor and -acceptor materials improves the likelihood that the excitons (electron–hole pairs) reach the interface, where they are separated and collected in respective contacts. Organic solar cell technology can be cheap, but the organic absorber materials need to be protected from humidity, moisture, and air, which restricts their industrial application [16].

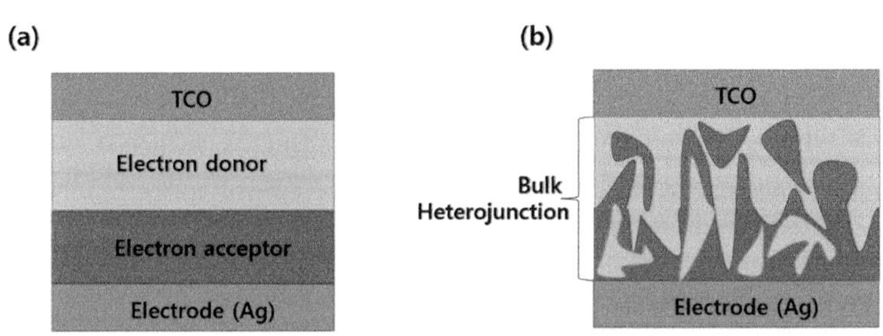

Figure 4.13 Schematic illustration of (a) the layered organic solar cells and (b) bulk heterojunction organic solar.

4.3.2 Quantum Dot Solar Cell

4.3.2.1 Quantum dots

The concept of nanotechnology was first introduced by Richard Feynman's talk in 1959, to the American Physical Society, "There's Plenty of Room at the Bottom." Since then, nanoscience and nanotechnology attracted enormous attention, especially in materials science and technology. Nanomaterials refer to materials with at least one dimension less than 100 nm. For their small size, nanomaterials show different chemical, physical and electrical properties from bulk materials. Quantum dots (QDs) are one type of nanomaterials with great potential for scientific and industrial applications. By definition, QDs are nanoparticles or nanocrystals of a semiconducting material, typically ranging between 2 and 10 nm in diameter. Due to their extra-small size, QDs exhibit different properties such as large extinction coefficient, luminescence properties, tunable bandgap, multiple exciton generation, and so on, which make them outstanding candidates for research explorations in optoelectronics, photovoltaics, biomedicine, and environment, and catalysts, etc. [17].

4.3.2.2 Quantum confinement effect

In QD semiconductors, because of dimension restrictions, the charge movement is confined to a small area, causing quantum confinement. Thus, quantum confinement is the spatial confinement of excitons in one or more dimensions within a material. The quantum confinement effect is observed when the size of the particle is too small to be comparable to the wavelength of the electron. The quantum confinement effect causes a large bandgap with observable discrete energy levels. By reducing the QD size the bandgap gradually increases, resulting in a blue shift in absorption and emission bands (Fig. 4.14).

4.3.2.3 QD-sensitized solar cells

Owing to different electrical and optical properties such as broad and tunable bandgap and low-cost fabrication and multiple exciton generation, QDs have attracted more attention as photoactive materials for PV applications. Figure 4.15 shows a schematic structure of quantum dot sensitized solar cells (QDSCs). QDSCs are composed of photoanode, QD sensitizer, electrolyte, and counter electrode. By receiving incident light, photo-generated electrons are injected into a wide-bandgap semiconductor and flow toward the electrode.

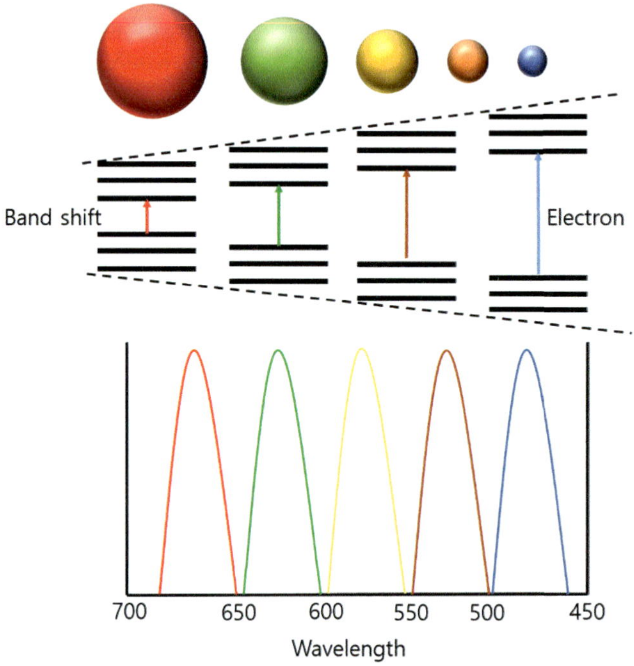

Figure 4.14 Quantum confinement effect, change of bandgap, and optical properties of QDs by size.

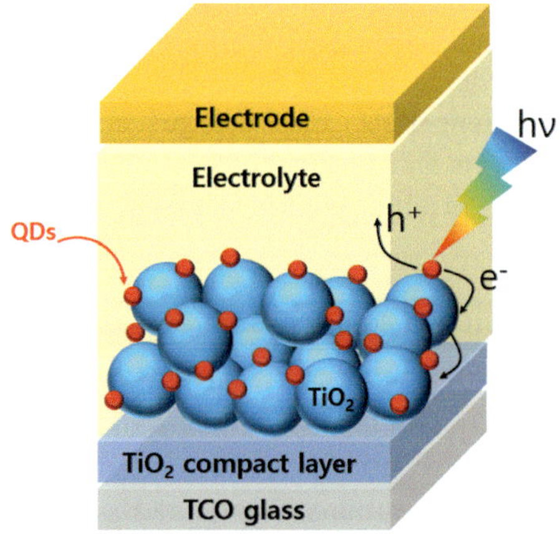

Figure 4.15 Schematic structure of QDSCs.

In QDSCs, the photoanode is composed of mesoporous nanoparticles of wide bandgap semiconductor material (e.g., TiO_2 and ZnO). The mesoporous nanoparticles provide a high surface area capable of large adsorption of the QD sensitizer. The common choices of QDs as sensitizers are CdSe, CdS, PbS, PbSe, HgS, or core–shell type CdS/CdSe or alloyed $CdSe_xTe_{1-x}$. However, because of the toxicity of Pb, Cd, and Hg, other QDs such as $CuInS_2$ and $CuInSe_2$ also have been widely experimented with.

Generally, the commonly used I^-/I^-_3 electrolyte used in DSSCs is not suitable for QDSCs, because the redox reaction can cause corrosion on the photoanode. Instead, an aqueous solution of polysulfide redox electrolyte showed promising performance in regeneration QD efficiency. However, still there are a few limitations regarding electrolytes. For example, the redox potential of polysulfide electrolytes has a highly negative value for lowering the photovoltage. Therefore, new types of electrolytes have been introduced, such as organic McMT$^-$/BMT electrolyte, $[(CH_3)_4N]_2S/[(CH_3)_4N]_2S_n$ as modified polysulfide electrolyte, etc. However, the overall performance of new electrolytes still cannot compete with conventional polysulfide electrolytes, thus the search for suitable electrolytes is a major challenge.

The counter electrode is responsible to collect the charges from the external circuit and cause oxidation in the electrolyte. For this role, the counter electrode should be catalytically active. For common polysulfide electrolytes, metallic chalcogenides are usually used as counter electrodes. Compounds such as CoS, Cu_2S, and $CuInS_2$ are catalytically active materials to reduce polysulfide (S_n^{2-}) electrolytes. Cu_2S-based electrodes showed the highest performance as counter electrodes in QDSCs. QDSCs will be discussed in detail in Chapter 6.

4.3.3 Organic–Inorganic Hybrid Solar Cells

4.3.3.1 Dye-sensitized solar cells

Dye-sensitized solar cells (DSSCs) have been classified as third-generation solar cells for their flexibility and transparency, which are the characteristics that do not exist in the first- and second-generation devices using single-crystalline silicon or thin films. The

DSSCs substantially imitate the photosynthesis process, in which a pigment absorbs light through chlorophyll, and photo-induced electron transfer produces fuel (i.e., starch). In the DSSCs, dye absorbs light, and charge generation yields the PV effect.

The efficiency of DSSC can be improved by optimizing the porosity of metal oxide electrode. The most common type of DSSC (n-type DSSC) contains a transparent electrode covered by a mesoporous layer of TiO_2 nanoparticles (mp-TiO_2), molecular dye absorbed on the mp-TiO_2 that receives sunlight, and an electrolyte solution that is injected into the cell to immerse TiO_2/dye, a Pt-based catalyst as the cathode. Electrodes are sealed together, and the electrolyte is a liquid conductor between the electrodes.

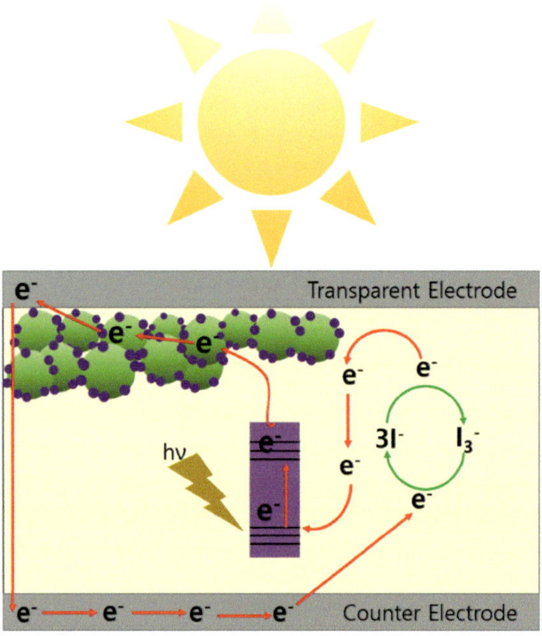

Figure 4.16 Operation processes of DSSCs.

The working principles of DSSCs are (i) light enters through a transparent electrode, (ii) dye absorbs light and excites an electron, (iii) photo-generated electrons injected into the conduction band of TiO_2, (iv) electrons flow through TiO_2 nanoparticles to transparent electrode, where electrons are collected and conducted through an external circuit, (v) electrons re-introduced into the cell through metal electrode (counter electrode) and into the electrolyte, (vi) at the end the electrolyte transports the electrons to the oxidized dye and regenerates the dye molecule. Figure 4.16 shows the operation processes of DSSCs [18–20]. In a p-type DSSC, the semiconductor is the p-type metal oxide (i.e., NiO), and instead of injecting electrons holes are generated and injected from the dye into the valence band of the p-type metal oxide. DSSCs will be explained in detail in Chapter 5.

4.3.3.2 Perovskite solar cells

Perovskite solar cells are one of the promising hybrid organic–inorganic solar cells that develop high and fast in the photovoltaics community. Operation methods are still new, but with more than a decade of the engineering of perovskite materials and improvement of devices performance, the power-conversion efficiency has significantly increased by more than 25% as of June 2021. Perovskite is a material with the same crystalline structure as calcium titanium oxide ($CaTiO_3$), the first perovskite crystal discovered by Russian mineralist Lev Perovskie in 1839. Generally, perovskite compounds have a chemical formula ABX_3, where 'A' and 'B' represent cations and X is an anion that bonds to both. The perovskite lattice arrangement is demonstrated in Fig. 4.17. Perovskite has a cubic structure, where large organic cations (A^+) are in corners of the cube. The center of the cube is occupied by inorganic cation (B^+) and the faces of the cube are occupied with negatively charged halogen anions (X^-) [21].

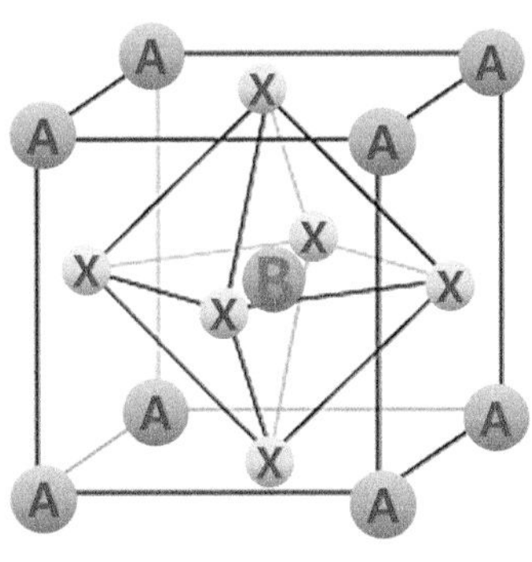

A = Cation - methylammonium ($CH_3NH_3^+$), formamidinium ($NH_2CHNH_2^+$), Cs^+
B = Inorganic cation - lead(II) (Pb^{2+}), Tin(II) (Sn^{2+})
X_3= Halogen anion – Br, Cl^-, I^-

Figure 4.17 Crystallographic structure as perovskite with the generic form of ABX_3 [21].

The first perovskite solar cells were based on solid-state DSSCs, with mesoporous TiO_2 as an electron-transport layer (ETL) scaffold. Many devices adopt the same architecture with TiO_2 mesoporous and also Al_2O_3 layer as interfacial engineering, but because of the high-temperature processing and UV instability of TiO_2, planar structure devices are similar to thin-film solar cells were introduced. Figure 4.18 shows the planar (either conventional or inverted) and the mesoporous (conventional) structures of perovskite solar cells [20].

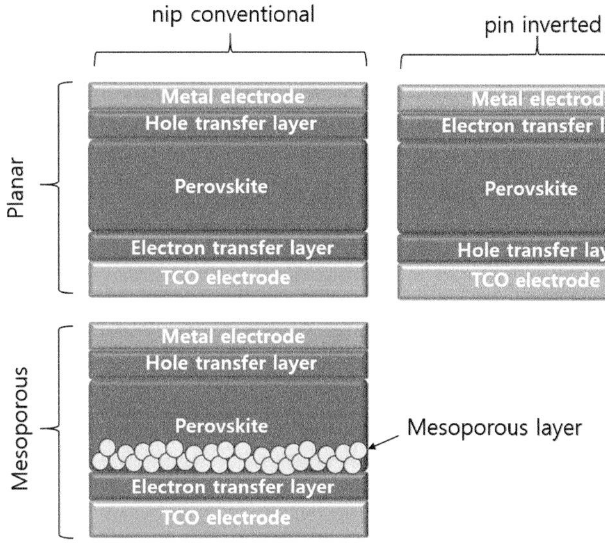

Figure 4.18 Structures of planar (conventional and inverted) and mesoporous (conventional) perovskite cells.

Most perovskite solar cells are based on the configuration of the TCO electrode/ETL/perovskite/HTL/metal electrode. TCO, ETL, and HTL represent transparent conductive oxide, electron-transport layer, and hole-transport layer, respectively. The HTL material is typically Spiro-OMeTAD or PEDOT:PSS and the ETL materials are mostly semiconductor metal oxides such as TiO_2 or SnO_2 or conductive organic PCBM. A well-optimized energy band design that considers interfacial interactions and interface engineering in perovskite solar cells is critical to the final device performance. In perovskite solar cells, the photoactive light-harvesting perovskite layer must be several hundreds of nanometers (thicker than organic

PV) with highly uniform grains. Extensive research work led to the high efficiency of over 25% using solution-based methods, showing that a large-scale fabrication of perovskite solar cells is very feasible. Perovskite solar cells will be described in detail in Chapter 8.

References

1. SAGA, T. (2010) *NPG Asia Mater.*, **2**, pp. 96–102.
2. Dale, B. and Rudenberg, H. G. (1960) High efficiency silicon solar cells, in *Proceedings of the 14th Annual Power Sources Conference*, p. 22.
3. Abdullah, M. F., Alghoul, M. A., Naser, H., Asim, N., et al. (2016) *Renew. Sustain. Energy Rev.*, **66**, pp. 380–398.
4. Räisänen, J., Hirvonen, J., and Anttila, A. (1981) *Solid-State Electron.*, **24**(4), pp. 333–336.
5. Trube, J., Fischer, M., Erfert, G., Li, C. C., et al. (2018) International technology roadmap for photovoltaic (ITRPV) 2019: VDMA photovoltaic equipment, *Solar Industry Reports,* **24**, pp.77–97.
6. Amorphous Silicon. https://en.wikipedia.org/wiki/Amorphous_silicon
7. Smets, A. H. M. and van de Sanden, M. C. M. (2007) *Phys. Rev. B*, **76**, pp. 073202.
8. Chopra, K. L., Paulson, P. D., and Dutta, V. (2004). *Prog. Photovolt.: Res. Appl.*, **12**, pp. 69–92.
9. Shah, A., Torres, P., Tscharner, R., Wyrsch, N., and Keppner, H. (1999) *Science*, **285**, pp. 692–698.
10. Poortmans, J. and Arkhipov, V. (eds.) (2006) *Thin Film Solar Cells: Fabrication, Characterization and Applications*, Vol. 18. John Wiley & Sons.
11. Hegedus, S. S. and Shafarman, W. N. (2004) *Prog. Photovolt.: Res. Appl.*, **12**, pp. 155–176.
12. Vigil-Galán, O., Courel, M., Andrade-Arvizu, J. A., Sánchez, Y., et al. (2015) *J. Mater. Sci.: Mater. Electron.*, **26**, pp. 5562–5573.
13. Kosyachenko, L. A. (ed.) (2011) *Solar Cells: Thin-Film Technologies*. BoD - Books on Demand: Norderstedt, Germany.
14. Dyndał, K., Sanetra, J., and Marszalek, K. W. (2021) Study of the second-generation of CdTe and CIGS thin film PV modules under natural sunlight conditions. *19th International Conference on Renewable Energies and Power Quality* (ICREPQ'21) Almeria (Spain).

15. Green, M. A. (2002) *Phys. E: Low-Dimens. Syst. Nanostructures*, **14**(1–2), pp. 65–70.
16. Nozik, A. J., Beard, M. C., Luther, J. M., Law, M., et al. (2010) *Chem. Rev.*, **110**(11), pp. 6873–6890.
17. Yan, J. and Saunders, B. R. (2014) *RSC Adv.*, **4**(82), pp. 43286–43314.
18. Mingsukang, M. A., Buraidah, M. H., Arof, A. K., and Das, N. (2017) *Nanostructured Solar Cells.* InTechOpen, pp. 1–9.
19. Dharmadasa, I. M. (2005) *Sol. Energy Mater. Sol. Cells*, **85**(2), pp. 293–300.
20. Hodes, G. (2013). *Science*, **342**(6156), pp. 317–318.
21. Yi, Z., Ladi, N. H., Shai, X., Li, H., Shen, Y., and Wang, M. (2019). *Nanoscale Adv.*, 1, 1276–1289.

Problems

4.1 What is the first generation of solar cells?

4.2 What are the components of a typical c-Si cell?

4.3 What is the difference between monocrystalline and polycrystalline solar panels?

4.4 What are the causes of optical losses in solar cells and what are the ways to reduce the optical loss?

4.5 One of the processing steps of Si solar cells is texturing. Explain the texturing process.

4.6 Explain the diffusion and ion implantation process in Si solar cells?

4.7 Briefly introduce thin-film solar cells?

4.8 Briefly explain the difference between c-Si, a-Si, and H-Si.

4.9 Briefly introduce the third generation of solar cells?

4.10 What is quantum confinement and its effect on the light absorption efficiency of materials?

4.11 A solar cell with an aperture area of 0.09 cm^2 can obtain a maximum power point (MPP) of 20 W under one sun illumination (1000 W/m^2), and its open-circuit voltage and current of 1.17 V and 2.07 mA. Calculate solar cell fill factor and efficiency?

Chapter 5

Organic Solar Cells

In 1954, Akamatu et al. first found high conductivity in some complexes between polycyclic aromatic compounds and halogens. Especially, the perylene–bromine complex is stable and has good electrical conductivity [1]. Afterward, organic semiconductors were used in the fabrication of organic light-emitting diodes (OLEDs) by Tang et al. in 1970 [2]. More importantly, organic solar cells (OSCs) were first realized by Ghosh et al. in the 1970s. They used tetracene as the active layer sandwiched between Al and Au metal electrode (i.e., Al/tetracene/Au structure), and obtained less than 10^{-3}% in PCE [2].

In 1985, Tang first reported bilayer OSCs utilizing two thin organic layers as an electron donor and an electron acceptor, yielding about 1% efficiency [3]. Two-layer OSCs significantly improved interface contacts and charge generation, resulting in a higher fill factor (FF). Such an architecture also is called a heterojunction that combines donor-acceptor materials. This work provided an important step in designing an efficient device architecture to obtain high performance of OSCs. Beyond that, OSCs present several advantages such as low-cost, lightweight, ease of fabrication, flexibility as well as tunability of their optical and electronic properties. The OSCs also can achieve colorful and semitransparent building-integrated PVs (BIPVs).

Next-Generation Solar Cells: Principles and Materials
Yoon-Bong Hahn, Tahmineh Mahmoudi, and Yousheng Wang

ISBN 978-981-4968-66-9 (Hardcover), 978-1-003-37238-7 (eBook)
www.jennystanford.com

5.1 Organic Semiconductors

Organic semiconductors are a class of material based on π-conjugated oligomers and polymers. Atoms in organic semiconductors are bonded by conjugated π-bonds with weak van der Waal's force, to provide inherent flexibility, lightweight, and low sublimation point. These characteristics indicate that organic semiconductors enable low-cost and flexible displays, lighting panels, semitransparent solar cells, or other optoelectronic devices through low-temperature solution processes.

In organic semiconductors, the splitting of the highest occupied molecular orbital (HOMO) and lowest unoccupied molecular orbital (LUMO) levels due to the interaction of adjacent chains along given directions yields the transfer integral to be used for the description of hole (electron) transport in these directions [4]. The HOMO and LUMO of organic semiconductors are similar to the valence band and conduction band in inorganic semiconductors, respectively. According to the electron wave function overlap of neighboring atoms, the π-bond system in organic semiconductors shows distinctive bond structures. As shown in Fig. 5.1, the HOMO and LUMO of organic semiconductors represent the hybridization between bonding (π-orbital) and anti-bonding (π*-orbital) states corresponding to different energy levels.

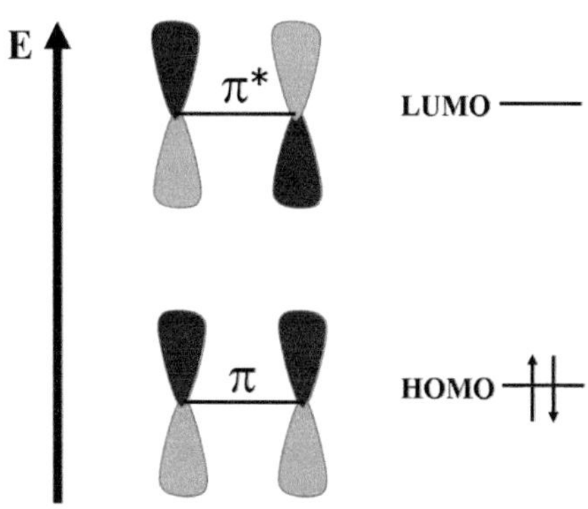

Figure 5.1 Illustration of bonding and anti-bonding interactions between the HOMO and LUMO levels of organic semiconductors.

5.2 Basic Operation Principles and Physical Mechanism

5.2.1 Absorption and Exciton

Upon the absorption of photons, electrons in the donor materials (or active layer) are excited from the HOMO to the LUMO as shown in Fig. 5.2 (step 1). This process is analogous to exciting electrons from the VB to CB in inorganic semiconductors. Interestingly, electrons and holes are free to move in inorganic semiconductors after absorbing photons. However, the exciton formation is bound to electron–hole pairs in organic semiconductors, which is mainly attributed to the smaller dielectric permittivity and localized electron and hole wavefunctions. Besides, the excitons have a relatively large binding energy of 0.1–1.4 eV, which is opposite to a much lower binding energy in inorganic semiconductors [5]. Thus, compared to free electron–hole pairs in inorganic semiconductors, the exciton (bound electron–hole pair) in organic semiconductors needs to overcome the binding energy and dissociate into free charges.

It is noteworthy that the absorption coefficient of organic semiconductors is usually high at $\sim 10^5$ cm^{-1}, which can have an efficient light-harvesting for a few hundreds of nanometers (100–200 nm) of the active layer [6]. However, organic semiconductors usually have narrow absorption spectra, leading to low absorption efficiency of photons in the long-wavelength region. Therefore, to absorb the light spectrum range as much as possible, semiconductor bandgap tuning is important.

5.2.2 Diffusion and Dissociation

As-formed excitons can further diffuse to the donor-acceptor interface and form charge-transfer states (Fig. 5.2, step 2), which is attributed to the interfacial electronic interactions at the heterojunction. The hole and electron are located at the HOMO of the electron donor and the LUMO of the electron acceptor under the state, respectively. However, they are still bound together by the electrostatic force. After overcoming the binding energy (around 0.1–0.5 eV) of charge-transfer excitons [7], the bound electron–hole pairs dissociate into free charge carriers, as shown in Fig. 5.2, step 3.

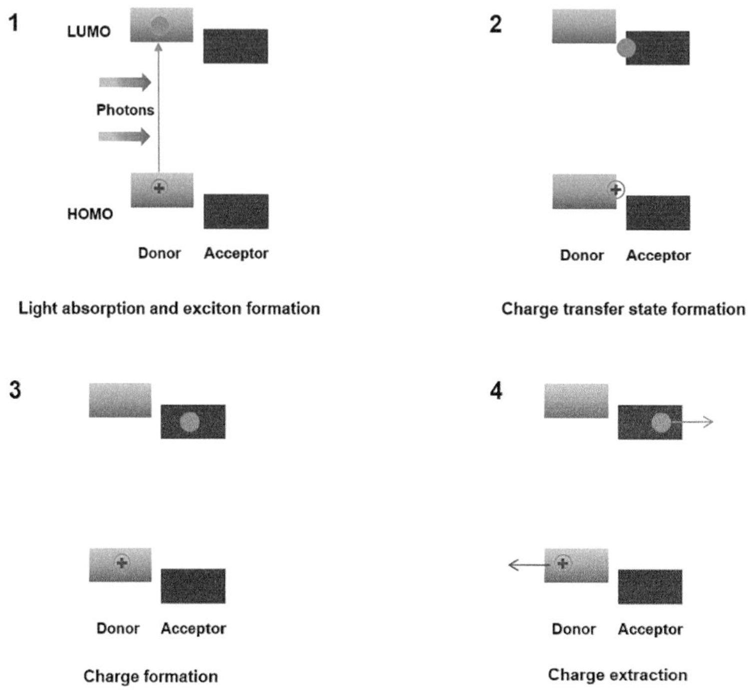

Figure 5.2 Illustration of working principles in donor and acceptor layers.

After dissociation of the bond between electron–hole pairs, charge carriers driven by the electric field move to the respective electrodes and are extracted by the external circuit (Fig. 5.2, step 4). An efficient exciton dissociation is vital to suppress electron–hole recombination. If the lifetime of excitons generated at a distance from the heterojunction is longer than exciton diffusion length (i.e., the distance that excitons can diffuse before recombination), exciton dissociation efficiency will be reduced. Therefore, the absorber active layer should be thin enough to guarantee that phase separation between the donor and acceptor is within the exciton diffusion length. However, a thin active layer will reduce light absorption efficiency, resulting in lower photocurrent density. To solve this issue, it is important to enlarge the interface contact area between the donor and acceptor. Thus, the bulk heterojunction and nanostructured active layers have been developed, which will be discussed in the next section.

5.3 Organic Solar Cell Configurations

5.3.1 Planar Solar Cells

The first planar OSCs were fabricated based on a single p-type organic film as an active layer sandwiched between anode and cathode. However, such a simplest solar cell configuration shows very low efficiencies ($\sim 10^{-3}\%$) due to the short exciton diffusion lengths in the single organic film, interfacial defects, and energy-level diagram mismatch. After that, a bilayer planar solar cell was developed to improve efficiency.

Figure 5.3a shows a typical configuration of a bilayer planar solar cell consists of an anode, an active layer composed of donor and acceptor layers, and a cathode. More importantly, a bilayer heterojunction is formed at the donor-acceptor interface, which facilitates charge formation and extraction. Although bilayer planar solar cells showed improved efficiency by around 1%, they still have certain challenges and limitations. For example, the limited interfacial area results in a low photocurrent; the short exciton diffusion length of organic materials limits the thickness of bilayers. If the thickness of the donor or acceptor layer is too high, the excitons generated far away from the heterojunction will lead to recombination before reaching the heterojunction. Also, to guarantee sufficient light absorption by organic materials, the thickness of donor and acceptor layers should be at least 100 nm, however, the exciton diffusion length of organic materials is only ~10 nm. For this reason, the bulk heterojunction concept is proposed to solve these problems.

5.3.2 Bulk Heterojunction Solar Cells

To overcome issues related to the limited exciton (tightly bound electron–hole pair) diffusion length in planar solar cells, a bulk heterojunction (BHJ) composed of donor-acceptor mixtures was explored (Fig. 5.3b). This concept was first illustrated by Hiramoto et al. by co-evaporation of donor and acceptor molecules, which emphasized the importance of large interfacial contacts between donor and acceptor for the effective charge photogeneration [8]. In 1995, Heeger et al. first reported efficient BHJ polymer solar cells

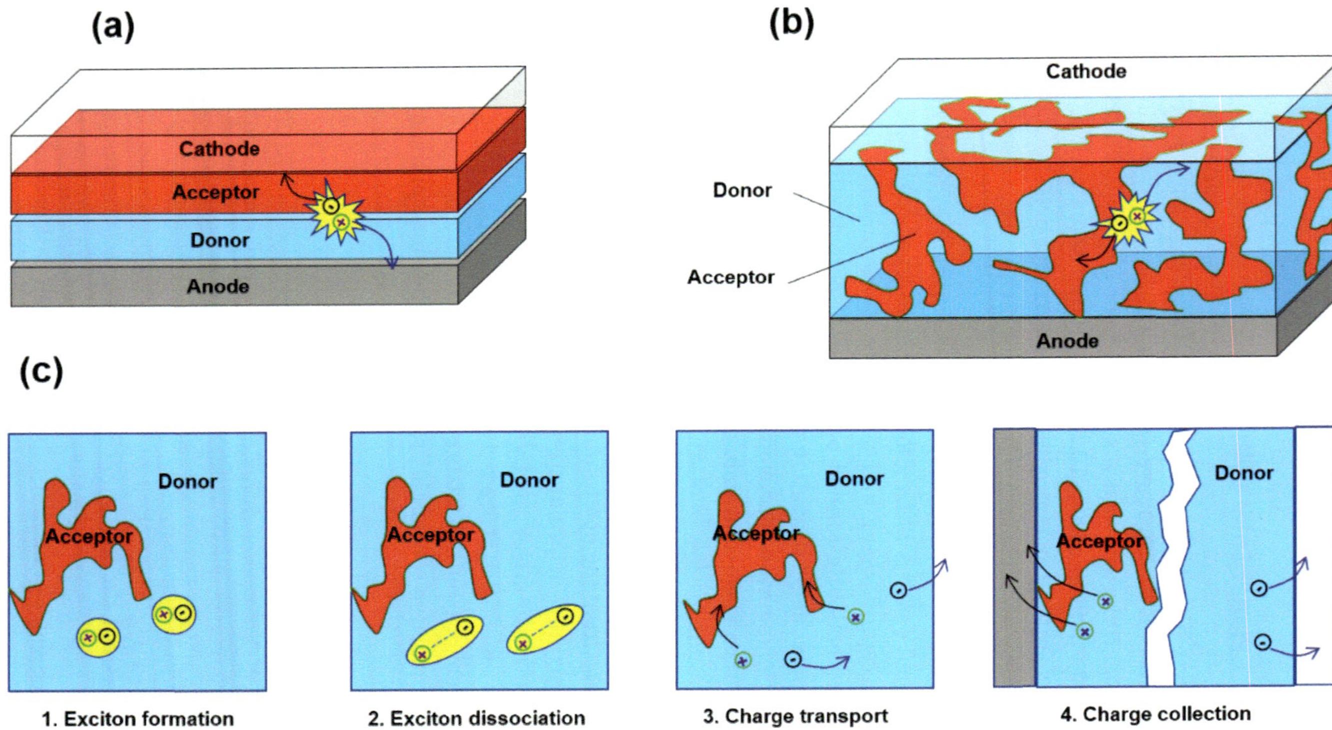

Figure 5.3 OSCs of (a) basic planar structure, (b) bulk heterojunction structure, and (c) corresponding work principles in bulk heterojunction structure.

by mixing polymer and fullerene [9]. In the same year, Halls et al. developed a polymer- and a polymer mixture–based BHJ solar cells and large-area photodetector [10]. Over the past three decades, the invention of the BHJ concept was a milestone in the OPVs field. Over the past two decades, numerous efforts have been dedicated to the development of BHJ solar cells which achieved a certified efficiency of 18.2% [11]. The BHJ solar cells have the opportunity to achieve over 20% or higher efficiencies in the near future. More importantly, there have been various methods, such as inkjet printing, screen printing, spray coating, roll-to-roll printing, flexographic printing, etc., which enable large-area solar modules.

As shown in Fig. 5.3b, a preferred morphology of BHJ should be a bi-continuous interpenetrating network with sufficient interfacial areas for efficient exciton dissociation. Usually, donor and acceptor domains are twice the size of the exciton diffusion length (~10 nm), which facilitates excitons to diffuse into the BHJ interface and thus obtains efficient exciton diffusion and charge separation [12]. The working principle consists of four steps (Fig. 5.3c): (1) light absorption to form exciton (electron–hole pairs); (2) charge separation and diffusion to the donor-acceptor interface; (3) hole-electron transportation; (4) charge collection by an external circuit. Usually, the donor material is divided into two categories: π-conjugated polymers and π-conjugated small molecules. By increasing the interfacial area between the electron donor and electron acceptor, BHJ solar cells overcome the issues of short exciton diffusion length, limited exciton lifetime, and charge separation that limits bilayer junctions. Until now, all kinds of BHJ solar cells have been explored, including polymer solar cells (polymer-fullerene or polymer-non-fullerene), all-polymer solar cells, ternary solar cells, and hybrid solar cells, which will be discussed in detail in the next sections.

5.3.3 Polymer Solar Cells

Polymer:fullerene bulk heterojunction. Most BHJ polymer solar cells use a conjugated semiconductor polymer as a photoactive electron donor and a non-photoactive fullerene derivative as an acceptor, i.e., polymer:fullerene BHJ solar cells, which is also called the fullerene generation of BHJ solar cells. In the first report of solution-processed BHJ solar cells, Yu et al. [9] used a semiconductor polymer poly(2-methoxy-5-(2′-ethyl-hexyloxy)-1,4-phenylene vinylene) (MEH-PPV) blending with a highly soluble fullerene derivative and phenyl-C_{61}-butyric acid methyl ester (PCBM) to achieve sufficient percolation of charges through both components. As a result, a 7.4% efficiency was achieved under 20 mW/cm^2 light intensity. This performance marked a significant increase over single polymer and planar solar cells.

In the fullerene generation of BHJ solar cells, PCBM is one of the best performing and most popular electron acceptors due to its excellent electron-transporting properties and the favorable bulk heterojunction morphology from optimal combinations of polymer donors and fullerene acceptors [27, 28]. Although MEH-PPV:PCBM system-based BHJ solar cells have improved efficiency, a higher PCE was limited due to low charge-transport mobility in PPV-type polymers. In 2002, Brabec et al. reported a BHJ solar cell based on a blend of poly-3(hexylthiophene) (P3HT) with PCBM. It is noteworthy that this type of solar cell has a high external quantum efficiency (EQE) with high photocurrent density [13]. By controlling the morphology of P3HT:PCBM BHJ structures using annealing temperature and solvent additives, the efficiency of the solar cells has increased substantially, approaching 5% [14]. As shown in Fig. 5.4a, thermal annealing can soften the P3HT matrix, which allows PCBM molecules to diffuse out of disordered P3HT clusters and form larger fullerene aggregates. Besides, the P3HT aggregates can be converted into P3HT crystallites in PCBM-free regions [15]. Thus, the improved efficiencies are mainly attributed to the effective phase separation and compositional gradient of P3HT and PCBM, which can facilitate efficient exciton dissociation and charge transport, respectively. By rational synthesis and design of donor and acceptor, the efficiencies of more than 10% in polymer:fullerene BHJ solar cells have been achieved.

Polymer:non-fullerene bulk heterojunction. Due to the drawbacks of fullerene acceptors including poor absorption in the visible region, limited synthetic tunability, and energy-level, non-fullerene acceptors (NFAs) have attracted considerable attention to fabricate high-performance BHJ solar cells. Compared to the fullerene acceptors, the n-type NFAs show distinct advantages,

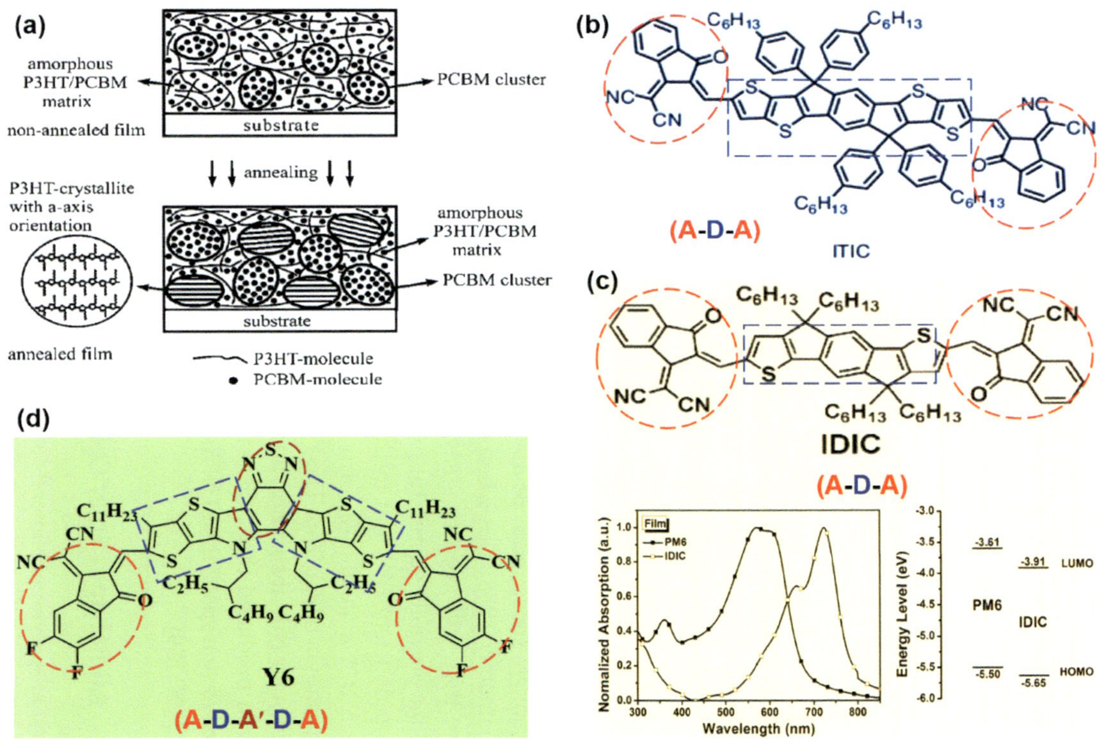

Figure 5.4 (a) Structural changes in P3HT:PCBM films before and after thermal annealing (Reprinted with permission from [15]. Copyright 2015. American Chemical Society). (b) Molecule structure of non-fullerene electron acceptor (ITIC) with acceptor-donor-acceptor (A-D-A) structure. (c) Non-fullerene electron acceptor (IDIC), UV–Vis spectra, and energy-level diagram. (d) A novel non-fullerene acceptor with a D-AA′D-A structure. Reprinted with permission from [19], Copyright 2019, Elsevier.

such as highly tunable optical absorption and energy levels, and strong absorption in the near-infrared (NIR) region. Zhan et al. first reported a new class of NFAs (ITIC, (3,9-bis(2-methylene-(3-(1,1-dicyanomethylene)-indanone))-5,5,11,11-tetrakis(4-hexylphenyl)-dithieno[2,3-d:2′,3′-d′]-s-indaceno[1,2-b:5,6-b′]dithiophene)), which successfully fabricated high-performance fullerene-free BHJ solar cells [16]. Such a type of acceptor broke the bottleneck of fullerene-based BHJ solar cells and boosted the rapid development of NFAs due to their excellent optical and electronic properties. In general, the desired NFAs should have HOMO and LUMO energy levels adjusted with respect to the energy level of the used donor material.

The most widely used NFAs are the narrow-bandgap n-type organic semiconductor materials with an acceptor-donor-acceptor (A-D-A) structure, such as ITIC and its derivatives [Fig. 5.4(b,c)]. When ITIC or its derivatives are mixed with a D-A structural conjugated polymer (e.g., PBDB-T [17] and its derivatives, PM6 [18]), the blend films not only obtain wide and complementary absorption spectra of 300–800 nm but also have a matchable energy-level alignment (Fig. 5.4c), resulting in high photocurrent density, voltage, and PCE.

Although those new NFAs based solar cells have achieved approximately 13% PCEs when combined with suitable polymeric donors, they still keep challenge toward higher PCEs over 20%. Based on previous successful experience, Zou et al. developed an electron-deficient (A′) unit in the middle of the central fused ring of the A-D-A molecule to form an A-DA′D-A framework acceptor (Y6), as shown in Fig. 5.4d. The Y6 has more π-π stacking properties and stronger NIR absorption and enhanced charge transfer [19]. It has been predicted that PCE over 20% can be achieved when the rational design spectrum of NFAs is 780–950 wavelengths (bandgap 1.6–1.3 eV) [20].

5.3.4 All-Polymer Solar Cells

The first all-polymer solar cells were developed by Halls et al. using a mixture of two PPV polymers, donor MEH-PPV and acceptor CN-PPV [10]. In the last two decades, various donor/acceptor polymers have been developed and applied in the fabrication of all-polymer solar cells. Some ordinary chemical structures can be found in the literature

[21, 22]. Compared to other types of OSCs, all-polymer solar cells present some unique advantages: (1) high absorption coefficients in the visible spectral region; (2) tunable energy levels; (3) superior flexibility in polymer-polymer blends; (4) robustness of BHJ film morphology; (5) high stability of the device to environmental and mechanical stresses. These excellent properties enable efficient and stable flexible large-scale modules and wearable-based electronics.

Until now, polymer PM6 has been considered one of the most efficient donors to achieve highly efficient polymer solar cells, which was first reported by Zhang et al. [23]. In general, PM6 has a suitable optical bandgap of ~1.8 eV and a deep HOMO level of ~5.45 eV, which can achieve a high V_{oc}. In recent years, most researchers have focused on designing and exploring new types of narrow-bandgap acceptor polymers to combine with PM6 for higher efficiencies. For example, Yan et al. reported a PCE of 15.2% for all-polymer solar cells by utilizing a regio-regular polymer acceptor PYF-T-o with a stronger conjugated polymerization site to mix with donor PM6 [24]. Note that a record PCE of 16.2% was reported by a mixture of ternary PM6:L15:MBTI in 2021 [25]. Thus, efficiencies over 20% for all-polymer solar cells can be predicted by the rational design of polymer structures and device configurations.

One of the critical challenges for OPVs is how to enhance environmental stability. A major barrier to the stability of OPVs using fullerenes or small-molecule non-fullerenes in the active layer is BHJ chemical sensitivity, photo/thermal microstructural instability, and external mechanical stresses, which are attributed to molecular brittleness, diffusion, and poor adhesion, as well as photo/thermal-induced oxidation and self-aggregation processes. However, compared to other BHJ OSCs, all-polymer solar cells present good stability due to superior mechanical and morphological stabilities against thermal annealing, light exposure, and donor/acceptor compositional deviations.

5.3.5 Ternary Polymer Solar Cells

According to previous knowledge and experience in binary polymer solar cells, ternary solar cells (TSCs) containing three components in an active layer with wide absorption spectra present a simple and effective strategy to enhance the efficiency of BHJ polymer solar cells.

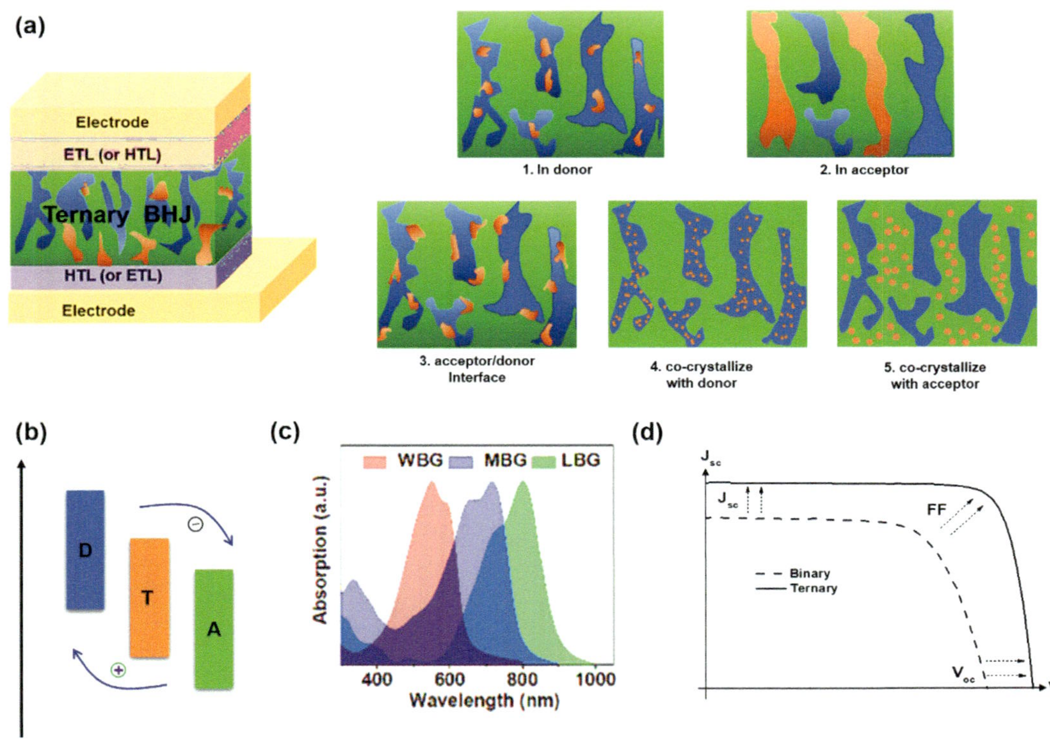

Figure 5.5 (a) Basic configuration of BHJ-TSCs and possible positions for the third component, (b) energy diagram level, and (c) UV–Vis spectra of donor, acceptor, and a third component. Reprinted with permission from [26], Copyright 2020, American Chemical Society. (d) Schematic illustration of *J–V* curves for binary and TSCs, presence of the third component may improve V_{oc}, J_{sc}, or FF.

Figure 5.5a shows a basic structure of TSCs with an acceptor, donor, and third component. Interestingly, the third component may locate at several different positions, such as mixture with donor (1) or acceptor (2), donor/acceptor interface (3), co-crystallize with donor (4), or acceptor (5), as shown in Fig. 5.5a. More importantly, the third component should have appropriate energy levels (Fig. 5.5b) and absorption spectra (Fig. 5.5c) for efficient charge transport and photon harvesting.

Besides, the third component also can be as morphology controller to enhance phase separation for efficient exciton dissociation and charge transport in ternary active layers. Thus, the higher V_{oc}, J_{sc}, and FF can be achieved when incorporating the third component than that of binary polymer solar cells, as shown in Fig. 5.5d. The enhanced V_{oc} is attributed to the formation of an energy cascade alignment. The third component can reduce trap density and increase charge mobility because of enhanced morphology and crystallization in the BHJ structure, resulting in improvement of FF. Note that high-performance TSCs may be achieved by further careful design of the three or more components in the BHJ structure in the near future.

5.3.6 Organic Tandem Solar Cells

Due to the tunable bandgap of organic semiconducting materials, organic tandem solar cells (OTSCs) are very popular for further improving the PCE of OPVs. Usually, a tandem architecture consists of two sub-cells, which can absorb a broader part of the light spectrum and reduce thermalization loss of photon energy. For organic semiconducting materials, it is easy to tune their light absorption wavelength and bandgaps by different donor-acceptor units or branched chains. Such properties enable highly efficient OTSCs with two or three sub-cells. Figure 5.6 shows the most basic OTSC configuration: normal and inverted structures, which involve stacking of two subcell units, one fronting anode, nominated as the bottom subcell, designed for capturing energetic photons in a short wavelength by means of wide-bandgap organics, and one directing the metal electrode denoted as top subcell and assembled for receiving photons with low energies occupying the long-wavelength region indulging low-bandgap absorbers. Two sub-cells are connected by a hole and electron-transport layers called the interconnection layer.

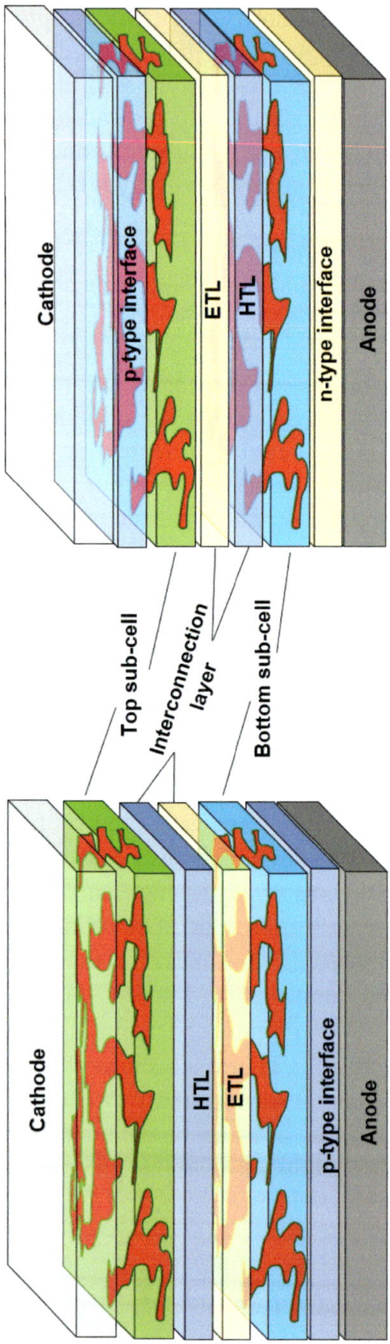

Figure 5.6 Schematic illustration of OTSCs with two different bandgap sub-cells: normal (left) and inverted configuration (right).

For a typical OTSC consisting of two sub-cells connected via an interconnection layer, working principles can be described briefly as follows: (1) absorption of photons and generation of free carriers; (2) collection of electrons from the bottom subcell and holes from the top subcell by the interconnection layer (and vice versa in an inverted device); (3) recombination of electron/holes in the interconnection layer; (4) collection of holes in the bottom subcell and electrons in the top subcell by the external circuit (and vice versa in an inverted device). One of the characteristics of a good OTSC is that the V_{oc} equals the sum of the V_{oc} values of the sub-cells and no V_{oc} loss occurs in the interconnection layer [29]. It is worth mentioning that OSCs also can combine with emerging perovskite solar cells (PSCs) to achieve high-performing organic-perovskite tandem solar cells, which will be discussed in detail in Chapter 11.

5.4 Charge Dynamics in Polymer Solar Cells

The understanding of charge dynamics in polymer solar cells is very important for further enhancing their efficiencies. The charge dynamics in polymer solar cells contain exciton dissociation, charge generation, and charge-recombination processes. Here, we will mainly focus on studying the charge-transfer (CT) electronic states formed at the D-A interfaces. The CT electronic states in polymer solar cells play a crucial role in exciton dissociation, charge generation, and charge recombination. Photo-induced excitons in the D and/or A components can dissociate at D-A interfaces, resulting in the formation of CT states. Over the past few decades, theoretical and experimental methods have been used to characterize CT states and to explain how factors such as the D and A components and exciton states influence the characteristics of the CT states.

5.4.1 Charge Dynamics Measurements

5.4.1.1 Transient absorption spectroscopy

It is important to choose an analysis technique for the charge dynamics characterization of polymer solar cells, which allows a good understanding of their working and energy loss mechanisms. The transient absorption spectroscopy (TAS) is the most powerful method for directly observing photovoltaic conversion events in polymer solar cells ranging from ~10^{-15} s for exciton generation to ~10^{-5} s for charge collection to the electrode [30]. Here, a typical example of P3HT/PCBM blend films will be analyzed by the TAS technique. Figure 5.7 shows transient absorption spectra (excited at 400 nm) of P3HT pristine and P3HT/PCBM blend films (broken curve) and decays of P3HT/PCBM blend film measured at 1030 nm and 480 nm (solid curves). A large absorption band can be observed at around 1000 nm in P3HT pristine film, which is attributed to polymer singlet excitons. The negative signal at around 500 nm is ascribed to photo-bleaching of the ground state, which indicates the generation of photo-excitons. However, the optical density (ΔOD) of the singlet exciton band in P3HT/PCBM blend films is reduced by almost half compared to the pristine P3HT film. Besides, for BHJ P3HT/PCBM blend film, new absorption bands can be observed at around 800 nm and 1600 nm due to the polymer polarons even at 0 ps. Note that the broadband of singlet excitons completely disappears at 0.2 ps, suggesting that there exists a rapid quenching pathway such as charge generation in BHJ P3HT/PCBM blend films.

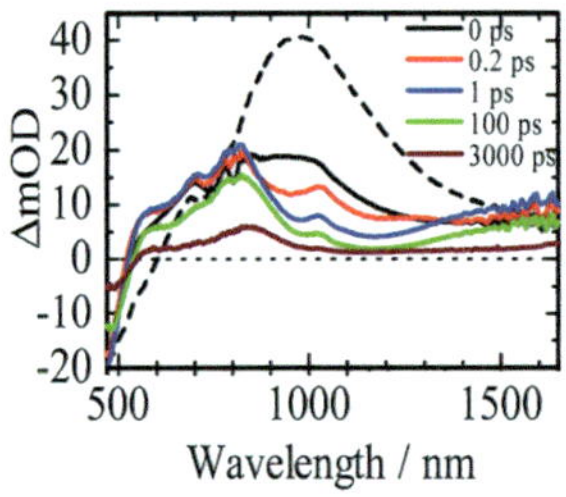

Figure 5.7 Femtosecond TAS of P3HT and P3HT/PCBM excited at 400 nm. Reprinted with permission from [31], Copyright 2010, American Chemical Society.

5.4.1.2 Transient photovoltage and photocurrent

Measurements of transient photovoltage (TPV) and transient photocurrent (TPC) are important in analyzing charge-carrier dynamics in polymer solar cells. The TPV measurement is used to analyze the recombination of charges by collecting the transient voltage decay of the device under open-circuit conditions. In addition, the dynamics of charge extraction efficiency can be obtained by the TPC measurement performed on the device under short-circuit conditions. Durrant et al. first used TPV and TPC measurements to study charge dynamics in polymer solar cells [32]. For TPV and TPC measurements, polymer solar cells are operated in open-circuit and short-circuit under solar light illumination and then excited by a small perturbation of pulsed laser light, respectively.

When the solar cells are illuminated under pulse light, there will be photon-generated electrons and holes, followed by the generation of photovoltage. After the pulsed light is off, the voltage decreased due to the charge recombination at defect sites. Furthermore, the decay of the charges follows pseudo-first-order kinetics and hence is given by an exponential function. Thus, the lifetime $\tau_{\Delta n}$ can be evaluated as a time constant of the exponential decay. Meanwhile, the lifetimes $\tau_{\Delta n}$ for different V_{OC} can be obtained by changing light intensities. The following relationship between $\tau_{\Delta n}$ and V_{OC} is given by

$$\tau_{\Delta n} = \tau_{\Delta n_0} \exp\left(-\frac{eV_{OC}}{\upsilon k_B T}\right) \quad (5.1)$$

where e, k_B, and T are the elementary charge, Boltzmann constant, and absolute temperature, respectively. $\tau_{\Delta n}$ and υ are obtained from an intersection and a slope in logarithmic plots of $\tau_{\Delta n}$ vs. V_{OC}, respectively. Under the short-circuit condition, transient photocurrent decay is ascribed to charge collection to electrodes. The following relationship between charge carrier density (n) and V_{OC} is given by

$$n = n_0 \exp\left(\frac{eV_{OC}}{mk_B T}\right) \quad (5.2)$$

where n_0 and m are obtained from an intersection and a slope in the logarithmic plot of n against V_{oc}, respectively [33].

5.4.2 Exciton Dissociation and Charge Generation

As shown in Fig. 5.8, the electron and hole can be transferred to the A and D, respectively, forming a coulombically bound electron–hole state at the interfaces, i.e., CT state. Note that the singlet exciton binding energy ($E_{exciton}$) can be overcome by the CT process. In general, the binding energy of the CT state (E_{CT}) is in the range of 0.1–0.5 eV, which is remarkably larger than the thermal energy at room temperature (~25 meV). The CT state is formed by excess thermal energy because of the energy difference between the $E_{exciton}$ and E_{CT}. Gao et al. pointed out that the CT state with excess thermal energy is denominated as a hot CT state. In this case, two processes may happen: the hot CT state can directly dissociate into the charge-separated states or thermally relax to the lowest-lying CT state and follow the formation of the charge-separated state, thus generating free charge carriers [34].

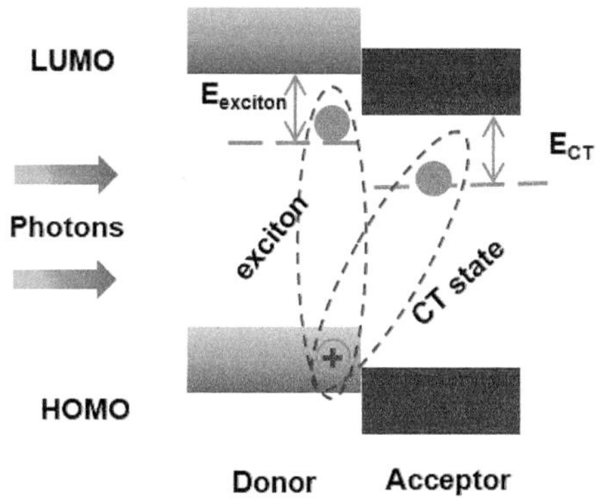

Figure 5.8 Schematic illustration of the charge-transfer state formation, exciton, and CT states with binding energies.

The CT states are generated with excess energy, thus the excess energy assists to dissociate the hot CT states directly to the free charge carriers before they thermalize to the ground (thermally relaxed) CT states. In this hot CT state theory, the initial kinetic competition is between dissociation and thermalization. The dissociation from this hot CT state is faster than the thermalization process around the order of several hundred femtoseconds (fs). Indeed, it was found that free carriers could be generated in the range of 50 fs based on the TAS measurements [35]. Furthermore, once relaxed to the ground CT states, the polaron pairs will primarily recombine as they do not have enough driving force to overcome the Coulombic attraction between each other under normal device operation conditions. Ohkita et al. found that the yield of the dissociated polarons varied by two orders depending on polythiophenes used, in spite of efficient PL quenching for all the blends. They observed a strong dependency of the yield of dissociated polarons on the energy difference between the singlet exciton and the dissociated polarons. Based on this observation, they proposed that the excess energy after the exciton dissociation provides extra kinetic energy for the CT state dissociation [36].

Interestingly, Vandewal et al. found that charge carriers are generated from the relaxed CT states, rather than hot CT states, for a range of material combinations [37]. Based on the experiments under different pre-biases, they found that the extracted free charge carriers depend on excitation energy. Such results suggest that free charge-carrier generation at different biases is not dependent on the excess energy. However, one augment still exists that the below-gap excitation energy does not produce the lowest-energy relaxed CT states exclusively. Even with below-gap excitation, most CT states can be relatively hot because of the formation of vibrationally excited CT states, attributed to the reorganization of the atoms after photon absorption. Unless the relaxed CT states are accessed exclusively, it is difficult to present that relaxed CT states are exclusive sources for charge-carrier generation [34].

5.4.3 Charge Recombination

The charge recombination becomes the dominant loss mechanism as the active layer thickness increases, reducing the V_{OC}, J_{SC}, and FF, thereby decreasing PCE. Therefore, determining the initial recombination mechanisms and their physical origin is essential for the future of polymer solar cells. As it is well known, the recombination lifetimes of the charges in the D/A domains restrict efficient carrier extraction to external circuits. The recombination can be reduced in optimized BHJ films, but it increases as the thickness of BHJ film increases because of the lower charge-carrier mobility in disordered D/A materials.

The V_{oc} values measured in polymer solar cells are usually lower than those in inorganic or perovskite photovoltaic devices. The lower values are mainly due to the fact that the CT state energy is usually lower than the optical gap, which corresponds to the local-exciton energy in the D or A material. It was found that the $\mathrm{V_{oc}}$ values correlate with the CT state energies in a large set of D-A systems. The voltage loss is also related to the non-radiative decay of the CT states to the ground state. For example, the non-radiative voltage loss (ΔV_{nr}) in polymer solar cells is generally ~0.4 eV, which is higher than that of high-quality GaAs solar cells of only 0.04 eV [38]. In general, the value of ΔV_{nr} can be obtained from the EQE of electroluminescence (EQE_{EL}) by

$$\Delta V_{nr} = \frac{k_B T}{q} \ln\left[\frac{1}{EQE_{EL}}\right] \tag{5.3}$$

5.5 Dye-Sensitized Solar Cells

In the early 1970s, it was found that electricity can be produced by irradiated organic dyes in the electrochemical reaction cell. Dye-sensitized solar cells (DSSCs) were considered an economic alternative to p-n junction PV devices. In 1972 at the University of California, for the first time, chlorophyll dye was injected into ZnO, and photons were converted to electricity through photoexcitation of dye. Later on, the absorption efficiency of DSSCs were improved by using porous semiconductors due to the enhancement of dye absorption on metal oxide, so that higher absorption of dye could enhance light-harvesting efficiency.

DSSC cells are also known as Grätzel cells and initially reported a power-conversion efficiency of 7%. The first DSSC was fabricated based on a transparent mesoporous TiO_2 film with a high surface area and a single layer of a CT dye with ideal spectral characteristics. The device could harvest up to 46% of the incident solar energy and the conversion efficiency of incident photons to electrical current was 7.1–7.9% under solar simulator light and 12% in daylight.

The dye sensitizers are mainly designed with –COOH, $–PO_3H_2$, and $–B(OH)_2$ functional groups for stable adsorption on a semiconductor substrate. Compared to conventional Si solar cells, where Si serves as photoelectrons source and charge separation to create current, in DSSC photosensitive dyes provide photoelectrons and semiconductors act as charge carriers. The theoretical power-conversion efficiency of DSSCs is about 20% [39]. This section provides a review of the DSSC structure, operating principle, and main issues such as low efficiency, scalability, and stability.

5.5.1 Structure of DSSC

Four key components of DSSC are the working electrode, dye (sensitizer), electrolyte (redox media), and counter electrode. Working and counter electrodes soaked with dye and electrolyte are sealed to each other with hot tape. The overall structure and operation principle are shown in Fig. 5.9.

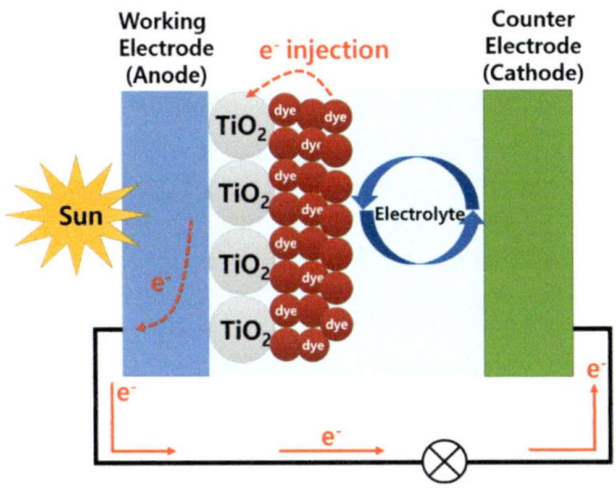

Figure 5.9 Overall structure and operation principle of DSSC.

5.5.1.1 Transparent conductive oxide (TCO) substrate

DSSCs are made of two TCO glass substrates in which semiconductor and catalyst are deposited. TCO is soda-lime glass coated with layers of conductive oxides, i.e., fluorine-doped tin oxide (FTO) or indium-doped tin oxide (ITO). The transmittance of ITO and FTO films are >80% and >75%, with a sheet resistance of 18 Ω/cm^2 and 8.5 Ω/cm^2, respectively. The TCO substrate should have more than 80% transparency to pass the optimal sunlight to the active material. In addition, the TCO must have high conductivity for efficient charge transport to reduce energy loss in DSSC devices.

5.5.1.2 Working electrode

The electrode is a solid conductor through which current enters or exits the medium. In an electrochemical cell, electrodes are called either an anode or a cathode. The anode is the electrode where electrons leave the device and oxidation occurs (signed by "−"), and the cathode is the electrode where electrons enter the device and reduction occurs (signed by "+"). For DSSC, the anode is the working electrode on which the semiconductor is coated and incident sunlight comes through it. The working electrode is fabricated by depositing a thin film of metal oxide semiconductor on an FTO or ITO substrate. Metal oxides are typically TiO_2, ZnO, SnO_2 (n-type), or NiO (p-type) with a wide energy bandgap of 3–3.2 eV. The most commonly used semiconductor is anatase TiO_2 with an energy bandgap of 3.2 eV, which is non-toxic, low cost, and high availability. TCO/TiO_2 working electrode is submerged in the photosensitive dye solution and the dye molecules attach to the TiO_2 surface. Porous TiO_2 with a high surface area absorbs a large number of dye molecules, increasing the light absorption at the semiconductor surface.

5.5.1.3 Dye (or photosensitizer)

Dye is the active component of a DSSC, which absorbs the incident light and generates photocurrent. There are general photophysical and electrochemical properties that the dye material should have. First, the dye should be a luminescent material having an absorption range of ultraviolet–visible (UV–Vis) and near-infrared (NIR) regions. Second, the HOMO and LUMO energy levels of dye material should be proportional to TiO_2 and electrolyte. Third, to fabricate a highly stable device, the water-induced instability should be minimized. To do this, hydrophobic dye material can be used to minimize device deterioration. Finally, to better distribute dye molecules on the surface of TiO_2 particles and prevent agglomeration, a co-absorbent and a fixed group are inserted between TiO_2 nanoparticles and the dye.

5.5.1.4 Electrolyte

An electrolyte in DSSCs functions as a medium for the charge transport between electrodes inside the cell in the form of ions as well as regeneration of dye and itself under working conditions. The electrolyte is an important component that affects the photocurrent efficiency and stability of DSSC devices. All PV parameters (J_{SC}, V_{OC}, and FF) are significantly affected by the electrolyte and its interaction with electrode interfaces. For example, the transport of the redox couple in the electrolyte affects J_{SC}, and the charge diffusion in the electrolyte and the resistance in the electrolyte/electrode interface significantly affect FF. The redox potential in electrolytes could make significant changes in V_{OC}. To act as a transportation medium for charge carriers, it is essential for an electrolyte must have the following characteristics:

- Besides the charge transport between working and counter electrodes, electrolytes should have a redox potential to regenerate oxidized dye into the ground state.

- The electrolyte should fast charge diffusion and conduction and create effective contact between the working and counter electrode.
- The electrolyte must have long-term chemical, thermal, electrochemical, and optical stability.
- The absorption spectrum of the electrolyte should not be in the range of the light absorber (i.e., dye). The redox couple (such as I^-/I^-_3, SCN^-/SCN_2^-, Br^-/Br_2^-, and Co(II)/Co(III)) in the electrolyte can react with injected electrons and affect the dark current.

Based on physical states, components, and formation mechanisms, the electrolytes can be classified into three groups: liquid, quasi-solid, and solid-state electrolytes. There are a few limitations with liquid electrolytes such as corrosion of electrodes and leakage factor. To avoid the drawbacks, quasi-solid and solid-state electrolytes were developed [40, 41].

5.5.1.5 Counter electrode (CE)

The counter electrode is a cathode on which platinum or carbon are coated. The counter electrode has three roles: (i) It completes the electric circuit by catalyzing the reduction process of electrolyte, and the oxidized dye accepts electrons and is reduced. (ii) As a charge collector, it collects electrons from the device circuit and returns them to the cell to aid in charge circulation. (iii) As a reflecting surface, it reflects the unabsorbed light to the cell to enhance light absorption efficiency.

5.5.2 Operating Principles of DSSC

The operating principle of DSSCs is different from that of the first- and second-generation solar cells and is highly related to an electrochemical process where different substances are responsible for electrochemical processes and light absorption and charge transport. Conversion of photons into the current in DSSC includes four steps of light absorption, charge injection, electron transportation, and charge collection. Figure 5.10 schematically illustrates the energy diagram of a DSSC and electrochemical processes [42].

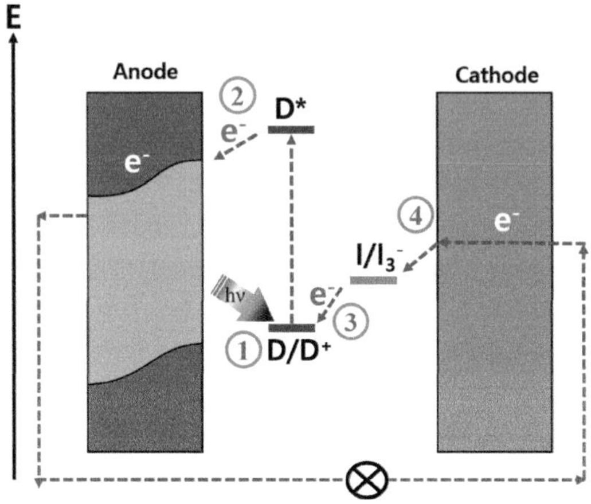

Figure 5.10 Operating process of DSSCs.

The operation processes occuring in a DSSC is as follows:

(1) Photoexcitation of dye molecule: firstly, dye is in the ground state (D). After light absorption, the dye molecule is excited to a higher-energy state (D*): $D \xrightarrow{hv} D^*$

(2) The excited dye molecule (D^*) is oxidized ($D^* \rightarrow D^+ + e^-$) and injects an electron into the conduction band of the semiconductor and transfers to the anode.

(3) The oxidized dye molecule (D^+) is reduced and regenerated by accepting an electron from the iodide in the electrolyte. $\left(D^+ + \frac{3}{2}I^- \rightarrow D + \frac{1}{2}I_3^-\right)$.

(4) The regeneration of iodide by reduction of triiodide on the cathode. $\left(\frac{1}{2}I_3^- + e^- \rightarrow \frac{3}{2}I^-\right)$.

5.5.3 Performance of DSSC

The DSSC device performance can be estimated by the following parameters: short-circuit current (J_{sc}, mA.cm^{-2}), open-circuit voltage (V_{oc}, V), FF (%), incident photon-to-current conversion efficiency (IPCE, %), maximum power output (P_{max}), efficiency (η, %). In a cell, when anode and cathode are short-circuited at a zero voltage, the current is zero mA; under open-circuit conditions, the voltage referred to as V_{oc} (V) presents the potential difference between the LUMO of the semiconductor and the redox potential of electrolyte [43].

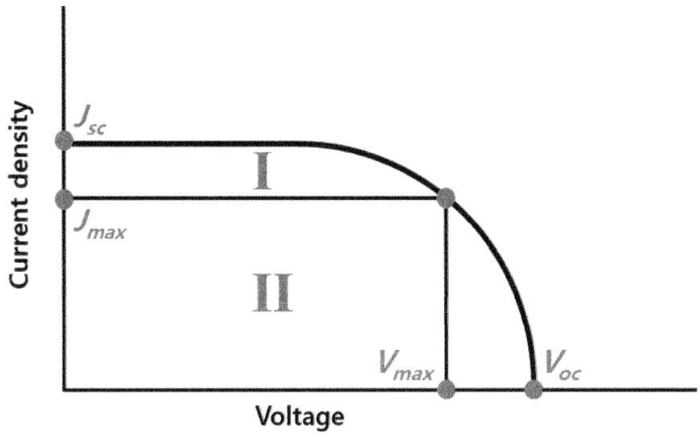

Figure 5.11 J-V plot of a solar cell.

P_{max} is the maximum power point and the ratio of maximum power output ($J_{max} \times V_{max}$) to the product ($V_{oc} \times J_{sc}$) is called *FF* (Fig. 5.11).

$$FF = \frac{\text{Area II}}{\text{Area I}} = \frac{J_{max} \times V_{max}}{J_{sc} \times V_{oc}} \quad (5.4)$$

Also, the overall efficiency (η) is the percentage of the photon-to-electron conversion.

$$\eta(\%) = \frac{J_{sc} \times V_{oc} \times FF}{P_{in}} \quad (5.5)$$

5.5.4 Limitations of DSSCs

DSSCs with promising performances have been reported over the years. However, due to the stability limitations of the devices, further modifications are required. There are two extrinsic and intrinsic limitations associated with DSSCs. For instance, in the dye regeneration process, a large amount of energy loss occurs due to an energy mismatch between the electrolyte and oxidized dye. To solve this problem, several electrolytes have been introduced. Also, the sheet resistance of TCO is another limiting factor for small cells. To improve the sheet resistance, the spacing between electrodes needs to be maintained in small module cells. To avoid the electrolyte or small molecules leakage, a sealing material like hot tape can be used. Another factor is the low conductivity of the TCO. ITO and FTO can be combined to improve the conductivity of TCO, thereby reducing the sheet resistance of TCO glass.

References

1. Akamatu, H., Inokuchi, H., and Matsunaga, Y. (1954) *Nature*, **173**, pp. 168–169.
2. Ghosh, A. K. and Feng, T. (1973) *J. Appl. Phys.*, **44**, pp. 2781–2788.
3. Tang, C. W. (1986) *Appl. Phys. Lett.*, **48**, pp. 183–185.
4. Bredas, J. L., Calbert, J. P., Filho, D. A. d. S., and Cornil, J. (2001) *PNAS*, **99**, pp. 5804–5809.
5. Mayer, A. C., Scully S. R., Hardin, B. E., Rowell, M. W., and McGehee, M. D. (2007) *J. Appl. Phys.*, **10**, pp. 28–33.
6. Brütting, W. (2006) *Physics of Organic Semiconductors.* Wiley-VCH Verlag GmbH & Co. KGaA, Weinheim, pp. 1–14.
7. Clarke, T. M. and Durrant, J. R. (2010) *Chem. Rev.*, **110**, pp. 6736–6767.
8. Hiramoto, M., Fujiwara, H., and Yokoyama, M. (1992) *J. Appl. Phys.*, **72**, pp. 3781.
9. Yu, G., Gao, J., Hummelen, J. C., Wudl, F., and Heeger, A. J. (1995) *Science*, **270**, pp. 1789–1791.
10. Halls, J. J. M., Walsh, C.A., Greenham, N. C., Marseglia, E. A., et al. (1995) *Nature*, **376**, pp. 498–500.
11. Green, M. A., Dunlop, E. D., Hohl-Ebinger, J., Yoshita, M., et al. (2021) *Prog. Photovolt. Res. Appl.*, **29**, pp. 3–15.
12. Li, G., Zhu, R., and Yang, Y. (2012) *Nat. Photon.*, **6**, pp. 153–161.
13. Schilinsky, P., Waldauf, C., and Brabec, C. J. (2002) *Appl. Phys. Lett.*, **81**, pp. 3885.
14. Kim, Y., Cook, S., Tuladhar, S. M., Choulis, S. A., et al. (2006) *Nat. Mater.*, **5**, pp. 197.
15. Lu, L. Y., Zheng, T. Y., Wu, Q. H., Schneider, A. M., et al. (2015) *Chem. Rev.*, **115**, pp. 12666–12731.
16. Lin, Y. Z., Wang, J. Y., Zhang, Z. G., Bai, H. T., et al. (2015). *Adv. Mater.*, **27**, pp. 1170–1174.
17. Zhao, W., Qian, D., Zhang, S., Li, S., et al. (2016) *Adv. Mater.*, **28**, pp. 4734–4739.
18. Fan, Q. P., Wang, Y., Zhang, M. J., Wu, B., et al. (2016) *Adv. Mater.*, **30**, pp. 1704546.
19. Yuan, J., Zhang, Y. Q., Zhou, L. Y., Zhang, G. C., et al. (2019). *Joule*, **3**, pp. 1140–1151.

20. Yuan, J., Zhang, H. T., Zhang, R., Wang, Y. M., et al. (2020) *Chem*, **6**, pp. 2147–2161.
21. Wang, G., Melkonyan, F. S., Facchetti, A., and Marks, T. J. (2019) *Angew. Chem. Int. Ed.*, **58**, pp. 4129–4142.
22. Facchetti, A. (2013) *Mater. Today*, **16**, pp. 123–132.
23. Zhang, M. J., Guo, X., Ma, W., Ade, H., and Hou, J. H. (2015) *Adv. Mater.*, **27**, pp. 4655–4660.
24. Yu, H., Pan, M., Sun, R., Agunawela, I., et al. (2021) *Angew. Chem.*, **133**, pp. 10225–10234.
25. Sun, H. L., Liu, B., Ma, Y. L., Lee, J. W. O., et al. (2021). *Adv. Mater.*, **33**, pp. 2102635.
26. Zhang, Y. and Li, G. (2020) *Acc. Res.*, **1**, pp. 158–171.
27. Mahmoudi, T., Wang, Y., and Hahn, Y.-B. (2018) *Nano Energy*, **47**, pp. 51–65.
28. Mahmoudi, T., Seo, S., Yang, H.-Y., Rho, W.-Y., et al. (2016) *Nano Energy*, **28**, pp. 179–187.
29. Ullah, F., Chen, C.-C., and Choy, W. C. H. (2021) *Adv. Energy Sustain. Res.*, **2**, pp. 2000050.
30. Ohkita, H., Kosaka, J., Guo, J., Benten, H., and Ito, S. (2011) *J. Photon. Energy*, **1**, pp. 011118.
31. Guo, J., Ohkita, H., Benten, H., and Ito, S. (2010) *J. Am. Chem. Soc.*, **132**, pp. 6154–6164.
32. Shuttle, C. G., O'Regan, B., Ballantyne, A. M., Nelson, J., et al. (2008) *Appl. Phys. Lett.*, **92**, pp. 093311.
33. Hiramoto, M. and Izawa, S. (Eds.) (2020) *Organic Solar Cells: Energetic and Nanostructural Design*. Springer Nature.
34. Gao, F. and Inganäs, O. (2014) *Phys. Chem. Chem. Phy.*, **16**, pp. 20291.
35. Grancini, G., Maiuri, M., Fazzi, D., Petrozza, A., et al. (2013) *Nat. Mater.*, **12**, pp. 29–33.
36. Ohkita, H., Cook, S., Astuti, Y., Duffy, W., et al. (2008) *J. Am. Chem. Soc.*, **130**, 3030–3042.
37. Vandewal, K., Albrecht, S., Hoke, E. T., Graham, K. R., et al. (2014) *Nat. Mater.*, **13**, pp. 63–68.
38. Coropceanu, V., Chen, X.-K., Wang, T., Zheng, Z., and Brédas, J.-L. (2019) *Nat. Rev. Mater.*, **4**, pp. 689–707.
39. Tributsch, H. and Calvin, M. (1971) *Photochem. Photobiol.*, **14**, pp. 95–112.

40. Nazeeruddin, K. Baranoff, E., and Gratzel, M. (2011) *Sol. Energy*, **85**, pp. 1172–1178.

41. Snaith, H. J. (2010) *Adv. Funct. Mater.*, **20**, pp. 13–19.

42. Wu, J., Lan, Z., Lin, J., Huang, M., et al. (2015) *Chem. Rev.*, **115**(5), pp. 2136–2173.

43. Sharma, K., Sharma, V., and Sharma, S. S. (2018) *Nanoscale Res. Lett.*, **13**(1), pp. 1–46.

Problems

5.1 Describe the operation principles of OSCs.

5.2 How many configuration types of OSCs are there? Describe their advantages and disadvantages.

5.3 What are the main advantages of NFAs in OSCs?

5.4 What is the design principle of TSCs?

5.5 Briefly describe the charge dynamics of OSCs and their characterization methods.

5.6 The electrolyte is an important component of DSSCs. Explain the essential characteristics of an electrolyte.

5.7 The mechanism of dye solar cells is based on photo-electrochemical processes. In the following energy diagram of a dye solar cell, describe all relevant electrochemical processes numbered 1 to 4.

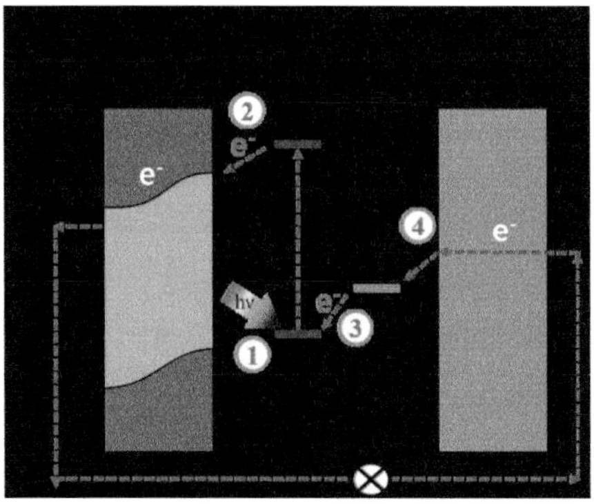

5.8 What is a charge-transport layer and what are the properties of a charge-transport layer?

5.9 Except for metal oxides, graphene-based materials are also used for charge transport in solar cells. Briefly explain the role of graphene-based materials as HTL and ETL.

Chapter 6

Quantum Dot Solar Cells

As a member of the third-generation solar power generation, quantum dot solar cells (QDSCs) have a theoretical efficiency of more than 44%, which is likely to exceed the S-Q limit of 33.7%. This high efficiency is mainly due to the excellent properties of quantum dots (QDs), such as resizable absorption, quantum limitation, high optical extinction coefficient, and multiple excitation effects. In addition, a facile and solution-processed synthesis method enables the cost-effective development of a wide-area solar cell module by a printing technique. This chapter first describes the basic concepts, characteristics, synthesis methods, and optoelectronic applications of QDs, and then explains the basic configuration, manufacturing method, and operating principles of QDSC. Finally, important strategies such as material and device interface optimization to improve the performance of QDSC are presented.

6.1 Physical Properties of Quantum Dot

6.1.1 What Are Quantum Dots?

QDs are quasi-zero-dimensional semiconductor nanocrystals, which are composed of 10^3–10^9 atoms and a comparable number of electrons [1]. Usually, the size of QDs in three spatial dimensions is smaller

Next-Generation Solar Cells: Principles and Materials
Yoon-Bong Hahn, Tahmineh Mahmoudi, and Yousheng Wang

ISBN 978-981-4968-66-9 (Hardcover), 978-1-003-37238-7 (eBook)
www.jennystanford.com

than the de Broglie wavelength of the excitons in corresponding bulk materials. Their typical sizes of QDs are in the range of 1–10 nm in diameter (10–50 atoms). Their size, shape, and characteristics can be precisely controlled by adjusting the reaction time and conditions of the synthesis method. QDs have unique optical, photochemical, semiconductor, and catalytic properties that are dependent on the spatial confinement of the electrons, thus making them useful for various devices application. In the early 1980s, the first methods for synthesizing QDs were achieved by Henglein and Rossetti et al. [2, 3]. Typical synthetic methods will be presented and discussed in the next section.

6.1.2 Synthesis of Quantum Dots

Due to their excellent optical and electronic properties, QDs have been widely studied for various device applications. Thus, the synthetic methods of high-quality QDs are of great importance. There are several typical routes for the synthesis of nanocrystals with controlled size and shape, such as hot injection [4, 5], successive ionic layer adsorption and reaction (SILAR) [6], and chemical bath deposition (CBD) [7]. The hot-injection method is very useful for synthesizing colloidal QDs (CQDs) with a highly monodisperse and narrow size distribution.

As shown in Fig. 6.1a, the formation of monodisperse CQDs typically involves two steps: rapid nucleation and slow growth. This method was first used to synthesize semiconductor nanocrystals by Murray et al. in 1993 [8]. The hot-injection method is based on the fast injection of organometallic precursors into a hot coordinating solvent to yield isolated nucleation and permit the controlled growth of nanocrystallites. By controlling the growth time, the monodisperse and/or different sizes of nanocrystals are successfully formed. Figure 6.1b shows the schematic process of nucleation and growth for monodisperse CQDs. It is noted that the nucleation and growth of QDs have to experience size change and Ostwald ripening in sequence. The factors, such as chain length of ligands, reaction temperature, and time, are important, and can remarkably affect the size, shape, and physical and chemical properties of QDs.

The SILAR is also an important method for achieving high surface coverage and size-controlled QDs, which has been widely used for the fabrication of DSSCs. As shown in Fig. 6.1c, the SILAR method needs to experience four important steps with several cycles to form desired QD films: adsorption, rinsing, reaction, and rinsing. In brief, an as-prepared substrate is immersed into the cationic and anionic solutions in sequence, and QDs are formed by chemical reactions. Note that each immersing step should rinse excessive ions. For example, the deposition process of CdS and CdSe QDs on the TiO_2 surface contains two steps: (1) the nanostructured TiO_2 electrodes are dipped into Cd^{2+} solution, and then (2) dipped into S^{2-} or Se^{2-} solutions. After dipping, the next step every time is rinsing with deionized water. To achieve desired surface coverage of CdS and CdSe QDs on the TiO_2 electrodes, several cycles of two steps are needed. Except for the deposition cycle, precursor type and concentration, temperature, and pH value also can affect the quality of QDs.

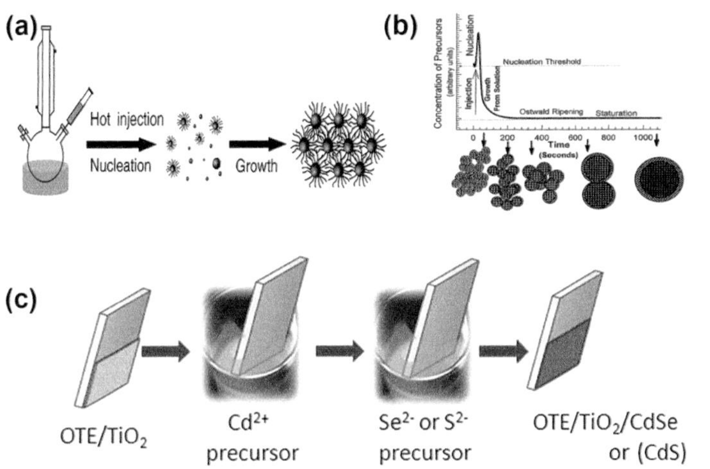

Figure 6.1 Typical synthesis methods of QDs: (a) hot injection and (b) schematic illustration of nucleation and growth for monodisperse QDs. Reprinted with permission from [4], Copyright 2015, American Chemical Society. (c) SILAR. Reprinted with permission from [6], Copyright 2013, American Chemical Society.

The CBD method is useful and cost-effective for large-scale preparations as it can obtain high surface coverage with a random size distribution of QDs on a target substrate. For a typical CBD method, as-mixed two solutions are reacted at a fixed temperature for a certain time. Yu et al. [7] reported an efficient one-step linked assisted CBD method to prepare CdTe or core/shell CdTe/CdS

QDs on TiO_2 photoelectrode. Certain synthesis methods mainly include immersing a TiO_2 substrate in a mixture solution composed of $Cd(CH_3COO)_2$, NaHTe, NaOH, thioglycolic acid, and water in an autoclave. The CdTe or core/shell CdTe/CdS QDs sensitized TiO_2 photoelectrode can be prepared by a hydrothermal reaction process according to reaction temperature and time. The CBD parameters such as solution concentration, complexing agents, pH, reaction temperature, and time have to be optimized to achieve multifunctional QD layers for various device applications.

6.1.3 Optical and Electronic Properties of Quantum Dots

Based on the quantum confinement effect, the size of QDs significantly influences their optical and electronic properties [9]. Thus, QDs endow a wide range of photo-response in the solar spectrum by adjusting their size. Absorption onset shift is dependent on QD size that affects its optical bandgap. The UV absorption onsets (i.e., 1s exciton peak) can be tuned from 1200 to 2100 nm by changing QD size from 3.3 to 8.1 nm as shown in Fig. 6.2a. The 1s exciton peak shifts toward a longer wavelength and/or lower energy. Figure 6.2b shows how the ground state wavefunction experiences increased confinement, which verifies the absorption onset shift in the lowest-energy transition. Figure 6.2c shows the relationship between QD radius and the position of the lowest-energy (1s) transition of PbSe QDs [10]. As another example, the bandgap of Cu_2ZnSnS_4 nanocrystals can be changed from 1.48 to 1.89 eV by controlling the crystal size from 10.5 to 3 nm [11].

It is worth noting that the doping strategy in quantum dots is a vital method to tune the optical and electronic properties of QDs. For example, Cu-doped InAs QDs show a blue-shifted absorption spectrum, indicting an n-type semiconductor property [12]. Adversely, a red-shifted absorption spectrum in Ag-doped InAs QDs suggests a p-type semiconductor property [13]. Thus, well-adjusted optical and electronic properties of QDs with different doping types enable various optoelectronic applications.

As discussed in Chapter 3, a semiconductor produces an electron–hole pair when absorbing photons with higher energy than its bandgap, but excessive photon energy cannot be completely converted into electrical energy because of thermalization through electron–photon coupling. The concept of multiple exciton generation (MEG) is important to utilize excessive photon energy in QD semiconductors. The MEG is a process that can create two or more electron/hole pairs by only one photo-absorption, which is attributed to the impact of ionization [14]. Furthermore, the MEG efficiency (η_{MEG}) can be defined as $\eta_{\text{MEG}} = E_g/E_{e/h}$, where E_g and $E_{e/h}$ are bandgap and generated energy of electron/hole pairs. As reported, the η_{MEG} of QDs is higher than that of the corresponding bulk semiconductor [15]. In addition, other processes such as hot-electron injection, plasmonic effect, and nanophotonic light trapping also can gain momentum to break the efficiency limitations of energy capture and conversion [16–18].

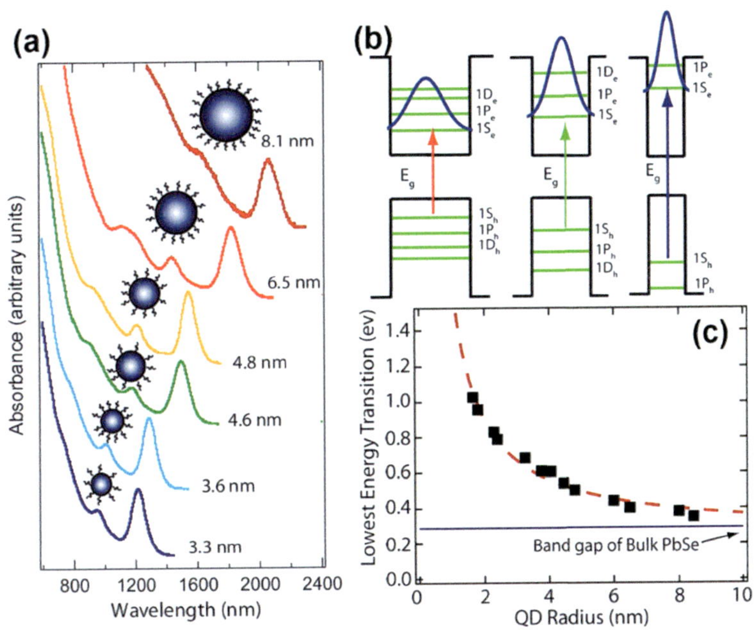

Figure 6.2 (a) UV–Vis absorption spectra of PbSe QDs in terms of nanosize range from 3.3 nm to 8.1 nm, (b) schematic illustration of the relationship between PbSe QD size and confinement caused by electronic wavefunction (QD size decreases from left to right), (c) the relationship between QD radius and the position of the lowest-energy (1S) transition of PbSe QDs. Reprinted with permission from [10], Copyright 2012, Elsevier.

6.1.4 Application of Quantum Dots

The advent of QDs is a milestone in the fields of nanotechnology so far. Since their first synthesis in the 1980s [19, 20], QDs have been widely used in biology and medicine, environment, photovoltaics, and optoelectronics. In the following, some typical applications of QDs will be presented and discussed. For example, Wu et al. used CdSe/ZnS core–shell nanocrystals to label cells. All labeling signals were specific for the desired targets and more photo-stable than organic dyes, indicating the QD-based probes are very effective in cellular imaging over organic dyes in multiplex target detection [21]. Some highly toxic heavy metal ions, such as Hg^{2+} and Cd^{2+}, present a severe threat to the environment and human life. Thus, it is important to develop an efficient and easy method for detecting heavy metal ions. It is one of the most effective detection methods by observing changes in fluorescence (FL) when QDs bind to heavy metal ions. CdSe QDs were successfully used for Hg^{2+} detection. Li et al. found that the FL intensity of CdSe/ZnS decreased when Hg^{2+} concentration increased [22]. You et al. used nitrogen and sulfur-doped carbon dots (N, S/C-dots) for Hg^{2+} detection. It was observed that a significant FL quenching and blue shift phenomenon when adding Hg^{2+} solution in N, S/C-dots [23]. QDs also can be catalysts for the conversion of CO_2 to high-order hydrocarbons and oxygenates. Wu et al. reported that functionalized graphene QDs enable direct selective CO_2 to CH_4 conversion by the electron-donating groups [24].

Except for the above QD applications, QDs are potential in the field of optoelectronics, especially in light-emitting diodes (LEDs), photodetectors, and photovoltaics. The QD-LEDs are desired for large-panel displays due to their excellent efficiency, reliability color purity, and cost-effective fabrication. Intensive efforts have produced green-, blue-, and red-emitting QD-LEDs. For example, Choi et al. successfully developed a ZnO core wrapped in a shell of single-layer graphene QDs to achieve white LEDs [25]. Recently, emerging hybrid perovskite quantum dots (PQDs) based LEDs have achieved impressive external quantum efficiency. Sargent et al. reported bright and stable LEDs by using stable matrix $CsPb(Br_xI_{1-x})_3$@$CsPbI_3$ QDs, leading to 18% external quantum efficiency (EQE) [26]. The solution processability and quantum confinement of colloidal

QDs have made them one of the promising materials for low-cost, high-efficiency, third-generation photovoltaic technology. In the next sections, we will present and discuss in-depth QD materials as absorbers and their basic PV working principles.

6.2 Quantum Dots Based Solar Cells

Size-tunable absorption, quantum confinement, high optical extinction coefficient, and multiple exciton generation in semiconductor QDs are suitable as light absorbers in solar cells and capable of achieving a theoretical efficiency of over 44% that overcomes the S-Q limit of 33.7%. As we discussed earlier, facile and solution-processed synthetic methods of QDs enable the cost-effective development of large-area solar cell modules by printing techniques. Thus, QDSCs have been considered one of the most meaningful third-generation photovoltaics. In the next sections, the basic configuration and operation principles of QDSC are described in detail.

6.2.1 Quantum Dots Solar Cell Configuration

To pursue a higher efficiency of QDSCs, various types of configurations have been explored: (a) QD/semiconductor heterojunction, (b) polymer–QD hybrid, (c) QD-sensitized, and (d) perovskite QD solar cells, as shown in Fig. 6.3. They are mainly based on the charge separation at the interface between excited short bandgap QDs and metal oxides (such as TiO_2 and ZnO_2). The light absorbers can be metal chalcogenides such as PbSe, CdSe, PbS, CdS, and Sb_2S_3, which have appropriate bandgaps. Note that hybrid organic/inorganic absorbers, such as inorganic QDs/polymeric hole conductors (P3HT, PBDB) [27], and perovskites [28] are also appropriate to develop highly efficient solar cells.

The QD/semiconductor heterojunction solar cells are mainly composed of (1) QDs light absorption layer with two electrodes (i.e., Schottky junction); (2) QDs absorption layer sandwiched between electron and hole-transport layers with two electrodes (i.e., depleted heterostructure). The polymer–QD hybrid solar cells have mixed polymer–QD composites as an absorber layer, and are also sandwiched between electron and hole-transport layers with two electrodes. The inorganic QD-sensitized or perovskite QD-sensitized solar cells can be achieved by depositing metal chalcogenide or PQDs on nanocrystalline metal oxide surfaces (typically TiO_2) by SILAR, CBD, or spin-coating methods. Note that perovskite QDs sensitized solar cells have shown higher efficiencies over 20% than the other three types of QD solar cells.

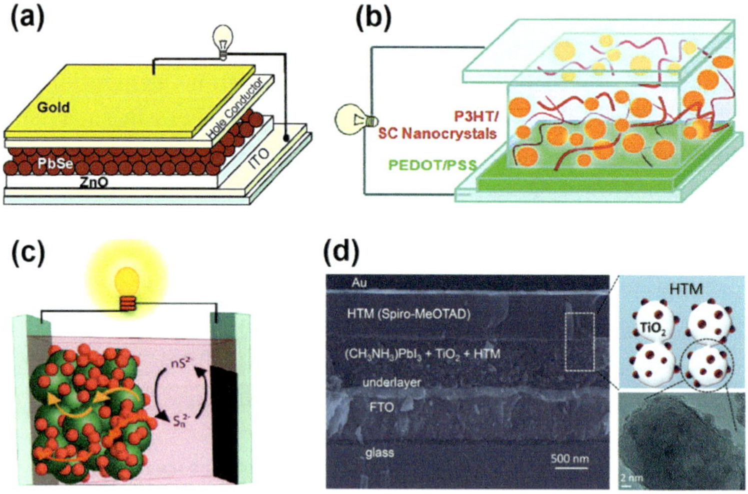

Figure 6.3 QD solar cell configurations: (a) QD–semiconductor heterojunction, (b) polymer–QD hybrid, (c) QD-sensitized. Reprinted with permission from [29], Copyright 2012, American Chemical Society. (d) Perovskite QD solar cells. Reprinted with permission from [30], Copyright 2012, Springer Nature.

6.2.2 Basic Operation Principles and Physical Mechanism

The advent of QDs has shown great potential to develop highly efficient photovoltaics with an opportunity to break the S-Q efficiency limit due to their size-tunable absorption, quantum confinement, high optical extinction coefficient, and multiple exciton generation effect. It is significant to understand basic operation principles and physical mechanisms. Figure 6.4 shows basic device structures and work mechanisms for various QDSCs. According to the Schottky-Mott theory, in a Schottky junction QD solar cell with p-type semiconducting properties as an absorber (Fig. 6.4a), the

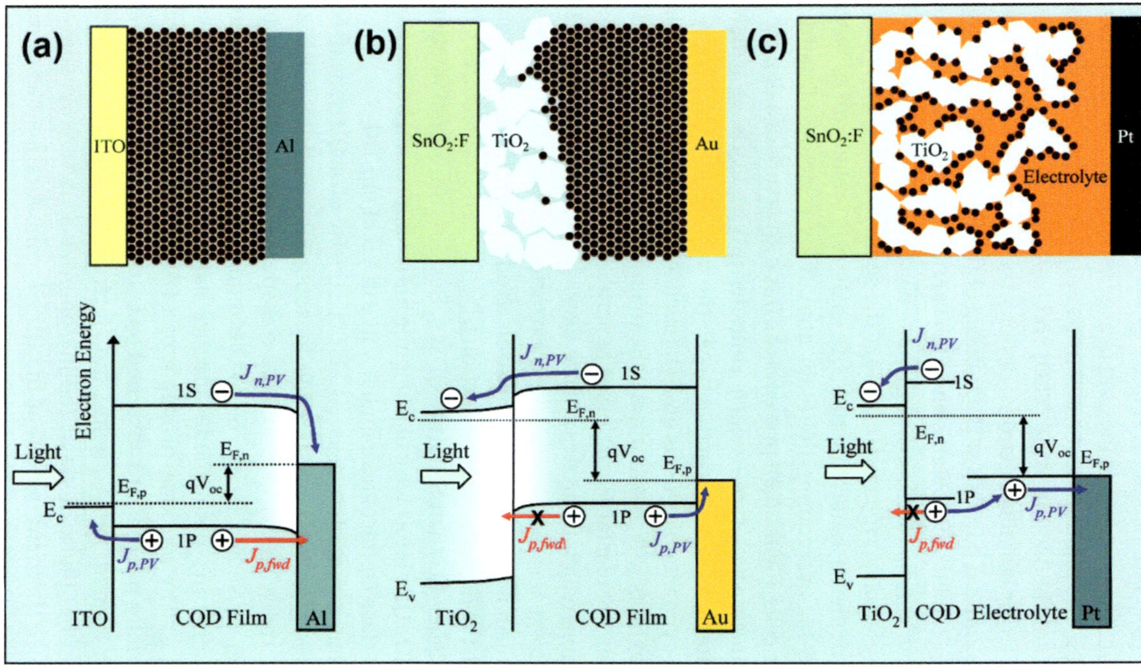

Figure 6.4 QD solar cells configurations (top) and corresponding energy diagrams (bottom): (a) Schottky junction, (b) depleted heterojunction, and (c) QDSSCs. Reprinted with permission from [31], Copyright 2010, American Chemical Society.

work function of metal contacts should be lower than that of the corresponding semiconductor, whereas the work function of the n-type semiconductor is the opposite.

The QD/metal contact is related to the Schottky barrier height, which significantly determines the charge separation and collection processes. The photo-induced electrons are first generated near the transparent conducting oxide (TCO) and have to pass through a long pathway to the metallic electrode. Besides, the V_{oc} of Schottky junction QDSCs is limited because of the pinning of the Fermi level resulting from the formed defect states and high carrier recombination at the metal-semiconductor interface. Thus, electron or hole-transport layers are usually prepared on/under the semiconductor QD layer.

The depleted heterojunction QD solar cell architectures were designed for overcoming the main limitations related to Schottky QD solar cells. Figure 6.4b presents the basic structure of a depleted heterojunction solar cell which contains p-type CQDs as a light absorption layer sandwiched between the metal contact and highly doped n-type metal oxide as an ETL (such as mp-TiO_2). As shown in band diagrams (Fig. 6.4, bottom), the main working principle is (1) a high density of charge carriers is generated in the depletion TiO_2–QDs region, which facilitates charge dissociation and transport; (2) the photo-generated electrons are extracted to the mp-TiO_2 layer, leaving the holes to flow to the metal electrode. According to the band diagram, the depleted heterojunction QD solar cell architectures can give a higher V_{oc} than Schottky QDSCs. Besides, the recombination at the TiO_2-QDs interface can be significantly suppressed because of well interface contact and a higher value of VB of mp-TiO_2.

QD-sensitized solar cells (QDSSCs) consist of a QD-sensitized photoanode having a metal oxide mesoporous structure (such as TiO_2), a redox electrolyte, or hole-transport layer, and a counter electrode, as shown in Fig. 6.4c. The redox electrolytes penetrate the mesoporous photoanode and are sandwiched between two electrodes. The working principle of QDSSC is similar to that of DSSC. Briefly, photons are absorbed by QDs and electron–hole pairs are then formed. Subsequently, the electrons can be injected into the CB of the photoanode of the mesoporous structure. As shown in Fig. 6.4c (bottom), note that the QDSSC configuration has a free barrier of hole transportation compared to other types of QD solar cell configuration.

6.3 Quantum Dot/Semiconductor Heterojunction Solar Cells

6.3.1 Schottky Junction Solar Cells

The metal-QD structure is the simplest semiconductor heterojunction solar cell. The QDs are directly deposited on the top of TCO glass and followed by evaporating or sputtering metal electrodes. The QD films not only act as absorbers but also play the role of charge-transport medium in the Schottky junction QD solar cells. In 2007, Sargent et al. first reported QD-based Schottky junction solar cells by using a p-type PbS or PbSe CQD film as absorbers and metal Al as top electrodes [32]. Afterward, improved efficiency of over 3% was achieved with new liganding strategies, barrier layers, and extending the response into the infrared [33–35]. On basis of ligand surface engineering, Loi et al. developed PbS QD Schottky solar cells with an efficiency of 5.2%, which is the highest reported until now [36].

Although the introduction of a buffer or passivation layer may slightly reduce the density of electronic trap states and recombination at the interface, the Schottky junction architectures have limitations in absolute device performance. Thus, it has to explore a new architecture of QD solar cells. In the following sections, we will present several novel and efficient QD solar cell architectures.

6.3.2 Depleted Planar Heterostructure Quantum Dot Solar Cells

The depleted heterojunction QD solar cells have two types of configurations, as shown in Fig. 6.5. Usually, the depleted heterojunction QD configuration with a dense ETL is named planar-depleted heterojunction QD solar cell. In the depleted planar QD solar cell architectures, TiO_2 or ZnO is the most commonly used wide-bandgap n-type semiconductor contact layer. The p-type CQD films (such as PbS) with desired thickness are deposited on the top of the ZnO layer, and then a thin layer of MoO_3 and metal electrodes are deposited on the CQD film in queue, as shown in Fig. 6.5a. The energy band alignment of the QD film and mesoporous oxide (such as TiO_2 and ZnO) layers should be tuned for injection of electrons to the wide-band-gap semiconductor without sacrificing the V_{oc}

[4]. Note that the p-type MoO_3 layer can act as a hole extraction layer, thereby improving hole transport.

On basis of materials innovation and device engineering, the performance of depleted planar heterojunction QD solar cells has been remarkably enhanced. To improve the performance of planar-depleted heterojunction QD solar cells, new ligand strategies to optimize the quality of QDs have been explored. For example, Tang et al. developed an atomic ligand strategy that used monovalent halide anions to facilitate electron transport and passivate surface defects of PbS CQD, resulting in a PCE of ~6% [37]. Zhang et al. developed a direct cation-exchange method to synthesize size-tunable PbSe QDs, followed by *in situ* chloride and cadmium passivation. As-synthesized PbSe QDs showed excellent air stability under ambient conditions and the PbSe QDs-based solar cells yielded a PCE of 6.2% [38]. However, the performance of planar-depleted heterojunction QD solar cells is still limited due to short carrier transport lengths in QDs. In other words, photo-induced electrons with a high thickness of the QD layer have to experience a long path to the ETL, but the low thickness of the QD layer reduces the photocurrent of the solar cell. A strategy for solving the above problem is to develop a bulk heterojunction QD architecture, which will be discussed in the next section.

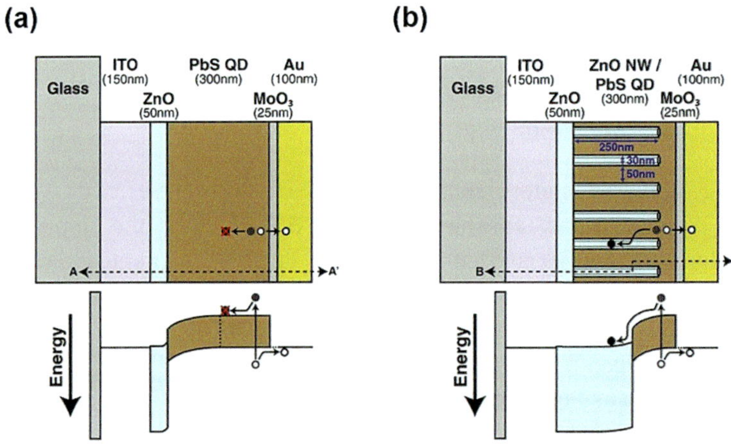

Figure 6.5 (a) Depleted planar and (b) bulk heterojunction quantum dot solar cells and corresponding energy diagrams. Reprinted with permission from [39], Copyright 2015, American Chemical Society.

6.3.3 Depleted Bulk Heterojunction Quantum Dot Solar Cells

The development of depleted bulk heterojunction QD architecture is one way to overcome the limitations in depleted planar device structures. The concept of bulk heterojunctions was first proposed in the field of organic photovoltaics, which emphasized the importance of large interfacial contacts between donor and acceptor for effective charge photogeneration. In bulk heterojunction QD solar cells, an interpenetrating layer can be formed by infiltrating QDs into the n-type mesoporous or nanowire, or nanorod metal oxides. As shown in Fig. 6.5b, ZnO nanowire arrays based on bulk heterojunction solar cells can reduce the distance of the electron-transport path compared to the planar heterojunction case. A typical example, Barkhouse, et al. deposited large-size TiO_2 nanoparticles (150–250 nm) as a porous layer on a dense small-sized TiO_2 (10–30 nm) layer so that PbS CQDs of 3.7 nm could well infiltrate into the porous TiO_2 layer. As a result, light absorption remarkably improved leading to higher photocurrent density [40]. Recently, Pradhan et al. used a mixed nanocrystal layer with ZnO and PbS QDs, which effectively reduced interfacial recombination and electron transfer barrier, resulting in over 10% efficiency [41].

6.4 Quantum Dots Sensitized Solar Cells

In 1991, a novel structure of dye-sensitized solar cells (DSSCs) was first introduced as a promising alternative to conventional p-n junction first-generation devices. DSSCs, composed of a natural dye or organic dye-sensitized photoanode, a liquid redox electrolyte (I^-/I_3^-), and a counter electrode, has attracted considerable attention owing to their lightweight, flexible, low-cost, and simple manufacturing processes and have a high potential for commercialization. To promote the DSSC, dyes were replaced by new types of light-absorbing semiconductors, i.e., quantum dots with unique optoelectronic properties, and thus QDSSCs were introduced. QDs with a diameter of <10 nm have advantageous features such as tunable bandgap by controlling size, multiple exciton generation (MEG), stability in atmospheric oxygen and water, high absorption coefficient, large dipole moments, etc. The MEG process can generate multiple electron–hole pairs with a high-energy photon instead of a single pair in a bulk material, which can generate higher photocurrents. However, research on QDSSCs has faced challenges such as finding the right charge carriers for different types of QDs with appropriate redox electrolytes without corrosion of QDs and a decisive combination of QDs with other layers for optimal photovoltaic properties. Several challenges to the fabrication of efficient and stable QDSSC devices have been overcome and solved.

6.4.1 Structure and Working Principles of QDSSC

The QDSSCs in which dyes are replaced by QDs have basic components of [42]:

- Working electrode: a mesoporous layer of a wide-bandgap semiconductor like TiO_2 or ZnO coated on transparent conductive FTO glass
- Sensitizer: a thin layer of QDs adsorbed on the surface of mesoporous semiconductors for light absorption
- Electrolyte: redox couple which can be either a solid or a liquid (I_3^-/I^-, S_x^{2-}/S^{2-} etc.)
- Counter electrode: a thin layer of platinum, graphene, gold, copper sulfate, etc. coated on FTO glass

A schematic structure and working principles of a QDSSC are illustrated in Fig. 6.6. The working principles are similar to the DSSCs as follows:

- Electron–hole pairs are formed by photoexcitation in QDs under photon illumination.
- An electron is transferred from the conduction band of QD to the conduction band of the wide-bandgap semiconductor such as TiO_2 and ZnO, leaving a hole in the valence band of the QD.
- The excited state of QD is restored by electron injection from the redox reaction in the electrolyte.
- The injected electron in TiO_2 travels from the photoanode to the counter electrode through the external load.
- The electron is received by the redox couple.

- A voltage is generated due to the difference between the conduction band energy level of the photoanode semiconductor and the electrolyte redox potential and electrical power is generated.

Compared to bulk materials, the zero-dimensional size of the QDs facilitates the electron injection at a faster rate than the recombination rate which allows electron transfer from the redox reaction before the electron has the chance to recombine with the hole.

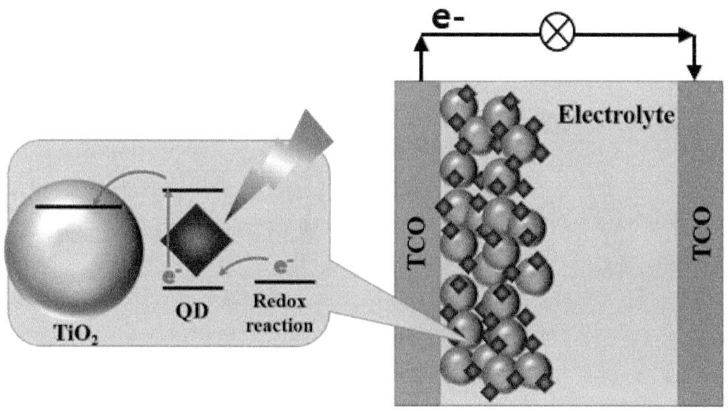

Figure 6.6 Schematic structure and working of QDSSC.

6.4.2 Components of QDSSC

6.4.2.1 Photoanode

The photoanode is the working electrode in QDSSCs and consists of semiconducting n-type metal oxides such as TiO_2, ZnO, and SnO_2 deposited on a transparent conductive oxide (FTO or ITO) glass. ITO films with ideal transparency and resistivity (of ~10^{-4} Ω cm) are widely used in optoelectronic devices. However, the low resistivity of ITO could decline via calcination in the QDSSC's fabrication process. Fabrication of the photoanode involves a high temperature (450 °C) process to improve the electrical properties of wide-bandgap metal oxide. At temperatures above 300°C, the sheet resistance of ITO film increases drastically leading to lower performance. FTO substrates with 70–80% transmittance and sheet resistance of <12 Ω are highly applicable in QDSSC applications.

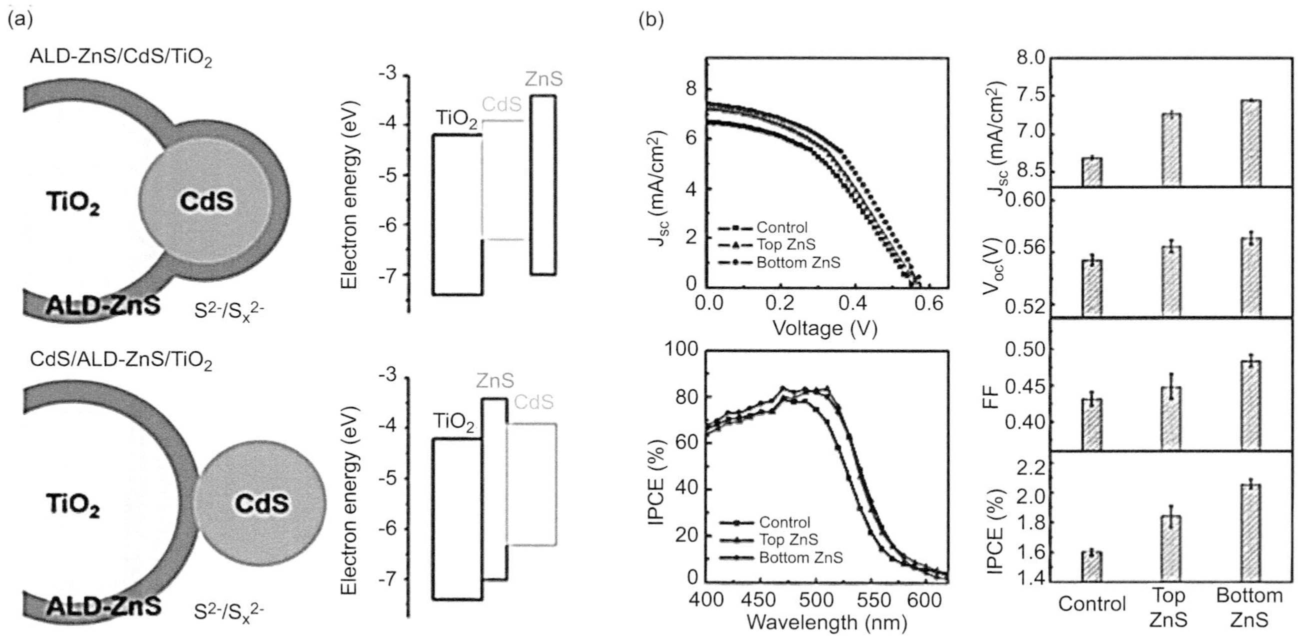

Figure 6.7 (a) Structures and band diagram of CdS-sensitized TiO_2 photoanode with top and bottom ZnS IPL. (b) *J–V* curves and photovoltaic parameters of the cells without and with ZnS IPL. Reprinted with permission from [44], Copyright 2021, Elsevier.

The main role of photoanode in QDSSCs is to support QDs molecules to transfer photo-excited electrons to a conductive substrate. A photoanode having sufficient surface area can provide supporting space more for QDs for efficient light harvesting. However, the rate of electron generation should be in balance with the device's current circulation to attain the highest conversion efficiency. High electron concentration could lead to recombination in interfaces. Hence, the electron-transport properties of photoanode are crucial for QDSSCs. In general, a combination of FTO and semiconductors such as TiO_2, ZnO, and SnO_2 are widely used as a photoanode to increase charge mobility in QDSSCs. Wide-bandgap TiO_2 is the most promising semiconductor of DSSCs, QDSSCs, and perovskite solar cells due to its low cost, non-toxicity, and biocompatibility. ZnO (E_g ~3.37 eV) with a large exciton binding energy of ~60 eV is also been used as TiO_2/ZnO hybrid arrays photoanodes to improve the performance of CdS/CdSe QDSSCs [43].

It is known that a photoanode based on TiO_2 nanostructure with a large surface area for QDs deposition improves the device performance of QDSSC. For example, CdS QDs distribution on a TiO_2 photoanode can be improved by ZnS interfacial passivation layer (IPL) deposited by an atomic layer deposition (ALD) [44]. The top ZnS IPL ($<\sim$0.5 nm) deposited after CdS QDs sensitization on the photoanode (i.e., ZnS/CdS/TiO_2) enhance light absorbance by acting as an anti-reflective coating, resulting in increased photocurrent (Fig. 6.7a, top). The bottom ZnS IPL deposited on the TiO_2 photoanode before QDs sensitization (i.e., CdS/ZnS/TiO_2) generates a uniform distribution of QD microaggregates by modification of the surface energy of TiO_2 (Fig. 6.7a, bottom), which improves the photo-generated charge-carrier injection from QDs to TiO_2 nanoparticles. Furthermore, ZnS IPLs serve as a recombination barrier layer, resulting in the enhancement of PCE by approximately 15% and 30% in cells with top and bottom ZnS IPLs, respectively (Fig. 6.7b) [44].

To enhance the light-harvesting efficiency in QDSSC, yttrium oxide–doped ZnO nanorods can be employed as photoanode. Yttrium oxide (Y_2O_3) enabled the growth of ZnO nanorods with a large surface area for QDs coating, leading to the enhancement of the light-harvesting ability of the QDs. The higher the Y_2O_3 concentration, the more QDs were loaded and significantly affected the chemical and surface deformation of the nanorods, improving the QDSSC performance by 44% compared to the un-doped ZnO nanorods [50]. Table 6.1 shows the comparison of photovoltaic parameters of QDSSCs according to various quantum dot-sensitized photoanodes.

Table 6.1 Quantum dots sensitized photoanodes and their effects on the photovoltaic performance of QDSSCs

Photoanode	QDs	J_{sc} (mA/cm^2)	V_{oc} (mV)	FF	PCE (%)	Ref.
TiO_2	CdS-dye CSC	6.09	710	0.64	2.77	50
TiO_2 TIR	CdS/CdSe/ZnS	5.89	690	0.66	2.71	51
TiO_2	CdSe	13.40	440	0.44	2.6	52
TiO_2 NFs-NTs	CdS/CdSe/ZnS	8.80	610	0.50	2.8	53
ZnONWs + ZnO MSs	CdS	9.01	511	0.51	2.35	54
ZnONRs/g-C_3N_4	CdS	11.1	650	0.34	2.43	55
TiO_2 NRs	CdS	3.68	590	0.49	3.14	56
ZnO NRs	CdS/CdSe/ZnS	9.93	610	0.52	3.14	57
TiO_2 NFs–ZnO NSs	CdS/CdSe/ZnS	13.64	511	0.44	3.05	58
TiO_2 NPs	ZCISe	11.11	590	0.63	4.13	59
TiO_2/ZnO NSs	CdS/CdSe	16.11	512	0.55	4.57	60
TiO_2 NRs	PbS/CdS	17.38	515	0.50	4.43	61
TiO_2 MSs	CdS/CdSe/ZnS	13.89	540	0.54	4.05	62
TiO_2 NTs	CdS/CdSe/ZnS	10.81	689	0.62	4.61	63
TiO_2 MPs	CdS/CdSe/ZnS	14.05	571	0.52	4.15	64
ZnO NWs	CdS/CdSe	17.30	627	0.38	4.15	65
ZnO TP	CdS/CdSe	13.85	722	0.42	4.24	66
ZnO NDs	CdS/CdSe	16.00	620	0.49	4.86	67

Photoanode	QDs	J_{sc} (mA/cm^2)	V_{oc} (mV)	FF	PCE (%)	Ref.
TiO_2NWsNWs	CdS/CdSe	17.98	465	0.50	4.20	68
TiO_2NWs–ZnO NSs	CdS/CdSe	16.11	512	0.55	4.57	69
TiO_2/ZnO NAs	CdS/CdSe	15.65	744	0.50	5.92	70
TiO_2 NPs	CdSe/ZnS	15.54	563	0.61	5.53	71
TiO_2 NSs	CdSe/ZnS/SiO_2	16.95	591	0.50	5.01	72
TiO_2 NRs	CdSe/ZnS	17.50	563	0.61	5.96	73
TiO_2 NWsTiO_2 NSs–ZnO NRs	CdS/CdSe	19.19	517	0.54	5.38	74
ZnO NDs/TiO_2 NPs	CdS/CdSe	15.34	659	0.53	5.36	75
ZnO NPs + ZnO MSs	CdS/CdSe/ZnS	17.13	560	0.53	5.08	76
ZnONRs + ZnO TPs	CdS/CdSe/ZnS	16.56	703	0.45	5.24	77
TZ3	CdS/CdSe/ZnS	16.60	622	0.59	6.05	78
TiO_2MPsNWs	CdS/CdSe	19.32	531	0.59	6.01	79

6.4.2.2 QD sensitizers

In QDSSCs, the photosensitizers are QDs, which are formed on the surface of metal oxide photoanode. Sensitizers should have a high absorption coefficient with a proper bandgap to enhance light-harvesting efficiency. The roles of the QD as a sensitizer are (1) to absorb a photon and generate an electron under illumination, and (2) to inject photo-generated electrons into the porous metal oxide photoanode.

For efficient electron injection, the LUMO energy level of QD should be higher than the CB of metal oxide (e.g., TiO_2). In QDSSCs, for high electron density at photoanode and photocurrent density (J_{sc}), the energy level of QD should match that of metal oxide semiconductors where the photo-generated electrons can be transferred easily. The method to fabricate a high-performance QDSSC can be categorized into *in situ* and *ex situ* methods.

- The *in situ* fabrication method, such as CBD and SILAR technique, is widely used to grow QDs on nanostructured metal oxide semiconductors, and results in dense coverage of QDs.

- The *ex situ* process is a linker-assisted assembly to deposit QDs on various photoanode materials and structures. For example, the functionalization of a metal oxide surface with bifunctional linker molecules can bind to QDs and form a link between QD and metal oxide photoanode.

Although *in situ* process is preferred over the *ex situ* technique due to the simple procedure, the precise control of QD size is challenging and the density of surface trap state in QDs is very high because limited size hinders crystal network formation. Extensive research has been conducted to enhance the QDSSCs efficiency using a variety of QDs (Table 6.2). Among all the QDs sensitizers, QDSSCs with Cd-based chalcogenide QDs such as CdS, CdSe, and CdTe exhibited higher stability, despite partial degradation under visible illumination. In addition, using a co-sensitizer with two or three QDs is a well-established method of expanding the light absorption range to improve cell performance.

With the recent development of QDs production, QD materials with exceptional properties can be designed compared to bulk materials. For example, $CsPbI_3$ QDs with a tunable bandgap in the range of 1.75–2.13 eV resulted in remarkable efficiency for QDSSCs with a PCE of 13.4% [82]. This outstanding performance of $CsPbI_3$ QDSSCs is attributed to high carrier mobility, high loading capacity of QDs, wide wavelength absorption range, low charge-recombination rate, and fast electron extraction from $CsPbI_3$ QD films. Another example is Zn–Cu–In–Se (ZCISe) QDs, where the Cu content may affect the optical properties of QDs, thus achieving the highest PCE of 12.57% with a Cu/In molar ratio of 0.7 [83]. Sequential deposition of ZCISe and CdSe QDs on mesoporous TiO_2 is another example, which significantly improves the photogeneration efficiency due to the wide absorption range of ZCISe-alloyed QDs and high extinction coefficient of CdSe QDs. A high PCE of 12.75% was achieved with V_{oc} of 0.752 V, J_{sc} of 27.39 mA cm^{-2}, and fill factor (FF) of 0.619 [84]. Table 6.2 summarizes the progressive improvement of QDSSCs with different QD sensitizers.

Table 6.2 Photovoltaic performance of QDSSCs with different QD sensitizers

QD	Photoanode	CEs	Electrolyte	PCE (%)	Ref.
Ag_2S/ZnS	TiO_2 NPs	Pt	I^-/I_3^-	1.76	85
CdSe/ZnS	TiO_2 NTs	Cu_2S	S^{2-}/S_x^{2-}	1.70	86
$CdS/Bi_2S_3/ZnS$	TiO_2 NPs	Cu_2S	S^{2-}/S_x^{2-}	2.52	87
CdSe/ZnS	TiO_2 NPs	Cu_2S	S^{2-}/S_x^{2-}	3.84	88
Cu_2ZnSnS_4/ZnS	TiO_2 NPs	Cu_2S	S^{2-}/S_x^{2-}	4.70	89
CdS/CdSe/ZnS	TiO_2 NPs	Au	S^{2-}/S_x^{2-}	4.22	90
CdS/CdSe	TiO_2 MSs	Pt	S^{2-}/S_x^{2-}	4.81	91
CdSe/ZnS	TiO_2 NPs	Cu_2S	S^{2-}/S_x^{2-}	5.42	92
$CuInSe_xS_{2x}/ZnS$	TiO_2 NPs	Cu_2S	S^{2-}/S_x^{2-}	5.51	93
CdS/CdSe/ CdS/ZnS	TiO_2 NTs	Cu_2S	S^{2-}/S_x^{2-}	5.47	94
CdTe/CdSeS	TiO_2	Cu_2S/ brass	S^{2-}/S_x^{2-}	7.00	95
$CuInS_2/ZnS$	TiO_2 NPs	Cu_2S	S^{2-}/S_x^{2-}	7.04	96
PbS/CdS/ZnS	TiO_2 NPs	Cu_2S	TiO_2 NPs	7.19	97
CuInSe/ZnS	TiO_2 NPs	Cu_2S	S^{2-}/S_x^{2-}	8.10	98
ZCIS	TiO_2	Cu_2S/ brass	S^{2-}/S_x^{2-}	8.47	99
$CdSe_xTe_{1x}$/ ZnS/SiO_2	TiO_2 NPs	Cu_2S	S^{2-}/S_x^{2-}	8.21	100
$CdSe_xTe_{1x}$/ $CdS/ZnS/SiO_2$	TiO_2 NPs	Cu_2S	S^{2-}/S_x^{2-}	9.48	101
ZnCuInSe/ZnS	TiO_2 NPs	Carbon	S^{2-}/S_x^{2-}	11.61	102
Zn–Cu–In–Se	TiO_2	MC/Ti	S^{2-}/S_x^{2-}	12.57	85

6.4.2.3 Redox electrolytes

An electrolyte is responsible for transporting charges between electrodes through mobile ions. The electrolytes can be classified into three main groups: liquid, solid, and quasi-solid electrolytes. In DSSCs the most used electrolyte is based on I^-/I_3^- redox couple.

The I^-/I_3^- redox electrolyte is corrosive to QDs and degrades the QDSSC device performance. Polysulfide electrolyte, a solution of sulfide salt and sulfur, is commonly used for QDSSCs for its high compatibility with QD materials and high charge extraction ability. However, using polysulfide electrolytes leads to the generation of QD oxides which cause poor device performance.

In general, the short-circuit current density of a solar cell is directly affected by the light-harvesting intensity and charge separation/transport efficiency. FF is affected by carrier transfer interfaces and V_{oc} is determined by energy matching between sensitizer and charge-transport material. In QDSSCs, FF can be affected by internal recombination, resistance, and concentration gradients in the electrolyte. V_{oc} depends on redox potential energy and photoanode Fermi level. Therefore, the electrolyte must have a negative redox potential compared to the QD oxidation potential to conduct high charge mobility in the medium. Redox electrolytes need to have high ionic mobility, fast charge-transport kinetics, and chemical and thermal stability under operating conditions. However, for the long-term application of QDSSCs, unwanted properties such as leakage, evaporation of electrolyte solvent, and photodegradation should be controlled.

It is also worth noting that a novel design using Ag plasmonic embedded titania and liquid sulfide electrolyte instead of sulfide gel electrolyte is effective for high-efficiency CdS QDSSCs. Gel-based cells have lower solar cell performance than liquid electrolytes, but the Ag/TiO_2/CdS electrode outperforms the TiO_2/CdS photoanode [103]. A crystal-based solid-state electrolyte made of plastic crystal succinonitrile and sodium sulfide (Na_2S) with ionic conductivity and charge-transfer ability enhances S_n^{2-}/S^{2-} mobility and improved the kinetics of $S_n^{2-} \rightleftharpoons S^{2-}$ interconversion [104]. Another novel approach is to utilize polyelectrolyte with superior water-absorbing and water-holding capacity. The high conductivity of the sodium polyacrylate gel and its strong affinity to the TiO_2 surface lead to a PCE of 8.54% [105]. Table 6.3 summarizes the effect of electrolytes on QDSSC performance.

Table 6.3 Photovoltaic parameters of QDSSCs with different electrolytes

Electrolyte	QDs	CE	V_{oc} (V)	J_{sc} (mA/cm^2)	FF	PCE (%)	Ref.
Spiro-OMeTAD	$Pb_5Sb_8S_{17}$	Au	0.48	11.9	0.31	1.76	106
Spiro-OMeTAD	PbS–CdS	Au	0.56	4.58	0.57	1.46	107
P_3HT	CdS	Au	0.67	4.31	0.50	1.42	108
Spiro-OMeTAD	PbS	Au	0.52	13.56	0.58	4.10	109
PAM-MBAf	CdS/CdSe	Cu_2S	0.61	12.80	0.55	4.30	110
[DHexBIm] [SCN]	CdS/CdSe	PbSe	0.60	12.58	0.56	4.26	111
PEGDMEh–Na_2S–S	CdSe/CdS	Carbon	0.72	17.84	0.42	5.45	112
P_3HT	PbS–CuS	Au	0.60	20.70	0.65	8.07	113
(PAAS) gel	CdSeTe	Cu_2S	0.664	20.63	0.62	8.54	114
(CMC-Na)	CdSeTe	Cu_2S	0.667	21.89	0.63	9.21	115
S^{2-}/S_x^{2-}	CdSeTe	Cu_2S	0.70	21.20	0.65	9.64	116
S^{2-}/S_x^{2-}	ZCISe	Cu_2S	0.63	26.01	0.58	9.56	117
S^{2-}/S_x^{2-}	CdSeTe	GHc-CuS	0.78	20.69	0.66	10.71	118
S^{2-}/S_x^{2-}	ZCISe	MC/Ti	0.74	25.25	0.62	11.61	119
S^{2-}/S_x^{2-}	Cu–In–Ga–Se	MC/Ti	0.74	25.01	0.62	11.49	120
S^{2-}/S_x^{2-}	$CdSe_{0.65}Te_{0.35}$	MC/Ti	0.80	20.68	0.68	11.16	121
S^{2-}/S_x^{2-}	ZCISe	N-MCd	0.76	25.53	0.63	12.23	122

6.4.3 Suppression of recombination in QDSSCs

In QDSSCs, charge recombination at the metal oxide/QDs/electrolyte interfaces is an important factor in lowering cell efficiency. An innovative method to suppress recombination is to utilize a TiO_2 compact layer at FTO/electrolyte interface. For example, the optimal thickness of the TiO_2 compact layer (~25 nm) on FTO increases the interfacial charge transfer, thereby increasing the PCE by about 24% compared to a reference cell [45]. Surface roughening and modification with CdS QDs on TiO_2 mesoporous film also reduce the recombination and increase the electron lifetime and transport rate, leading to higher light harvesting and reduced charge-transfer resistance [46].

In ZnO nanorod-based QDSSC, the nanorod surface can be roughened by ZnO nanoparticles (i.e., ZnO NR/ZnO NP), where the nanorod acts as a backbone of the photoanode and nanoparticles increase surface area to deposit more QDs [47]. The layered structure of photoanode in ZnO nanorod-based CdSe/CdS/PbS QDSSCs with copper sulfide (CuS) counter electrode showed higher PCEs with a high rate of photo injected electrons and low recombination within the electrolyte [48]. To impede the recombination in QD metal oxide surface, an alumina (Al_2O_3) barrier (i.e., ZnO NRs/Al_2O_3) can be used, forming ZnO NRs/Al_2O_3/CdS/CdSe structure. The Al_2O_3 in the ZnO NRs/Al_2O_3 core–shell acts as a barrier, helps with more QDs deposition, increases electron lifetime, and reduces recombination loss, thereby improving V_{oc} due to higher conduction band edge [49].

References

1. van der Wiel, W. G. (2003) *Rev. Mod. Phys.*, **75**, pp. 1–22.
2. Henglein, A. (1982) *J. Phys. Chem.*, **86**, pp. 2291.
3. Rossetti, R., Nakahara, S., and Brus, L. E. (1983) *J. Chem. Phys.*, **79**, pp. 1086.
4. Carey, G. H., Abdelhady, A. L., Ning, Z., Thon, S. M., et al. (2015). *Chem. Rev.*, **115**, pp. 12732–12763.
5. Murray, C. B., Kagan, C. R., and Bawendi, M. G. (2000) *Annu. Rev. Mater. Sci.*, **30**, pp. 545.
6. Kamat, P. V. (2013) *J. Phys. Chem. Lett.*, **4**, pp. 908–918.
7. Yu, X.-Y., Lei, B.-X., Kuang, D.-B., and Su, C.-Y. (2011) *Chem. Sci.*, **2**, pp. 1396–1400.
8. Murray, C. B., Norris, D. J., and Bawendi, M. G. (1993) *J. Am. Chem. Soc.*, **115**, pp. 8706–8715.
9. Alivisatos, A. P. (1996) *Science*, **271**, pp. 933–937.
10. Semonin, O. E., Luther, J. M., and Beard, M. C. (2012) *Mater. Today*, **15**, pp. 508–515.
11. Liu, W. C., Guo, B. L., Wu, X. S., Zhang, F. M., et al. (2013) *J. Mater. Chem. A*, 1, pp. 3182–3186.
12. Amit, Y., Eshet, H., Faust, A., Patllola, A., et al. (2013) *J. Phys. Chem. C*, **117**, pp. 13688–13696.
13. Amit, Y., Li, Y., Frenkel, A. I., and Banin, U. (2015) *ACS Nano*, **9**, pp. 10790–10800.

14. Lin, Z., Franceschetti, A., and Lusk, M. T. (2011) *ACS Nano*, **5**, pp. 2503–2511.
15. Pijpers, J., Ulbricht, R., Tielrooij, K., Osherov, A., et al. (2009) *Nature Physics*, **5**, pp. 811–814.
16. Tisdale, W. A., Williams, K. J., Timp, B. A., Norris, D. J., et al. (2010) *Science*, **328**, pp. 1543–1547.
17. Paz-Soldan, D., Lee, A., Thon, S. M., Adachi, M. M., et al. (2013) *Nano Lett.*, **13**, pp. 1502–1508.
18. Kim, S., Kim, J. K., Gao, J., Song, J. H., et al. (2013) *ACS Appl. Mater. Interfaces*, **5**, pp. 3803–3808.
19. Ekimov, A. I. and Onushchenko, A. A. (1981) *JETP Lett.*, **34**, pp. 345–349.
20. Rossetti, R., Nakahara, S., and Brus, L. E. (1983) *J. Chem. Phys.*, **79**, pp. 1086–1088.
21. Wu, X., Liu, H., Liu, J., Haley, K. N., et al. (2003) *Nat. Biotech.*, **21**, pp. 41–46.
22. Li, H., Zhang, Y., Wang, X., Xiong, D., and Bai, Y. (2007) *Mater. Lett.*, **61**, pp. 1474–1477.
23. Li, L., Yu, B., and You, T. (2015) *Biosens. Bioelectron.*, **74**, pp. 263–269.
24. Zhang, T., Li, W., Huang, K., Guo, H., et al. (2021) *Nat. Comm.*, **12**, pp. 5265.
25. Son, D. I., Kwon, B. W., Park, D. H., Seo, W.-S., et al. (2012) *Nat. Nanotech.*, **7**, pp. 465–471.
26. Liu, Y., Dong, Y., Zhu, T., Ma, D., et al. (2021) *J. Am. Chem. Soc.*, **143**, pp. 15606–15615.
27. Xue, Y., Yang, F., Yuan, J. Y., Zhang, Y. N., et al. (2019) *ACS Energy Lett.*, **4**, pp. 2850–2858.
28. Zhao, Q., Hazarika, A., Chen, X. H., Harvey, et al. (2019) *Nat. Comm.*, **10**, pp. 2842.
29. Kamat, P. V. (2012) *Acc. Chem. Res.*, **45**, pp. 1906–1915.
30. Kim, H. S., Lee, C. R., Im, J. H., Lee, K. B., et al. (2012) *Sci. Rep.*, **2**, pp. 591.
31. Pattantyus-Abraham, A. G., Kramer, I. J., Barkhouse, A. R., Wang, X., et al. (2010). *ACS Nano*, **4**, pp. 3374–3380.
32. Clifford, J. P., Johnston, K. W., Levina, L., and Sargent, E. H. (2007) *Appl. Phys. Lett.*, **91**, pp. 253117.
33. Koleilat, G. I., Levina, L., Shukla, H., Myrskog, S. H., et al. (2008) *ACS Nano*, **2**, pp. 833–840.

34. Choi, M.-J., Oh, J., Yoo, J.-K., Choi, J., et al. (2014) *Science*, **7**, pp. 3052–3060.
35. Johnston, K. W., Pattantyus-Abraham, A. G., Clifford, J. P., Myrskog, S. H., et al. (2008) *Appl. Phys. Lett.*, **92**, pp. 151115.
36. Piliego, C., Protesescu, L., Bisri, S. Z., Kovalenko, M. V., and Loi, M. A. (2013) *Energy Environ. Sci.*, **6**, pp. 3054–3059.
37. Tang, J., Kemp, K. W., Hoogland, S., Jeong, K. S., Liu, H., et al. (2011). *Nat. Mater.*, **10**, pp. 765–771.
38. Zhang, J. B., Gao, J. B., Church, C. P., Miller, E. M., et al. (2014) *Nano Lett.*, **14**, pp. 6010–6015.
39. Carey, G. H., Abdelhady, A. L., Ning, Z., Thon, S. M., et al. (2015) *Chem. Rev.*, **23**, pp. 12732–12763.
40. Barkhouse, D. A. R., Debnath, R., Kramer, I. J., Zhitomirsky, D., et al. (2011) *Adv. Mater.*, **23**, pp. 3134–3138.
41. Pradhan, S., Stavrinadis, A., Gupta, S., and Konstantatos, G. (2017) *ACS Appl. Mater. Interfaces*, **9**, pp. 27390–27395.
42. Surana, K., Mehra, R. M., Bhattacharya, B., Rhee, H. W., et al. (2015) *Renew. Sustain. Energy Rev.*, **52**, pp. 1083–1092.
43. Feng, H. L., Wu, W. Q., Rao, H. S., Wan, Q., et al. (2015) *ACS Appl. Mater. Interfaces*, **7**(9), pp. 5199–5205.
44. Jung, E. S., Basit, M. A., Abbas, M. A., Ali, I., et al. (2020) *Sol. Energy Mater. Sol. Cells*, **218**, pp. 110753.
45. Choi, H., Nahm, C., Kim, J., Moon, J., et al. (2012) *Curr. Appl. Phys.*, **12**, pp. 737.
46. Jeong, M.-S., Son, M.-K., Kim, S.-K., Park, S., et al. (2014) *Electron. Mater. Lett.*, **10**, pp. 621–626.
47. Kim, S.-K., Raj, C. J., and Kim, H.-J. (2014) *Electron. Mater. Lett.*, **10**, pp. 1137–1142.
48. Raj, J. C., Karthick, S. N., Park, S., Hemalatha, K. V., et al. (2014) *J. Power Sources*, 248, pp. 439–446.
49. Soo-Kyoung, K., Son, M. K., Park, S., Jeong, M. S., et al. (2014) *Mater. Chem. Phys.*, **143**, pp. 1404–1409.
50. Son, M. Y., Seo, H., Soo-Kyoung, K., Park, S., et al. (2014) *Phys. Status Solidi A*, **211**, pp. 1726–1731.
51. Buatong, N. Tang, I. M., and Pon-On, W. (2015) *Nanoscale Res. Lett.*, **10**, pp. 146.
52. Yun, H. J., Paik, T., Diroll, B., Edley, M. E., and Baxter, J. B. C. B. (2016) *ACS Appl. Mater. Interfaces*, **8**, pp. 14692–14700.

53. Han, H., Sudhagar, P., Song, T., Jeon, Y., et al. (2013) *Chem. Commun.*, **49**, pp. 2810–2812.

54. Chetia, T. R., Barpuzary, D., and Qureshi, M. (2014) *Phys. Chem. Chem. Phys.*, **16**, pp. 9625–9633.

55. Chetia, T. R., Ansari, M. S., and Qureshi, M. (2016) *J. Mater. Chem. A*, **4**, pp. 5528–5541.

56. Wan, J., Liu, R., Tong, Y., Chen, S., et al. (2016) *Nanoscale Res. Lett.* **10**, pp. 1.

57. Tian, J., Zhang, Q., Uchaker, E., Liang, Z., et al. (2013) *J. Mater. Chem. A*, **1**, pp. 6770–6775.

58. Cao, Y., Dong, Y. J., Chen, H. Y., Kuang, D. B., and Su, C. Y. (2016) *RSC Adv.*, **6**, pp. 78202–78209.

59. Shen, G., Du, Z., Pan, Z., Du, J., and Zhong, X. (2018) *ACS Omega*, **3**, pp. 1102–1109.

60. Fen, H. L., Wu, W. Q., Rao, H. S., Wan, Q., et al. (2015) *ACS Appl. Mater. Interfaces*, **7**, pp. 5199–5205.

61. Gonfa, B. A., Mee Rahn, K., Delegan, N., Tavares, A. C., et al. (2015) *Nanoscale*, **7**, pp. 10039–10049.

62. Wu, D., Shi, X., Dong, H., Zhu, F., et al. (2014) *J. Mater. Chem. A*, **2**, pp. 16276–16284.

63. Huang, H., Pan, L., Lim, C. K., Gong, H., et al. (2013) *Small*, **9**, pp. 3153–3160.

64. Du, X., He, X., Zhao, L., Chen, H., et al. (2016) *J. Power Sources*, **332**, pp. 1–7.

65. Seol, M., Kim, H., Tak, Y., and Yong, K. (2010) *Chem. Commun.*, **46**, pp. 5521–5523.

66. Cheng, H.-M., Huang, K.-Y., Lee, K.-M., Yu, P., et al. (2012) *Phys. Chem. Chem. Phys.*, **14**, pp. 13539–13548.

67. Chetia, T. R., Ansari, M. S., and Qureshi, M. (2015) *ACS Appl. Mater. Interfaces*, **7**, pp. 13266–13279.

68. Rao, H.-S., Wu, W., Liu, Q. Y., Xu, Y. F., et al. (2014) *Nano Energy*, **8**, pp. 1–8.

69. Feng, H. L., Wu, W., Rao, Q. H. S., Wan, Q., et al. (2015) *ACS Appl. Mater. Interfaces*, **7**, pp. 5199–5205.

70. Zhao, H., Huang, F., Hou, J., Liu, Z., et al. (2016) *ACS Appl. Mater. Interfaces*, **8**, pp. 26675–26682.

71. Du, Z., Zhang, H,. Bao, H., and Zhong, X. (2014) *J. Mater. Chem. A*, **2**, pp. 13033–13040.

72. Zhou, H., Li, L., Jiang, D., Lu, Y., and Pan, K. (2016) *RSC Adv.*, **6**, pp. 67968–67975.
73. Zhang, W., Zeng, X., Wang, H., Fang, R., et al. (2016) *RSC Adv.*, **6**, pp. 33713–33722.
74. Feng, H. L., Wu, W., Rao, Q. H. S., Li, L. B., et al. (2015) *J. Mater. Chem. A*, **3**, pp. 14826–14832.
75. Jin, B. B., Wang, Y. F., and Zeng, H. (2016) *Chem. Phys. Lett.*, **660**, pp. 76–80.
76. Tian, J., Lv, L., Wang, X., Fei, C., et al. (2014) *J. Phys. Chem. C*, **118**, pp. 16611–16617.
77. Zhu, Z., Qiu, J., Yan, K., and Yang, S. (2013) *ACS Appl. Mater. Interfaces*, **5**, pp. 4000–4005.
78. Wu, Q., Hou, J., Zhao, H., Liu, Z., et al. (2018) *Dalton Trans.*, **47**, pp. 2214–2221.
79. Xu, Y. F., Wu, W. Q., Rao, H. S., Chen, H. Y., et al. (2015) *Nano Energy*, **11**, pp. 621–630.
80. Soo-Kyoung, K., Gopi, C. V. V. M., Rao, S. S., Punnoose, D., and Hee-Je, K. (2016) *Appl. Surf. Sci.* **365**, pp. 136–142.
81. Jeong, M. S., Son, M. K., Soo-Kyoung, K., Park, S., et al. (2014) *Thin Solid Films*, **570**, pp. 310–314.
82. Sanehira, E. M., Marshall, A. R., Christians, J. A., Harvey, S. P., et al. (2017) *Sci. Adv.*, **10**, pp. 4204.
83. Zhang, L., Pan, Z., Wang, W., Du, J., et al. (2017) *J. Mater. Chem. A*, **5**, pp. 21442–21451.
84. Wang, W., Feng, W., Du, J., Xue, W., et al. (2018) *Adv. Mater.*, **30**, pp. 1705746.
85. Wang, Y., Zhang, Q., Li, Y., and Wang, H. (2015) *Nanoscale*, **7**, pp. 6185–6192.
86. Salant, A., Shalom, M., Hod, I., Faust, A., et al. (2010) *ACS Nano*, **4**, pp. 5962–5968.
87. Mu, L., Liu, C., Jia, J., Zhou, X., and Lin, Y. (2013) *J. Mater. Chem. A*, **1**, pp. 8353–8357.
88. Pedro, V. G., Xu, X., Mora-Sero, I., and Bisquert, J. (2010) *ACS Nano*, **4**, pp. 5783–5790.
89. Lee, S. H., Jung, C., Jun, Y., and Kim, S. W. (2015) *Opt. Mater.*, **49**, pp. 230–234.
90. Lee, Y. L. and Lo, Y. S. (2009) *Adv. Funct. Mater.*, **19**, pp. 604–609.

91. Yu, X. Y., Liao, J. Y., Qiu, K. Q., Kuang. D. B., and Su, C. Y. (2011) *ACS Nano*, **5**, pp. 9494–9500.

92. Zhang, H., Cheng, K., Hou, Y. M., Fang, Z., et al. (2012) *Chem. Commun.*, **48**, pp. 11235–11237.

93. McDaniel, H., Fuke, N., Makarov, N. S., Pietryga, J. M., and Klimov, V. I. (2013) *Nat. Commun.*, **4**, pp. 2887.

94. Choi, Y., Seol, M., Kim, W., and Yong, K. (2014) *J. Phys. Chem. C*, **118**, pp. 5664–5670.

95. Kim, J.-Y., Yang, J., Yu, J. H., Baek, W., et al. (2015) *ACS Nano*, **9**, pp. 11286–11295.

96. Jiao, S., Wang, J., Shen, Q., Li, Y., and Zhong, X. (2016) *J. Mater. Chem. A*, **4**, pp. 7214–7221.

97. Bai, B., Kou, D., Zhou, W., Zhou, Z., et al. (2016) *J. Power Sources*, pp. **318**, 35–40.

98. Yue, L., Rao, H., Du, J., Pan, Z., et al. (2018) *RSC Adv.*, **8**, pp. 3637–3645.

99. Zhao, K., Pan, Z., Mora-Sero, I., Canovas, E., et al. (2015) *J. Am. Chem. Soc.*, **137**, pp. 5602–5609.

100. Yang, J., Wang, J., Zhao, K., Izuishi, T., et al.. (2015) *J. Phys. Chem. C*, **119**, pp. 28800–28808.

101. Du, J., Du, Z. L., Hu, J.-S., Pan, Z. X., et al. (2016) *J. Am. Chem. Soc.*, **138**, pp. 4201–4209.

102. Zhang, L., Pan, Z., Wang, W., Du, J., et al. (2017) *J. Mater. Chem. A*, **5**, pp. 21442–21451.

103. Kumar, N. P., Deepa, M., and Srivastava, A. K. (2015) *Phys. Chem. Chem. Phys.*, **17**, pp. 10040–10052.

104. Duan, J., Tang, Q., He, B., and Chen, H. (2015) *RSC Adv.*, **5**, pp. 33463–33467.

105. Feng, W., Li, Y., Du, J., Wang, W., and Zhong, X. (2016) *J. Mater. Chem. A*, **4**, pp. 1461–1468.

106. Chang, Y. C., Suriyawong, N., Aragaw, B. A., Shi, J. B., et al. (2016) *J. Power Sources*, **312**, pp. 86–92.

107. Lee, H., Leventis, H. C., Moon, S. J., Chen, P., et al. (2009) *Adv. Funct. Mater.*, **19**, pp. 2735–2742.

108. Qian, J., Liu, Q. S., Li, G., Jiang, K. J., et al. (2011) *Chem. Commun.*, **47**, pp. 6461–6463.

109. Zhang, Z., Shi, C., Chen, J., Xiao, G., and Li, L. (2017) *Appl. Surf. Sci.*, **410**, pp. 8–13.

110. Yu, Z., Zhang, Q., Qin, D., Luo, Y., et al. (2010) *Electrochem. Commun.*, **12**, pp. 1776–1779.

111. Dang, R., Wang, Y., Zeng, J., Huang, Z., et al. (2017) *J. Mater. Chem. A*, **5**, pp. 13526–13534.

112. Kim, H., Hwang, I., and Yong, K. (2014) *ACS Appl. Mater. Interfaces*, **6**, pp. 11245–11253.

113. Park, J. P., Heo, J. H., Im, S.-H., and Kim, S.-W. (2016) *J. Mater. Chem. A*, **4**, pp. 785–790.

114. Feng, W., Li, Y., Du, J., Wang, W., and Zhong, X. (2016) *J. Mater. Chem. A*, **4**, pp. 1461–1468.

115. Feng, W., Zhao, L., Du, J., Li, Y., and Zhong, X. (2016) *J. Mater. Chem. A*, **4**, pp. 14849–14856.

116. Pan, Z. and Zhong, X. (2016) *J. Mater. Chem. A*, **4**, pp. 18976–18982.

117. Wang, W., Jiang, G., Yu, J., Wang, W., et al. (2017) *ACS Appl. Mater. Interfaces*, **9**, pp. 22549–22559.

118. Zhang, H., Yang, C., Du, Z., Pan, D., and Zhong, X. (2017) *J. Mater. Chem. A*, **5**, pp. 1614–1622.

119. Du, J., Du, Z. L., Hu, J.-S., Pan, Z. X., et al. (2016) *J. Am. Chem. Soc.*, **138**, pp. 4201–4209.

120. Peng, W., Du, J., Pan, Z., Nakazawa, N., et al. (2017) *ACS Appl. Mater. Interfaces*, **9**, pp. 5328–5336.

121. Du, Z., Pan, Z., Fabregat-Santiago, F., Zhao, K., et al. (2016) *J. Phys. Chem. Lett.*, **7**, pp. 3103–3111.

122. Jiao, S., Du, J., Du, Z., Long, D., et al. (2017) *J. Phys. Chem. Lett.*, **8**, pp. 559–564.

Problems

6.1 What are QDs and their properties?

6.2 Describe the synthesis methods of QDs and their advantages.

6.3 Describe briefly the application areas of QDs.

6.4 Describe briefly quantum dot solar cell configurations and their advantages.

6.5 Explain the MEG.

6.6 What are quantum confinement effects, and why does bandgap energy of the QDs vary with their size?

Chapter 7

Organic–Inorganic Hybrid Solar Cells

A typical hybrid solar cell comprises a large area of inorganic material and an organic conductive absorber, which can be an organic hole conducting material or a blend of organic electron and hole conductors. Two techniques are used to interface inorganic with organic photovoltaic materials. The first method is to deposit a suspension of stabilized inorganic nanoparticles or nanorods in polymer solution; the second method is to deposit porous inorganic architectures, subsequently intercalating organic materials [1]. Although the inorganic nanoparticles used in the first method can be formed in a large range of shapes, sizes, and materials, which is beneficial to the photovoltaic application because of size-dependent properties resulting from the quantum confinement effects, this type of device is still limited by the discontinuous charge-transport process via hopping of electrons between the nanoparticles. In these types of solar cells, nanostructured metal oxides are first fabricated via various methods and then filtered with organic materials. Since the mobilities of metal oxides are significantly higher than those of organic materials, it offers an ultrafast pathway for charge-carrier transport. The charge-transport dynamics of devices is further improved by modifying the surface of metal oxides to reduce the charge-carrier recombination as well as by controlling the spacing of nanostructures close to the exciton diffusion length.

Next-Generation Solar Cells: Principles and Materials
Yoon-Bong Hahn, Tahmineh Mahmoudi, and Yousheng Wang

ISBN 978-981-4968-66-9 (Hardcover), 978-1-003-37238-7 (eBook)
www.jennystanford.com

7.1 Graphene-Based Hybrid Solar Cells

Graphene has played the role of game changer for conductive transparent devices indebted to its unique two-dimensional (2D) structures and gained an exceptional opportunity to be employed in the energy industry. In the past two decades, graphene has been merged with the concept of PV material and exhibited a significant role as a transparent electrode, hole/electron-transport material, and interfacial buffer layer in solar cell devices. For example, Mahmoudi et al. developed a microwave-assisted simple method to fabricate Ag nanoparticles-decorated rGO to obtain a highly conductive, charge selective rGO composite. The Ag-rGO composite showed high conductivity of 9×10^6 S/m (one order of magnitude greater than pristine graphene, i.e., 1.59×10^5 S/m) and was utilized as charge-transfer material in the active layer of PCBM/P3HT. As a highly conductive p-type material, Ag-rGO composite showed an efficient charge separation in interfaces of rGO component and BHJ phase, as well as improved charge transfer, and photocurrent generation, resulting in improved PCE (Fig. 7.1). The Ag-rGO-based device showed a PCE of 4.23%, which

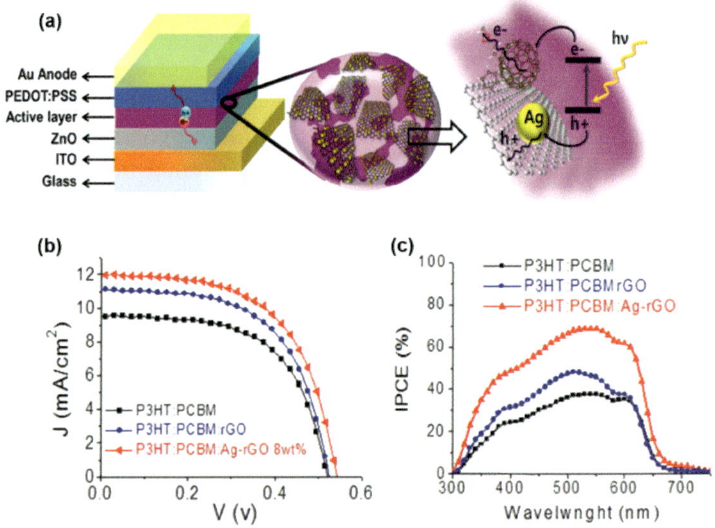

Figure 7.1 (a) Schematic structure of BHJ hybrid solar cells with Ag-rGO in the active layer, as conductive and charge selective material, (b) current–voltage plots, and (c) and IPCE spectra of OSCs with P3HT:PCBM, P3HT:PCBM:rGO and P3HT:PCBM:Ag-rGO as the active layer. Reprinted with permission from [2], Copyright 2016, Elsevier.

was about 42% higher than the control device [2]. Generally, compared to other types of solar cells discussed in the previous sections, the hybrid solar cells show low efficiencies.

7.2 Polymer–Quantum Dot Hybrid Solar Cells

A bulk heterojunction (BHJ) photovoltaic cell, consisting of quantum dots and conjugated polymers, is a promising device to fabricate a high-performance solar cell. Polymer–QD hybrid devices have the potential to take advantage of material properties (such as bandgap tunability, multiple exciton generation, and high mobility), flexibility, and the high molar extinction coefficient of conductive polymers. Despite such advantages, the PCE (~6%) of the polymer–QD hybrid devices is lower than that of all-organic or all-inorganic solar cells. For commercial applications of conjugated polymer–QD hybrid devices, the main challenges such as controlling interfacial structure, efficient charge transfer, and charge transport need to be overcome. In this section, the recent development of bulk heterojunction solar cells made of conjugated polymer–QD blends is described by highlighting the issues of research that need to be further developed.

7.2.1 Material Aspects

Conjugated polymers. The discovery of conjugated polymers opened a whole new area of material for the application of optics and PV electronics. Unlike saturated polymers, which are insulators with limited application in electronics, conjugated polymers are conducting or semiconducting materials with significant application in optoelectronic devices. Compared to inorganic semiconductors, conjugated polymers require cost-effective, solution-based, and simple processing technologies. Moreover, the mechanical flexibility of conjugated polymers enables the fabrication of flexible devices and molecular engineering of polymers enables the modification of energy bandgap levels, transparency, and major carrier type. However, the complex synthetic process of conjugated polymers and the scalability of processes are considerable factors for mass production and commercialization. Some conjugated conductive polymers used for BHJ solar cells are listed in Table 7.1.

Table 7.1 Conjugated conductive polymers for BHJ solar cells

Abbreviation	Structure	Name
P3HT	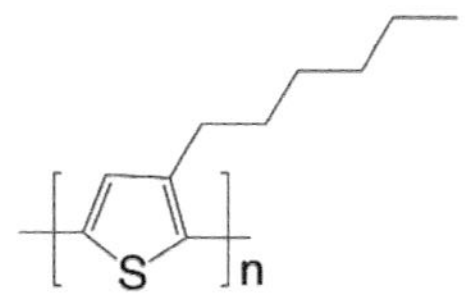	Poly(3-hexylthiophene)
PCPDTBT	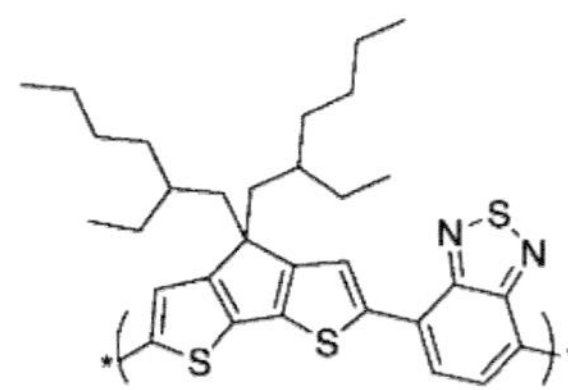	Poly[2,6-(4,4-bis-(2-ethyhexyl)-4H-cyclopenta[2,1-b;3,4-b′]-dithiophene)-alt-4,7-(2,1,3 benzothiadiazole)]
PDTPBT	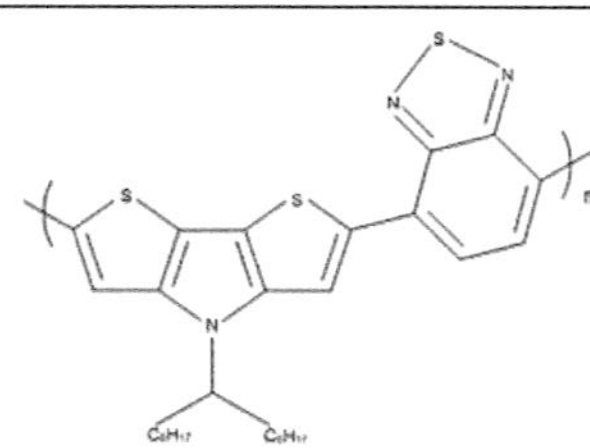	Poly(2,6-(N-(1-octylnonyl)dithieno[3,2-b:20,30-d]pyrrole)-alt-4,7-(2,1,3-benzothiadiazole))

PTB7	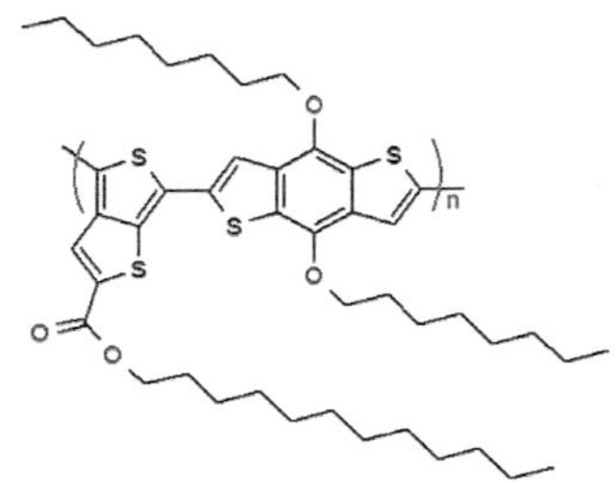	Poly(4,8-bis (octyloxy) benzo (1,2-b:4,5-b′ dithiophene-2,6-diyl) (2-((dodecyloxy) carbonyl) thieno(3,4-b) thiophenediyl))
MEH:PPV	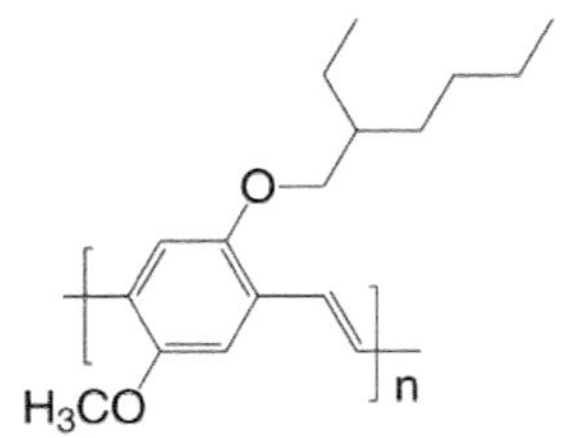	Poly[2-methoxy, 5-(2′-ethyl-hexyloxy)-p-phenylenevinylene)]
MDMO-PPV	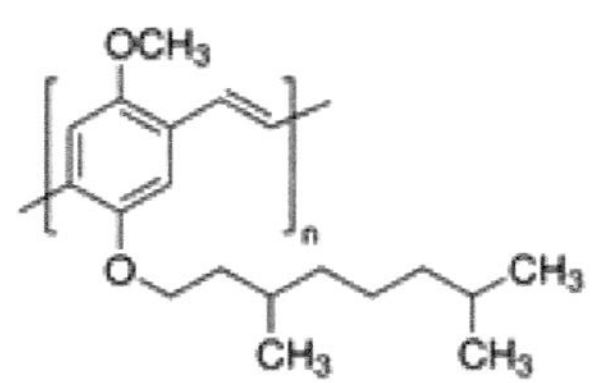	Poly[2-methoxy-5-(3′,7′-dimethyloctyloxy)-1,4-phenylenevinylene]
PPV	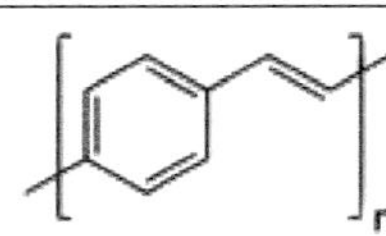	Poly(p-phenylene vinylene)

Quantum dots. Quantum dots are mainly double compounds such as PbS, PbSe, CdS, CdSe, InAs, and InP. As mentioned in Chapter 6, the unique properties arise from their small radius (smaller than the exciton Bohr radius), which is the main cause of quantum confinement, discreet energy states, and wide bandgap, compared to bulk materials. This effect is more significant as the particle size becomes smaller, meaning that the bandgap of QDs can be tuned to match the device energy diagram and harvest different regions of the solar spectrum.

Another significant property of QDs is efficient multiple exciton generation (MEG). In conventional PV theory, an incident photon with matching energy produces a single exciton, and if the photon had higher energy than the bandgap of the semiconductor, the extra energy is converted into heat by thermalization of "hot carriers." In the MEG effects, this excess energy generates additional electron–hole pairs and increases solar cell efficiency. In bulk materials, MEG could be observed through extremely high-energy photons, but QDs show the MEG effect through lower-energy photons within the solar spectrum. The mechanism of MEG is attributed to the discrete energy levels and carrier confinement effect of QDs. QDs can have different shapes and sizes, which is consequently an important factor in the performance of the BHJ hybrid solar cell. Various structures, (i.e., nanorods, tetrapods, and hyperbranched) enable the development of a well-matched interface between QD crystal and polymer in BHJ solar cells. For example, the P3HT-$CuInS_2$ blended hybrid solar cell exhibits superior performance when using pyramidal as opposed to elongated QDs [3].

7.2.2 Hybrid Bulk Heterojunction Solar Cells with Large Bandgap Nanocrystals

The active layer of hybrid solar cells is similar to that of organic solar cells. In organic solar cells, the active layer includes a polymer as a donor and PCBM as an acceptor. The disadvantage of this structure is the low mobility of PCBM which affects electron transfer and charge collection. By using semiconductor nanocrystals (NCs)/QDs as an acceptor (instead of PCBM), the polymer–QD hybrid thin film has the advantages of absorbance of conjugated polymers and high mobility of crystalline QDs, which is promising to obtain cheap, stable, and efficient solar cells. The common materials and their energy levels for polymer–QD BHJ solar cells are shown in Fig. 7.2. In semiconductor QDs, the bandgap depends on the dimensions of the nanocrystal, while the energy band levels in polymers can vary depending on the chain regularity.

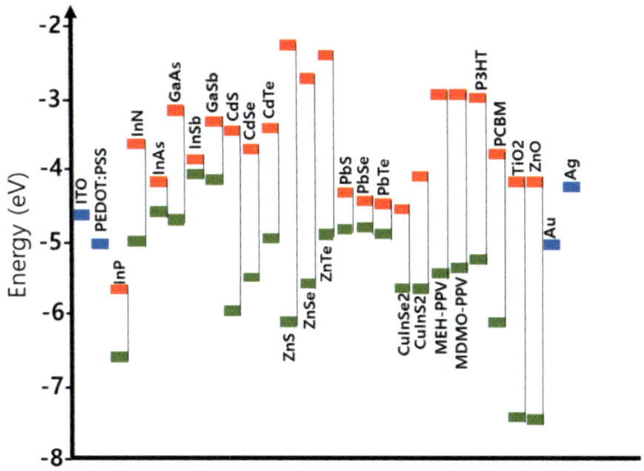

Figure 7.2 The energy levels of polymer and metal oxide QDs used in BHJ hybrid solar cells. The gray area is the electronic bandgap of each material.

For the first time, a hybrid solar cell (HSC) with indium tin oxide (ITO)/P3HT:CdSe/Al configuration along with CdSe as an electron transfer material was reported. Photo-generated electrons are injected from the CB of the polymer into the CB of the CdSe and holes are injected from CdSe to the polymer. Blend of CdSe:P3HT showed a complementary absorption in the visible range. By controlling the diameter of CdSe the onset of the absorption spectrum could be extended up to ~720 nm, resulting in a PCE of 1.7% [4]. The charge-transfer dynamics between the polymer and CdSe of the hybrid film showed that the electron transfer in the hybrid film is very fast (<65 fs) and the device performance is mainly dependent on the morphology of CdSe [5]. By replacing P3HT with poly(*p*-phenylenevinylene) derivative (OC1C10-PPV), and using less amount of volatile solvent (1,2,4-trichlorobenzene) instead of chloroform, a higher PCE of 2.8% was achieved [6]. The presence of long-chain organic ligands used to prevent QD aggregation in the active layer results in uncontrollable phase separation and limits the charge transfer between the polymer and QD, leading to poor device performance. To address this problem and to control the morphology of polymer/CdSe NCs, P3HT with amino (–NH_2) functional group is used. The amino group in P3HT covers the CdSe surface and forms a stable composite, resulting in a significant improvement of PCE [7]. Ligand chemistry engineering showed a substantial increase of PCE by over 12% in solution-processed PbS QD solar cells. It is also reported that hybrid solar cells based on P3HT:CdS QDs with ITO/PEDOT:PSS/P3HT:CdS/BCP/Mg:Ag configuration yields a PCE of 4%, attributed to fast charge separation and collection [8].

7.2.3 Hybrid Bulk Heterojunction Solar Cells with Low-Bandgap Nanocrystals

The HSCs based on polymer and wide-bandgap semiconductor QDs showed up to 4% PCE by optimizing the morphology of the hybrid layer and stabilizer engineering. However, the absorption range of HSCs is limited to visible light where the donor material determines the onset of the spectrum response, which restricts the HSC for the long-wavelength IR range. Recently, PbS [9], PbSe [10], and PbS_xSe_{1-x} [11] alloy have shown promising performance for their size-tunable bandgap, high mobility, multiple charge generation, and solution processing methods. By using PbS or PbSe in HSCs, both short and long-wavelength (>700 nm) photons can be used efficiently. Hybrid SC based on PbS and PDTPBT blend showed performance improvement by post-deposition ligand exchange using 1,2-ethanedithiol [12]. It is also reported that the PbS-polymer hybrid is close to organic-organic mixing, thereby improving charge separation efficiency and improving HSC device performance by suppressing morphology traps and defect traps [13].

7.2.4 Limiting Factors of Polymer–QD Hybrid Solar Cells

Interfacial properties. Stabilizers such as oleic acid and oleylamine with long organic chains are generally used to effectively disperse QDs and control QD size as well as protect against unwanted chemical reactions. However, these stabilizers are insulator organics, and when QD films blend with conjugated polymers, the presence of stabilizers prevents exciton dissociation and thus inhibits charge transport [15]. Long-chain stabilizers can be replaced by shorter chain conductive aryl-type with π-conjugated chains. However, the suitability of these stabilizers to polymers is a major consideration with respect to the electron matching and ligand direction of the active layer. The presence of polymer stabilizers on the surface of QDs greatly affects the electronic properties of the semiconductor by surface functionalization. Nuclear magnetic resonance (NMR) studies show there is partial bounding between stabilizers and QDs, which involves the reactive sites of QD and causes quenching of the luminescence of QDs by nearly 50% [15].

Crystallinity. The crystallinity of the polymer can affect the optical properties of QDs. Poor local crystallinity of donor and acceptor materials is another limiting factor for heterojunction solar cells. Lack of high crystallinity causes energy states that cannot be used for quick charge separation and causes polaron dissociation. It is worth noting that the addition of an electron or hole to a conducting polymer chain has been shown to cause the chain to distort, creating a large polaron.

Traps. Impurities and imperfections of the crystal, dangling bonds at the surface of QDs act as carrier trap sites. Deep traps (with deep energy levels) have a dramatically negative effect on the charge transfer and collection in hybrid solar cells. In the presence of deep traps, as the charge carriers cannot overcome the trap site energy, the extended extraction time may increase the recombination rate. Studies based on transient absorption spectroscopy (TAS) and photo-induced absorption (PIA) show that in smaller QDs, the high surface-to-volume ratio exhibits a higher trap density, thus affecting PV parameters. High-intensity dependence of PV parameters in CdSe QD-P3HT blend solar cells was observed by varying the QD sizes. For example, the CdSe QD-P3HT blend demonstrated significantly improved performance with larger QDs compared to smaller QDs [16].

7.2.5 Interfacial Engineering in Polymer–QD Hybrid Solar Cells

Capping ligands of stabilizers on the surface of QDs maintain colloidal stability in solvents and play an important role in the electronic properties of QDs. Ligands also determine the shape and size distribution of QDs during the fabrication process. They can passivate the trap sites, tune the PL intensity, and alter the energy levels of the QDs [17]. Recombination kinetics of QD–polymer blends and back-electron transfer from the QDs to the conductive polymer (a loss mechanism) are highly dependent on the type of organic stabilizer attached to the QD surface [18].

The use of stabilizers on QD surface increased spacing between QDs, which hinders electronic charge transport. To solve this problem, several methods such as replacing longer-chain organic stabilizers with shorter ones and post-plasma treatment of stabilizers. It is known that plasma treatment can cause oxidation on the surface of QDs and interfere with charge transport [19]. Moreover, the presence of organic stabilizers can affect the film morphology of the heterojunction layer. Control of surface chemistry of quantum dots can help adjust the molecular interface between QDs and organic semiconductor materials. Altering the long-chain stabilizer with small-molecule pyridine on CdSe QDs in QD–polymer blend hybrid solar cells showed an improvement in the charge transfer. Post-deposition treatment of stabilizer-QDs with organic materials with both thiol and carboxylic acid functional groups, such as 1,2-ethanedithiol (EDT), 3-mercaptopropionic acid (MPA), and malonic acid (MA), showed effective removal of the long-chain ligands [20] and resulted in substantial improvement of device performance. Studies reveal that not only energy level, work function, and carrier mobility of QDs but also charge recombination, trap density, and exciton diffusion length in QD films are substantially affected by the capping ligands.

7.3 Charge-Transport Materials for Hybrid Solar Cells

For all types of device architecture, most hybrid solar cells contain at least one metal oxide layer because metal oxides are useful to improve the device stability compared to organic alternatives. While the theoretical highest efficiency of hybrid solar cells has not yet been achieved, most studies mainly focus on improving the stability of the solar cells, which is a significant challenge for commercialization. In addition, high concern about improving the stability of solar cells leads to interest in the possibility of oxide-based hole-transport layer (HTL) and ETL because the inorganic charge-transport layers can lead to higher stability under ambient conditions compared to organic alternatives such as PEDOT: PSS, PCBM, and Spiro-OMeTAD.

A charge-transport layer should have (i) an energy-level alignment adjusted with the active layer for effective charge extraction, (ii) good charge mobility for efficient charge transport, (iii) low optical absorption for not interfering with incoming light, and (iv) low interfacial defects to lower recombination losses at interfaces. In general, most metal oxides satisfy the first three conditions, but there is the most problematic part in achieving low interfacial defects and recombination losses at interfaces. It is noteworthy that interfacial defects may vary greatly depending on the preparation and deposition conditions of ETL and HTL.

7.3.1 Metal Oxide–Based Charge-Transport Materials

7.3.1.1 Electron-transport materials

A metal oxide ETL can be formed by various methods such as sol–gel, spin-coating nanoparticles, doctor-blading, atomic layer deposition (ALD), sputtering, e-beam deposition, etc. Deposited film is often subjected to post-treatment processes such as annealing, UV–ozone treatment, and oxygen plasma treatment. The type of treatments depends on metal oxide precursors and deposition methods. Generally, metal oxide films require high-temperature annealing to obtain a high-quality crystalline layer. For transparent conductive electrodes, either fluorine-doped tin oxide (FTO) or ITO coated glasses can be used. FTO has a higher tolerance at a high processing temperature, while ITO is used in low-temperature processes because the conductivity of ITO decreases when treated at a high temperature. Flexible ITO substrates can only be used below 150°C. Post-deposition treatment may include UV–ozone treatment to remove residual organics and enhance the wettability of metal oxide. High-temperature processing of metal oxide film limits substrate selection in both conventional and inverted devices.

Cathode interface layers for both conventional and inverted OSCs require semiconductors with low work function to match energy levels for efficient charge extraction, transportation, and high compatibility with electrodes and active layers to reduce interfacial losses. In inverted OSCs, the ETL should be transparent and stable.

Zinc Oxide (ZnO). ZnO is an n-type semiconductor and has characteristics such as low cost, ease of fabrication, high stability, high mobility, and non-toxicity, making it one of the best ETL materials. Furthermore, ZnO has a low work function (WF) of about 4.30 eV, which is well-matched with a LUMO level of organic acceptors such as PCBM. The optical and electrical properties of ZnO can be engineered according to changes in crystalline structure, morphology, doping, and composition. Among various deposition processes such as electrochemical deposition, sol–gel synthesis, hydrothermal growth, and ALD, the sol–gel method is a promising process to achieve a high performance of the ZnO-based cell.

In inverted organic hybrid solar cells with sol–gel ZnO, poly (3-hexylthiophene) (P3HT)-based system exhibited PCEs $< 5\%$. With a smaller bandgap polymer poly ([9-(1-octylnonyl)-9H-carbazole-2, 7-diyl]- 2,5-thiophenediyl-2, 1,3-benzothiadiazole-4,7-iyl-2,5-thiophenediyl) (PCDTBT) PCEs increased over 6%. The development of active layer materials and processing methods improved the PCE to 7–8%, and more than 10% PCEs were obtained by sol–gel ZnO film and a new polymer (PffBT4T-2OD), a record high efficiency for single-junction OSCs [21, 22].

Titanium Oxide (TiO_x). Another n-type metal oxide with high optical transparency, high electron mobility, and stability is TiO_2. TiO_2 is the most widely used ETL material for perovskite solar cells. It has a wide bandgap of 3.6 eV which allows the transition of the visible light into the active layer. TiO_2 has lower electron mobility than perovskite ($0.14\ cm^2\ V^{-1}\ s^{-1}$), resulting in charge accumulation at interfaces. Therefore, the introduction of TiO_2 nanoparticles improves the contact surface, thereby lowering the charge concentration of the interface and efficiently extracting charges from the active layer.

In both inverted and conventional organic solar cells, TiO_x is an effective ETL that may be deposited on a TCO substrate through sol–gel or spin coating. In conventional OSCs, aside from the ETL role, TiO_2 acts as an optical spacer and redistributes light within the active layer. Between the active layer and metal contact, TiO_x ETL acts as an insulation and scavenging layer, preventing oxygen and moisture from penetrating the organic active layer. In the

OSC devices using TiO_x ETL, stability under ambient conditions is more than doubled compared to devices using other types of ETLs. Similarly, in inverted OSCs, when TiO_x is deposited between ITO and active layer, it significantly affects the film structures and interfacial properties of the fabricated device, ultimately improving the device performance. Adjustment of the film thickness and morphology of the TiO_x layer helps to improve and optimize the device efficiency of inverted OSCs [23].

As an alternative, the chemical doping of TiO_x is another effective approach that significantly improves device performance by increasing the carrier density of TiO_x and reducing the work function of the TCO electrode. Moreover, the use of thin film of metal oxides or polymer layers as surface engineering can help to passivate the trap states on the surface of TiO_x and effectively improve the PCE of OSCs [24].

Niobium Oxide (Nb_2O_5). Niobium oxide (Nb_2O_5 or NbO_x) is an n-type semiconductor with a wide bandgap (over 3 eV) and high transparency in the visible region. It can act as an electron transfer layer for SCs. Nb_2O_5 ETL deposited by sol–gel process is introduced as a binary oxide ETL in OPVs to enhance device efficiency of both conventional and inverted organic solar cells. The disadvantage of Nb_2O_5 is that it has a conduction band ($\approx$3.7 eV) higher than the LUMO level of the organic acceptor (PCBM), which can reduce electron extraction in fullerene-based devices. Due to its similar CB level and better electrical properties, Nb_2O_5 has been used in DSSC and perovskite solar cells to replace conventional TiO_2 ETL. Combining Nb_2O_5 with compact TiO_2 in perovskite solar cells resulted in improved device stability with less hysteresis. Nb_2O_5 modification of FTO can tune the work function from 4.6 to 4.3 eV without a significant change in surface roughness and optical transparency, which could reduce energy-level offset and increase V_{oc}. To improve the conductivity of Nb_2O_5 ETL, fine control of doping and deposition parameters is an effective way to improve the conductivity without adding impurities to the material. UV–ozone treatment may improve the wettability and hydrophilicity of Nb_2O_5 film, which is helpful for the growth of active layer film on the Nb_2O_5 coated TCO substrate [27, 28].

Tin Oxide (SnO_2). SnO_2 is a promising substitute for TiO_2 ETL due to its good optical and electronic properties. SnO_2 has higher charge mobility (421.70 $cm^2/V \cdot s$) than TiO_2, which can ensure efficient electron transport. SnO_2 has a wider energy bandgap than TiO_2, which has better transmittance, lower photon energy loss, and higher UV resistance. By combining with transparent conductive oxide electrodes, SnO_x has a smaller refractive index gradient than TiO_2, thereby reducing the reflection of incident light. In addition, the processing temperature of SnO_2 is significantly lower than that of other oxides (<200 °C), which makes it compatible with flexible devices with low fabrication costs for solar cells. Using SnO_x as ETL in inverted OSCs shows a high PCE with high thermal and ambient stability. Unlike other oxides, ALD deposited SnO_x does not need UV activation treatment to enhance electron extraction. Accurate control of surface structures of SnO_x can improve OSC performance. Optimization of the particle size film morphology along with high electrical properties of SnO_2 effectively helps to grow high-quality active layers. It is worth noting that tin oxide–based semiconductor materials are a new class of ETL for organic–inorganic hybrid SCs with high PCE and long-term stability.

7.3.1.2 Hole-transport materials

Organic p-type hole transfer materials, such as PEDOT:PSS, Spiro-OMeTAD, or PTAA, are widely used in organic and perovskite solar cells and produce high-performance devices. However, instability in O_2/H_2O and heat limits the long-term stability of solar cells. Inorganic materials especially metal oxides, naturally have better thermal stability and charge-transport properties, and can be used as charge-transport materials. Various inorganic materials such as CuO, CuSCN, Cu_2O, CoO, and NiO_x have been used for the HTL. Among them, NiO_x is one of the most efficient hole-transport materials for solar cells [25, 26].

Nickel oxide (NiO_x). NiO_x is a p-type semiconductor and one of the most promising hole-transport/electron blocking materials. NiO_x has a cubic structure, with a lattice parameter of 0.42 nm. Pure stoichiometric nickel oxide (i.e., NiO) is an insulator with low electrical conductivity ($<10^{-11}$ S/cm). However, NiO_x is in a non-stoichiometric state with excess oxygen and Ni vacancies, resulting in p-type behavior. NiO_x is highly transparent near the UV–Vis range, has a work function of ~5.4 eV, and has a wide bandgap in the range of 3.6 to 4.0 eV, making it suitable for p-type transparent conductive oxide (p-TCO).

NiO_x HTL can be deposited by various methods such as chemical vapor deposition, sputtering, chemical bath deposition, spin coating of nanoparticles, sol–gel deposition, etc. Like other metal oxides, the overall concentration of nickel vacancies is determined according to the deposition method, conditions, and post-treatment. Moreover, the surface properties of NiO_x film have a strong influence on the morphology of the active layer and the charge collection or recombination at interfaces. In an early work, solution-processed NiO_x thin films were employed in planar perovskite solar cells (PSCs) having a configuration of NiO_x/$MAPbI_3$/PCBM/BCP, resulting in PCE of 7.8%. Film quality improvement was achieved through sputtering NiO, and then PCE of 9.8% was obtained. The higher performance with NiO_x HTL than PEDOT:PSS was attributed to the higher VBE (valence-band energy) and higher air stability of the NiO_x. By reducing the non-radiative recombination on NiO_x surface in NiO/perovskite high PCE over 17% was achieved.

In general, the quality of NiO_x HTL plays an important role in reaching optimal device performance. The electrical properties of wide-bandgap metal oxide semiconductors can be highly improved by doping. In NiOx the p-type conductivity is rather low ($\sim 10^{-13}$ S cm^{-1}) because of Ni vacancies. Doping NiO_x with other metal ions such as Zn, Cu, Li, and Cs reduces the ionization energy of the Ni vacancies and subsequently increases the hole conductivity in HTL films. Doping directly affects charge-carrier collection efficiency due to changes in carrier concentration, mobility, and energy-level alignments. When the dopant has multiple valence states (e.g., copper with two states of Cu^+ or Cu^{2+}), the doping effect becomes complicated. Doping of NiO_x HTL with Mg^{2+} shows a small change in conductivity, but a significant shift in the work function, which improves charge collection [29].

Other oxides. Other metal oxides have also been studied for HTL applications, but few are reported because they are less efficient than commonly used NiO_x materials. Other oxide HTLs include Cu_2O, WO_3-water-free PEDOT:PSS, indium-doped $CuCrO_2$, $NiCo_2O_4$, Co_3O_4, etc. Unlike NiO_x, the impact of the surface optimization and interfacial modifications for these HTLs have not been explored.

Cuprous oxide (Cu_2O) and copper oxide (CuO) are well-known p-type semiconductors used as HTL in perovskite solar cells. The low VB energy levels of Cu_2O and CuO match well with that of $MAPbI_3$, leading to minimal energy loss. Theoretical studies expect that solar cells with Cu_2O as HTL could outperform other inorganic/organic PSCs with an efficiency of more than 24%.

7.3.2 Graphene-Based Charge-Transport Materials

The role of the charge-transport layer is to reduce the energy barrier between the active layer and the electrodes for better charge extraction/transportation from the active layer to the electrodes. Therefore, charge-transport layers enhance electrical contact at the interface of the active layer electrode, helping charge extraction as well as transport and collection. The charge-transport layers should be selective to allow a specific type of charge, while blocking the other type, as well as protecting the cell device from moisture and/or oxygen penetration, thus improving cell efficiency and stability.

7.3.2.1 Tuning of the work function of graphene

The WF of a material is the minimum energy required to extract an electron from the surface of a solid. WF of a material determines the relative position of the Fermi levels. In a device, at the heterojunction contact with the help of aligned WF, one can facilitate selective charge (electron or hole) transport. Graphene's WF and its tune-ability are important factors in applying graphene as a conductive charge selection material. Graphene is an important conductive material in many modern optoelectronic devices including solar cells, organic light-emitting diodes, liquid crystal displays, touch screens, etc. However, graphene is a zero-bandgap material, and unlike semiconductor materials, its conductivity cannot be turned off. Thus, tuning the WF of graphene opens new opportunities for graphene-based materials to be employed in optoelectronic and nano-electronic devices. For example, in PV devices, the choice of an electron injecting/transporting material with a WF that matches the energy level of the electron-donating layer prevents charge accumulation at interfaces. Facilitating the charge transport reduces accumulation and recombination and improves device performance, so the ability to change the WF of graphene-based materials to match the valence band and/or conduction band of the active materials is important. To tune the WF of graphene, several methods are suggested, such as introduction of a chemically bonded dopant, element or small organic molecule, metallic nanoparticle decorated on the graphene surface, absorption of gas molecules, and use of aromatic materials. One of the most helpful methods to change the WF of graphene or cause bandgap opening in graphene is chemical doping/functionalization or decoration with noble metals [30].

Graphene quantum dots (GQDs) and nanosize graphene oxide (GO) with good charge mobility and tunable bandgap have shown promising applications for light-emitting diodes and PV devices. GQDs have narrow size distribution and a high ratio of oxygen/carbon and oxygenated functional groups and show superb dispersion in polar/protic solvents. Several studies showed that GQDs could be an efficient alternative to the conventional PEDOT:PSS as HTL for OPV cells.

7.3.2.2 Electron-transport materials

In solar cells, graphene ETL controls the extraction of electrons from the active layer and transports them to the reactive electrode, blocks hole carriers, and suppresses recombination, thereby improving the overall efficiency. For ETL application, the WF of graphene should have an energy band matching the conduction band of the active layer material for effective electrons extraction and transportation. Good electron-transport materials provide good ohmic contact at the active layer/electrode interface and are chemically stable and moisture-resistant properties improves the stability of the devices.

Among graphene-based ETLs, the composites made of rGO and metal oxide semiconductors (e.g., rGO-SnO_2) in n-i-p PSCs helped to increase the electron mobility and suppressed charge recombination, thereby improving the PCE [31]. Since rGO-metal oxide ETL has excellent electron extraction performance due to the alignment of energy levels between the active material and the electrode, the energy barrier is lowered because the WF of graphene is between the FTO electrode and the metal oxide conduction band. Introduction of graphene into the ETL provides conductive electron pathways, which suppress carrier recombination and extend the carrier lifetime, resulting in a high photocurrent density (J_{sc}). ZnO with a wide bandgap is an effective hole-blocking layer, and the graphene/ZnO and N-graphene/ZnO composites match the energy levels with the photoactive layer to enhance rapid electron extraction and reduce the interfacial resistance with minimum recombination.

Integration of GQDs in ETLs in n-i-p PSCs to form composites, such as GQD-TiO_2, GQD-ZnO, and GQD-SnO_2, helps to reduce contact resistance at the photoactive layer/metal oxide interface, and reduce the series resistance and improve the electron transportation at the interface, thereby improving the J_{sc} and V_{oc} of the device [32, 33]. The presence of GQDs suppresses carrier recombination and significantly lowers the leakage current, which increases the shunt resistance and FF. Therefore, the inclusion of GQDs in the conventional ETL materials improves not only device efficiency but also the long-term stability of devices, which are both crucial for commercialization.

7.3.2.3 Hole-transport materials

Carbon-based conductive materials, such as graphene and its derivatives, have been developed as a competent HTL alternative for PV devices due to their low cost, solution process capability, excellent optical and electrical properties as well as chemical stability and flexibility. Because of these characteristics, graphene is an exceptional candidate for electronic devices application. Pristine graphene/graphite without any functional groups or doping is a conductive metal-like material made of carbon. Such material has little function in electronic devices. However, GO and reduced graphene oxide (rGO) allow tailoring the functional groups, metal nanostructures, metal oxides, etc. to be customized to the rGO surface and tuned its electrical and chemical properties for a wide range of applications. Theoretical studies suggested that tailored graphene-based materials have an outstanding performance in PV devices.

For example, ammonia-treated GO, boron-doped rGO, rGO/CuSCN, and rGO/NiO were developed as an HTL in planar perovskite SCs. Doping and functionalization of GO enhance the energy-level bands at the interface between HTL and the perovskite layer, thus performing as effective hole-transport material, resulting in relatively high-performance SCs.

Moreover, GO and rGO can be incorporated into conventional HTLs like PEDOT:PSS and Spiro-OMeTAD. The composition rises the work function of the composite HTLs and improves the energy-level matching between HTL and active material. The presence of rGO in an organic HTL increases the built-in potential and electrical conductivity for efficient hole extraction and transport, with minimal recombination. Moreover, graphene is chemically inert with high hydrophobicity, which prevents moisture and oxygen molecules from diffusing into the device. Hence, as a promising method, the incorporation of graphene-based materials in conventional charge-transport materials not only improves the efficiency of PSCs but also protects the active layer from moisture and suppresses the migration of ions through the active layer, thus enhancing devices stability.

References

1. Mahmoudi, T., Wang, Y., and Hahn, Y.-B. (2018) *Nano Energy*, **47**, pp. 51–65.
2. Mahmoudi, T., Seo, S., Yang, H.-Y., Rho, et al. (2016). *Nano Energy*, **28**, pp. 179–187.
3. Radychev, N., Scheunemann, D., Kruszynska, M., Frevert, K., et al. (2012) *Org. Electron.*, **13**, pp. 3154–3164.
4. Huynh, W. U., Dittmer, J. J., and Alivisatos, A. P. (2002) *Science*, **295**, pp. 2425–2427.
5. Couderc, E., Greaney, M. J., Brutchey, R. L., and Bradforth, S. E. (2013) *J. Am. Chem. Soc.*, **135**, pp. 18418–18426.
6. Sun, B., Marx, E., and Greenham, N. C. (2003) *Nano Lett.* **3**, pp. 961–963.
7. Liu, J., Tanaka, T., Sivula, K., Alivisatos, A. P., and Fréchet, J. M. (2004) *J. Am. Chem. Soc.*, **126**, pp. 6550–6551.
8. Ren, S., Chang, L. Y., Lim, S. K., Zhao, J., et al. (2011) *Nano Lett.*, **11**, pp. 3998–4002.
9. Blachowicz, T. and Ehrmann, A. (2020) *Appl. Sci.*, **10**, pp. 1743.
10. Zhang, X., Zhang, Y., Wu, H., Yan, L., et al. (2016) *RSC Adv.*, **6**, pp. 17029–17035.
11. Song, T., Cheng, H., Fu, C., He, B., et al. (2016) *Nanotechnology*, **27**, pp. 165202.
12. Seo, J., Cho, M. J., Lee, D., Cartwright, A. N., and Prasad, P. N. (2011) *Adv. Mater.*, **23**, pp. 3984–3988.
13. Piliego, C., Manca, M., Kroon, R., Yarema, M., et al. (2012) *J. Mater. Chem.*, **22**, pp. 24411–24416.

14. Fu, W., Wang, L., Zhang, Y., Ma, R., et al. (2014) *ACS Appl. Mater. Interfaces*, **6**, pp. 19154–19160.
15. Munro, A. M. and Ginger, D. S. (2008) *Nano Lett.*, **8**, pp. 2585–2590.
16. Colbert, A. E., Wu, W., Janke, E. M., Ma, F., and Ginger, D. S. (2015) *J. Phys. Chem. C*, **119**, pp. 24733–24739.
17. Kalyuzhny, G. and Murray, R. W. (2005) *J. Phys. Chem. B*, **109**, pp. 7012–7021.
18. Albero, J., Clifford, J. N., and Palomares, E. (2014) *Chem. Rev.*, **263**, pp. 53–64.
19. Nam, M., Lee, T., Kim, S., Kim, S., et al. (2014) *Org. Electron.*, **15**, pp. 391–398.
20. Colbert, A. E., Wu, W., Janke, E. M., Ma, F., and Ginger, D. S. (2015) *J. Phys. Chem. C*, **119**, pp. 24733–24739.
21. Kundu, S., Gollu, S. R., Sharma, R., Srinivas, G., et al. (2013) *Org. Electron.*, **14**(11), pp. 3083–3088.
22. You, J., Dou, L., Yoshimura, K., Kato, T., et al. (2013) *Nat. Commun.*, **4**(1), pp. 1–10.
23. Chen, S., Small, C. E., Amb, C. M., Subbiah, J., et al. (2012). *Adv. Energy Mater.*, **2**(11), pp. 1333–1337.
24. Li, P., Sun, C., Jiu, T., Wang, G., et al. (2014) *ACS Appl. Mater. Interfaces*, **6**(6), pp. 4074–4080.
25. Dkhil, S. B., Duché, D., Gaceur, M., Thakur, A. K., et al. (2014) *Adv. Energy Mater.*, **4**(18), pp. 1400805.
26. Agnihotri, P., Sahu, S., and Tiwari, S. (2017) Recent advances and perspectives in electron transport layer of organic solar cells for efficient solar energy harvesting. In: *2017 International Conference on Energy, Communication, Data Analytics and Soft Computing* (ICECDS), pp. 1568–1573.
27. Fernandes, S. L., Albano, L. G. S., Affonço, L. J., Silva, J. H. D. D., et al. (2019) *Front. Chem.*, **7**, pp. 50.
28. Guo, Y., Tao, J., Jiang, J., Zhang, J., et al. (2018) *Sol. Energy Mater. Sol. Cells*, **188**, pp. 66–72.
29. Hossain, M. I., Alharbi, F. H., and Tabet, N. (2015) *Sol. Energy*, **120**, pp. 370–380.
30. Garg, R., Dutta, N. K., and Choudhury, N. R. (2014) *Nanomaterials*, **4**(2), pp. 267–300.
31. Liu, R., Qui, R., Zou, T., Liu, C., et al. (2019) *Nanotechnology*, **30**, pp. 075202.

32. Khorshidi, E., Rezaei, B., Irannejad, N., Adhami, S., et al. (2020) *Electrochim. Acta*, **337**, pp. 135822.

33. Pang, S., Zhang, C., Zhang, H., Dong, H., et al. (2020) *Appl. Surf. Sci.*, **507**, pp. 145099.

Problems

7.1 Explain what are the limiting factors in QD–polymer hybrid solar cells.

7.2 What is the role of capping ligands of stabilizers in QD–polymer SCs?

7.3 Illustrate the working principles of a QDSSC.

7.4 The approaches to fabricate high-performance QDs for solar cell application can be categorized into *in situ* and *ex situ* methods, Briefly explain the two processes.

7.5 What are the required properties of charge-transport layer?

Chapter 8

Perovskite Solar Cells

Perovskite solar cells (PSCs) based on Earth-abundant hybrid organic metal halide perovskites (OMHPs) with the formula of ABX_3 and their intriguing optical and electronic properties, such as high carrier mobility and absorption coefficient, tunable bandgap, long charge diffusion lengths, low exciton binding energy, and broad light absorption wavelength, etc., have the superb potential for the fabrication of cost-benefited, simple solution-processed and high-performance next-generation solar cells.

The advent of PSCs has opened up great PV developments due to impressive power-conversion efficiency (PCE) beyond 25% in single junction solar cells and over 29% in tandem solar cells. The extraordinary high efficiencies are mainly attributed to several predominant breakthroughs, such as the optimization of perovskite films with additives and the innovative design of device configurations with ideal interfacial materials. It is noteworthy that single PSCs with a PCE of 30% can be achieved through light management and defect control strategies according to previous experiences of the Si and GaAs solar cells. PSCs are considered the most potential candidates that are competitive with commercial silicon-based solar cells.

This chapter discusses in detail the characteristics of OMHP, types of perovskite crystal structure, perovskite compositional engineering, history, and basic working principles of PSC.

Next-Generation Solar Cells: Principles and Materials
Yoon-Bong Hahn, Tahmineh Mahmoudi, and Yousheng Wang

ISBN 978-981-4968-66-9 (Hardcover), 978-1-003-37238-7 (eBook)
www.jennystanford.com

8.1 What Are Perovskites and Their Properties?

A perovskite is a material that has the same crystal structure as the naturally occurring mineral calcium titanium oxide ($CaTiO_3$). It was first discovered by German mineralogist Gustav Rose in 1839 and was named after Russian mineralogist L. A. Perovski. Generally, a perovskite has a three-dimensional (3D) structure ABX_3, where the cation A is stabilized in a cubo-octahedral cage formed by the 12 nearest X anions, and the cation B coordinates with the six nearest X anions to form octahedral geometry. Hybrid OMHPs usually are a broad class of materials adopting a general chemical formula of ABX_3 (A = monovalent cation, such as FA^+, MA^+, Cs^+; B = metallic divalent cations, such as Pb^{2+} and Sn^{2+}; X = halide anions such as I^-, Br^-, and Cl^-) as shown in Fig. 4.17 and Fig. 8.1a.

8.1.1 Organic–Inorganic Hybrid Perovskites

In 2009, Miyasaka et al. were the first to develop OMHPs as the sensitizer in DSSCs [1]. Until 2012, Park et al. [2] and Snaith et al. [3] had reported solid-state TiO_2-based $MAPbI_3$ and meso-superstructure Al_2O_3-based $MAPbI_2Cl$ hybrid solar cells, generated 9.7% and 10.9% efficiency, respectively. Since 2013, PSCs have become one of the hottest research areas in the photovoltaic community. The OMHPs present excellent optical and electronic properties, such as high absorption coefficient (>10^5 cm^{-1}), ultra-long carrier diffusion length, moderate carrier mobility, tunable bandgap, low exciton binding energy, and broad light absorption wavelength, etc. For example, mixed halide perovskite compositions of $FAPbI_yBr_{3-y}$ were developed, where their optical and electronic properties can be tuned by controlling the value y. Figure 8.2(a,b) shows the UV–Vis absorption spectra and photoluminescence spectra for different y values. A systematic shift of the absorption band edge to longer wavelengths is observed when the y value is increased. The photoluminescence emission also is red shifted when y is increased, which is well-matched with UV–Vis performance. All of these properties provide evidence that OMHPs can be considered a next-generation optoelectronic semiconductor material.

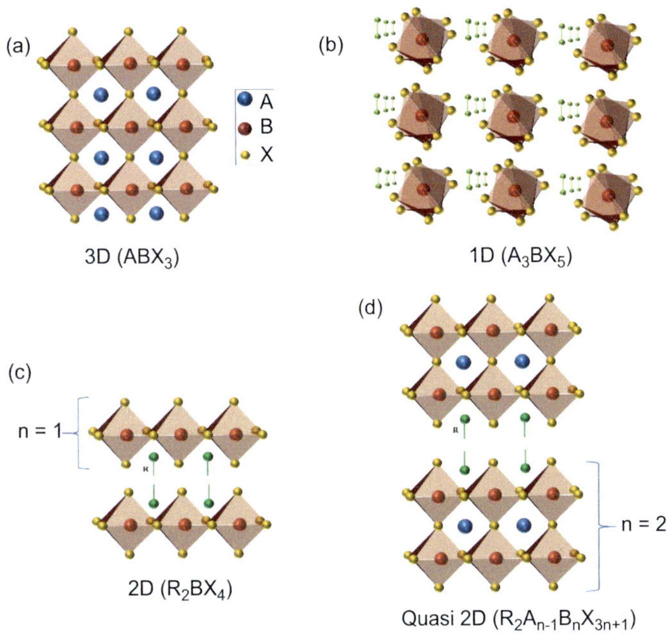

Figure 8.1 (a) Illustration of typical 3D hybrid perovskite structure ABX_3 (A = organic cation, B = metal cation and X = halide), (b) 1D structure A_3BX_5, (c) 2D structure R_2BX_3, and (d) Quasi-2D structure $R_2A_{n-1}B_nX3_{n+1}$.

8.1.2 Low-Dimensional Perovskites

The 3D crystal OMHPs present poor stability when expose to ambient conditions. The instability is mainly attributed to organic compounds and phase transformation. However, OMHPs endow with dimensional tailoring properties for more stable low-dimensional structures. For example, Ruddlesden–Popper phase-based 2D perovskites have been widely studied for solar cell applications owing to their high environmental stability [5, 6]. Figure 8.1b shows the stable 1D structure which is corner-sharing,

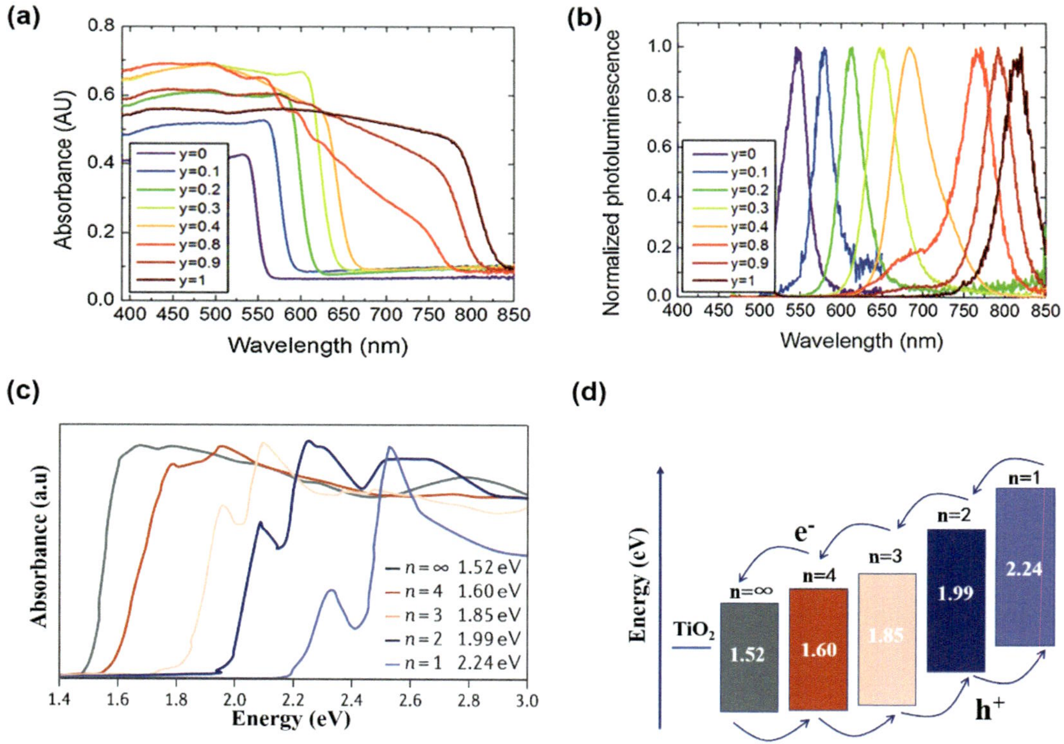

Figure 8.2 (a) UV–Vis absorption spectra and (b) photoluminescence spectra for $FAPbI_yBr_{3-y}$ compositional films. Reprinted with permission from [4], Copyright 2015, Elsevier. (c) Bandgaps of 2D (n = 1) and quasi-2D $BA_2MA_{n-1}Pb_nI_{3n+1}$ (BA^+: butylammonium) for different n values and (d) corresponding to their alignment of the energy levels and bandgaps. Reprinted with permission from [8], Copyright 2015, American Chemical Society.

edge-sharing, or face-sharing to form a 1D nanowire surrounded by organic cations. As shown in Fig. 8.1(c,d), 2D perovskites have the general formula $R_2A_{n-1}B_nX_{3n+1}$, where R is a bulky organic cation that plays the role of spacer between the inorganic framework. The 1D and 2D are unsuitable as light absorbers due to poor optical and electronic properties, but they can be selected as passivate materials to reduce defects and protect 3D crystals in photovoltaics or other optoelectronic devices. Note that the quasi-2D by controlling the value of '*n*' can obtain desired perovskite crystals (Fig. 8.1d). As shown in Fig. 8.2(c,d), the quasi-$BA_2MA_{n-1}Pb_nI_{3n+1}$ can give various bandgaps and optical properties depending on the number of layers. When $n \leq 2$, the 2D perovskites have a high bandgap and high exciton binding energy up to hundreds of meV [7]. Thus, the 2D perovskites are difficult as light absorbers and unsuitable for PVs. However, they exhibit strong excitonic behavior attributed to quantum confinement effects, which can yield high photoluminescence and make them suitable for LED applications. By contrast, when $n \geq 3$, the perovskites show a low exciton binding energy, smaller bandgap, and appropriate light absorption region, which present promising both efficiency and stability in PSCs [8].

8.1.3 All-Inorganic Perovskites

All-inorganic halide perovskites can be formed when organic cation site A is replaced by inorganic cation Cs in ABX_3, such as $CsPbI_3$, $CsPbBr_3$, and $CsPbCl_3$. Note that the cubic α-phase $CsPbI_3$ is the most prominent candidate for photovoltaic applications due to the appropriate E_g of 1.73 eV. However, the cubic α-phase $CsPbI_3$ is only stable at high temperature, and it rapidly transforms into a yellow orthorhombic non-perovskite δ-phase with E_g of 2.82 eV, resulting in poor photovoltaic properties at room temperature [9]. Although $CsPbBr_3$ presents the best phase stability among all Cs-based all-inorganic perovskites, the wide E_g of 2.3 eV limits the photocurrent and thus makes it difficult to achieve a higher PCE in solar cells. As reported, partial substitution of iodine (I) with bromide (Br) can build mixed-anion $CsPbI_2Br$ (E_g = 1.92 eV) and $CsPbIBr_2$ (E_g = 2.05 eV). However, such higher bandgaps severally limit the photocurrent, leading to lower PCEs [10]. Therefore, in-depth experimental and theoretical studies on the chemical and physical properties of inorganic perovskites are needed to further improve the efficiency and the long-term stability of inorganic photovoltaics under operational conditions.

8.2 Perovskite Composition Engineering

As discussed above, the hybrid perovskites have a general chemical formula of ABX_3. To establish a stable and cubic phase, the tolerance factor (t) of perovskites should be in a range of $0.8 \leq t \leq 1.0$ (orange area, Fig. 8.3a). The value of t can be calculated by the Goldschmidt equation $t = (r_A + r_X)/\sqrt{2}(r_{Pb} + r_X)$, where r is the ionic radius. Figure 8.3a shows t values for several A-site cations-based lead triiodide perovskites ($APbI_3$). It can be seen that methylammonium lead iodide ($CH_3NH_3PbI_3$, $MAPbI_3$), can form a stable cubic structure under ambient conditions. However, formamidinium lead iodide [$HC(NH_2)_2PbI_3$, $FAPbI_3$] and cesium lead iodide ($CsPbI_3$) phases are only stable at temperatures above 150°C and 308°C, though their tolerance factors are very near to 1.0 and 0.8, respectively.

The bare $FAPbI_3$ and $CsPbI_3$ undergo the phase transition from black metastable α-phase [cubic, Fig. 8.3(d–f)] to yellow and more stable non-perovskite δ-phase [Fig. 8.3(b,c)] under ambient conditions. Although the $FAPbI_3$ and $CsPbI_3$ have ideal bandgaps of 1.48 eV and 1.73 eV, respectively, it is difficult to fabricate stable pure perovskite films under ambient conditions. Thus, a great way to find that mixing FA and Cs cations can produce a black and highly stable perovskite phase (e.g., $FA_xCs_{1-x}PbI_3$) provides enhanced photovoltaic performance and stability. Note that the stable α-phase $FA_{1-x}Cs_xPbI_3$ perovskite phases have lower formation energy than that of bare $FAPbI_3$ and $CsPbI_3$ [11].

Furthermore, high-quality perovskite structures should have perfect lattice (Fig. 8.3g), however, the bare OMHPs-based films present high chemical defects (intrinsic and interface defects), such as A-site vacancy, interstitial, Frenkel defect, Schottky defect and grain boundaries (GBs) as illustrated in Fig. 8.3(h–l), respectively. The mechanism and formation of perovskite crystal defects will be further presented and discussed in Chapter 10. Thus, it is possible to use ideal ions to modify such lattice defects by doping strategy. The doping areas can be classified as vacancies, interstitials, and GBs in terms of the size of the doping ion. From previous works, the intentional doping of MHPs (and/or compositions) by mixed cations or anions at desired locations can not only control their optical, electronic, and magnetic properties, but also improve the structure of perovskite crystals [12–23].

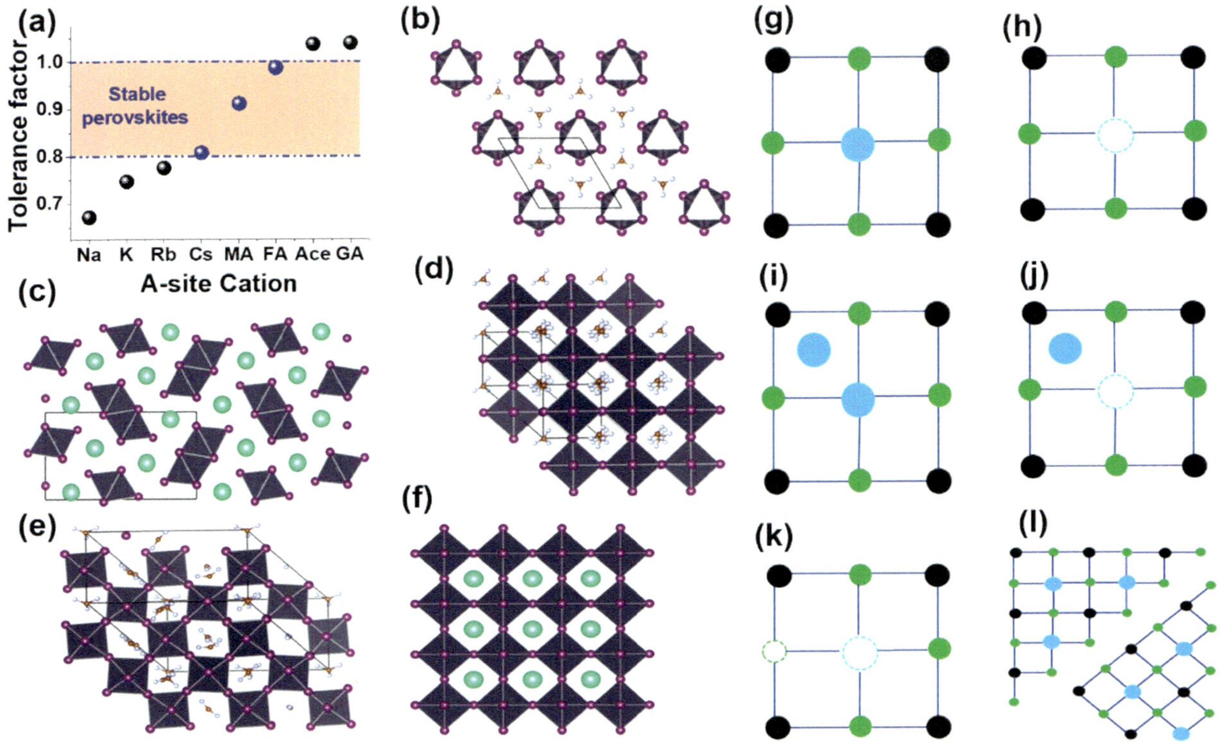

Figure 8.3 (a) Tolerance factor for $APbI_3$ with various A-site cations. (b, c) non-perovskite δ-phase of $FAPbI_3$ and $CsPbI_3$; (d, e) α- and β-phase of $FAPbI_3$ and (f) α-phase of $CsPbI_3$. Reprinted with permission from [11], Copyright 2017, American Chemical Society. (g–l) Defects of perovskite crystal lattice: perfect lattice, A-site vacancy, interstitial, Frenkel defect, Schottky defect, and grain boundary, respectively.

8.2.1 A-Site Doping

The motive for using the A-site doping strategy is to form a stable perovskite film with less formation energy and defects by selecting more stable inorganic elements or cations of different sizes. For example, a structurally stable perovskite composition of $(FAPbI_3)_{1-x}(MAPbBr_3)_x$ was developed for the fabrication of highly efficient and stable PSCs by Seok et al. [24]; a novel method to obtain stable and black α-$FA_{1-x}MA_xPbI_3$ film by treating δ-$FAPbI_3$ with methylammonium chloride (MACl) was reported by Xie et al. [25]. Such novel methods not only improve the balance of electron–hole in $FAPbI_3$ perovskite but also enable high-performance and long-term stable inverted-structure PSCs; Wang et al. developed high-quality and stable FAMA-perovskite-Cu:NiO composite films by incorporating larger FA cation and Cu:NiO nanoparticles, resulting in highly crystallized dense morphology having large-size crystals and fewer GBs [Fig. 8.4(a–c)]. More importantly, FA doped perovskite composite films also significantly enhance photovoltaic parameters (Fig. 8.4d) and stability of PSCs, which are attributed to the improvement of optical and electronic properties, suppression of non-radiative recombination, reduction of trap density, and formation of stable one-dimensional δ-$FAPbI_3$ at GBs [16]. Besides, the improved efficiency and stability of FAMA-based PSCs are also ascribed to the highly crystalline nature of FAMA perovskite film, stronger interaction between higher dipole MA-cation and PbI_6 octahedral cage [26], and substitutability of more acidic MA with partial FA cation [27].

Except for FAMA-perovskites, multi-cation A-site doped metal halide perovskites (MHPs) also were explored to further enhance the efficiency and operational stability of PSCs [12, 17, 28, 29]. A pioneering work related to quadruple cations (FAMACsRb) OMHPs-based highly efficient and stable PSCs was reported by Saliba et al. [12]. Although $RbPbI_3$ shows no black phases, Rb is very close to the Goldschmidt tolerance factor limit which makes it a desirable candidate for integration into perovskite lattice. Thus, the FAMACsRb perovskites-based cells not only obtained a *PCE* of 21.8% with a record V_{oc} of 1.24 V but also showed remarkable thermal- and photo-stability over 500 h at 85°C under continuous one-sun illumination in a nitrogen atmosphere.

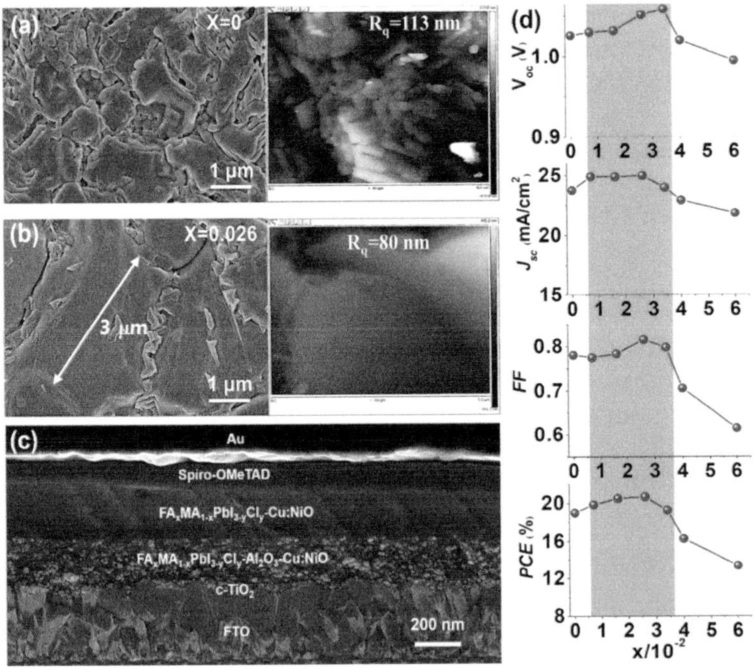

Figure 8.4 (a) Surface morphologies SEM (left) and atomic force microscopy (AFM) (right) images of $FA_xMA_{1-x}PbI_{3-y}Cl_y$-Cu:NiO composite film with (a) $x = 0$ and (b) $x = 0.026$; (c) Cross-sectional SEM image of a composite-based cell with FTO/c-TiO_2/mp-(Al_2O_3/Cu:NiO)/$FA_xMA_{1-x}PbI_{3-y}Cl_y$-Cu:NiO/Spiro-OMeTAD/Au structure ($x = 0.026$); (d) Effect of A-site doping with FA content (x) on the photovoltaic performances of the composite-based cells. Reprinted with permission from [16], Copyright 2020, WILEY-VCH.

Theoretical studies suggest that potassium (K) cation doped MHPs can effectively reduce the observable *J–V* hysteresis (mismatch *J–V* curves for both forward and reverse scans) in PSCs, which is mainly ascribed to suppression of Frenkel defects by K energetically preferring the interstitial site [28, 29]. Also, K cation doped MHPs have lower formation energy than that pristine MHPs (such as $MAPbI_3$). Wang et al. developed quadruple-cation K-doped CsFAMA perovskite compositions (K_δ-CsFAMA) to achieve highly efficient and stable NiO_x-based inverted PSCs and perovskite solar modules (PSMs) [17]. The K_δ-CsFAMA perovskite films show a uniform surface with bigger crystals over 1 μm and smaller roughness of 8.5 nm, as clearly shown in SEM and AFM images [Fig. 8.5(a,b)].

Besides, compared to CsFAMA perovskite films with disordered growth of smaller crystals, the K_δ-CsFAMA perovskite-based films present vertically aligned perovskite crystal growth with reduced GBs [Fig. 8.5(c,d)]. Furthermore, the content of K significantly affects the photovoltaic performance of inverted PSCs (Fig. 8.5e). It is also concluded that the incorporation of a small amount of K cation in CsFAMA-based perovskites can effectively reduce the perovskite thin-film defects and suppress the non-radiative recombination and photo-induced ion migration of PSCs.

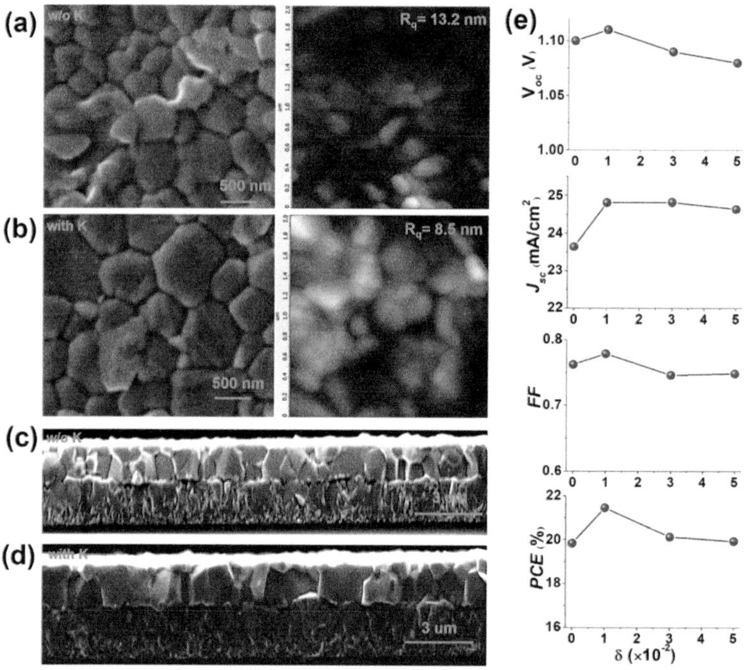

Figure 8.5 SEM and AFM images of (a) triple-cation CsFAMA perovskite and (b) quadruple-cation K_δ-CsFAMA perovskite; SEM images of (c) FTO/NiO_x/Al_2O_3/CsFAMA perovskite and (d) FTO/NiO_x/PTAA/Al_2O_3/K_δ-CsFAMA perovskite ($\delta = 0.01$); (e) normalized photovoltaic parameters as a function of δ. Reprinted with permission from [17], Copyright 2021, Elsevier.

8.2.2 B-Site Doping

In general, the B-site ions in OMHPs can be metallic divalent cations for establishing effective perovskite structures, however, the most efficient and stable photovoltaic perovskite materials are based on Pb^{2+} ion until now. Note that the ion radius of Sn^{2+} (1.18 Å) is very close to that of Pb^{2+} (1.19 Å), namely, $APbX_3$, and $ASnX_3$ perovskite structures have almost the same tolerance factors and ns^2 electronic structures. Compared to $APbX_3$ perovskites, however, the higher energy of Sn 5s orbitals in $ASnX_3$ perovskites can increase metal contribution to the valence band edge states, resulting in a lower bandgap [30]. Therefore, $ASnX_3$ perovskites are highly desired for the fabrication of perovskite–perovskite tandem solar cells. Unluckily, $ASnX_3$ perovskites are not stable due to the fast oxidation of Sn^{2+} to Sn^{4+} under an oxygen atmosphere, which becomes a burden to fabricate stable $ASnX_3$ perovskite films. Thus, the B-site doped Pb-OMHPs become an important strategy that not only tunes their optical and electrical properties but also improves the perovskite crystalline growth and stability. In general, Pb sites or vacancies can be substituted or occupied when the ionic radius and electronegativity of doping ions are similar to Pb^{2+}. Otherwise, the interstitial doping is dominant when the ionic radius of doping ions is smaller than Pb^{2+}, and GBs doping is dominant when the ionic radius of doping ions is larger than Pb^{2+} [31, 32]. Admittedly, compared to A-site and X-site doping, partial B-site doping is more difficult because of the higher formation energy of metallic divalent cation-based perovskites [33]. Therefore, it is significant to control the doping contents for fabricating efficient and stable perovskites.

Although pristine $ASnX_3$ perovskites are not stable under ambient conditions, Pb–Sn mixed perovskites become one of the most effective light absorbers for low-toxicity and high-performance PSCs. As can be seen in Fig. 8.6(a,b), the bandgap and absorption wavelength of Pb–Sn mixed perovskites can be remarkably adjusted by Sn-doping. Interestingly, to further enhance the performance and stability of Pb–Sn mixed PSCs, Li et al. developed vertically aligned 2-dimensional/3-dimensional (2D/3D) Pb–Sn mixed heterostructures by one-step vacuum-assisted blade-coating 4-fluorophenethylamine (FPEAI)-modified $(MAPbI_3)_{0.75}(FASnI_3)_{0.25}$ perovskite precursors [20]. The grazing incidence wide-angle X-ray scattering (GIWAXS) images and PL spectra provide strong evidence that (FPEAI)-modified Pb–Sn perovskite films have vertically aligned perovskite crystal growth and formation of 2D perovskites in GBs [Fig. 8.6(c–f)]. The (FPEAI)-modified Pb–Sn PSCs showed

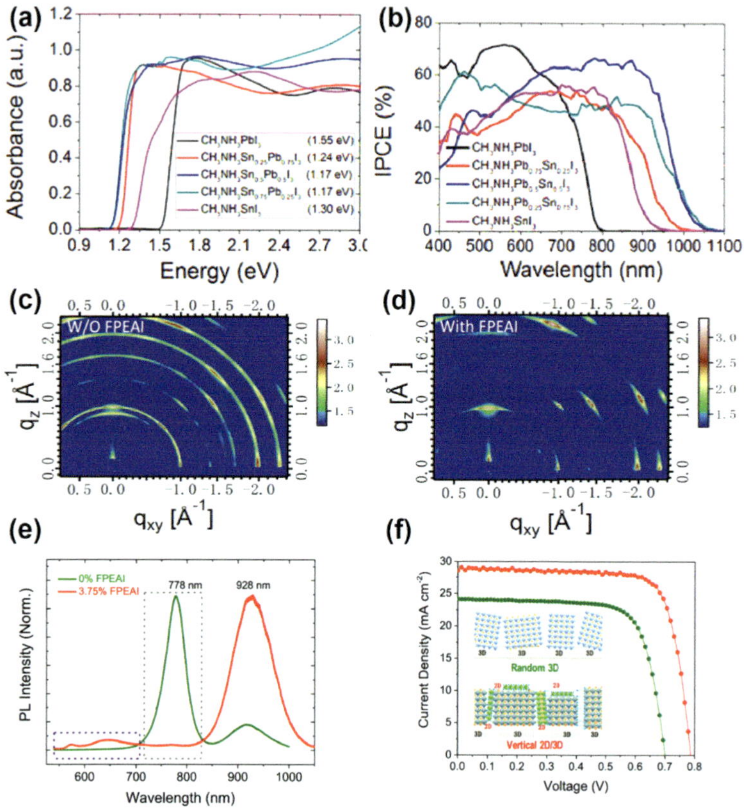

Figure 8.6 (a) Absorption spectra of the $CH_3NH_3Sn_{1-x}Pb_xI_3$ perovskites; (b) IPCE spectra of $CH_3NH_3Sn_{1-x}Pb_xI_3$ perovskites-based solar cells. Reprinted with permission from [19], Copyright 2014, American Chemical Society. GISAXS spectra of Pb–Sn mixed perovskite films (c) w/o FPEAI and (d) FPEAI treatment, (e) normalized PL spectra of Pb–Sn mixed perovskite films w/o FPEAI (green), and FPEAI (red) treatment, (f) *J–V* curves and schematic illustration of Pb–Sn mixed perovskite crystals w/o FPEAI (green) and FPEAI (red) treatment. Reprinted with permission from [20], Copyright 2020, American Chemical Society.

excellent photovoltaic performance (Fig. 8.6f) and air stability. In addition, Tan et al. reported all-perovskite tandem solar cells with a certified efficiency of 24.2% by using surface-anchoring zwitterionic antioxidants [21]. The formamidine sulfinic acid (FSA) additive as an antioxidation reactant can prevent Sn^{2+} from being oxidized to Sn^{4+} in the Pb–Sn perovskite film. These great works present an important milestone to fabricate highly efficient and stable Pb–Sn

PSCs. Except for Sn-doped Pb-perovskites, other B-site doing in a small amount also can improve the performance and stability of PSCs, such as improvements of perovskite crystal quality, structural stability, and passivation of trap states at GBs by doping Eu^{3+} [34], Sr^{2+} [35], Cu^{2+} [36], Co^{2+} [37], etc.

8.2.3 X-Site Doping

As discussed, A- and B-site doping strategies not only tune optical and electrical properties but also stabilize structures of OMHPs. X-site doping is another strategy in OMHPs for fabricating efficient and stable perovskite photovoltaics. The X-site generally can be halide ions such as I^-, Br^-, and Cl^-. Note that the hybrid lead triiodide perovskites ($APbI_3$) are one of the most efficient pure X-site halide photovoltaic materials than other perovskites, such as $APbBr_3$ and $APbCl_3$. More importantly, Br- or Cl-doped OMHPs not only can suppress the formation of halide vacancies via lattice strain relaxation but also improve perovskite film quality with fewer GBs and lower trap density. For example, Noh et al. systematically studied the structural, optical, and electronic properties of bromide-doped iodide perovskite system $MAPb(I_{1-x}Br_x)_3$ by changing the content of bromide [22]. It was found that the $MAPb(I_{1-x}Br_x)_3$ structure can be transformed from an unstable tetragonal phase ($x = 0$–0.13, Fig. 8.7b, top) to a stable cubic phase ($x \geq 0.2$, Fig. 8.7b, bottom), which is ascribed to the reduction of lattice parameter with an increase in the amount of Br (Fig. 8.7c). Thus, the cubic phase $MAPb(I_{1-x}Br_x)_3$ ($x \geq 0.2$) based solar cells are more stable against high humidity than the tetragonal phase ($x = 0 - 0.13$) based devices.

More interestingly, the Cl-doped OMHPs not only can improve crystal coalescence and growth but also suppress defects and lattice distortions. Kim et al. reported that MACl induced perovskite intermediate phase stabilization for highly efficient $FAPbI_3$ PSCs with a certified PCE of 23.48% [38]. According to density functional theory (DFT) calculations, the cationic chloride significantly stabilizes the $FAPbI_3$ structure thermodynamically and enhances the interaction between FA and I, leading to improvement of the physical and electronic properties, and stability of $FAPbI_3$. In addition, another interesting work was reported regarding wide-bandgap triple-halide alloyed perovskites (i.e., Cl, Br, and I) which were used

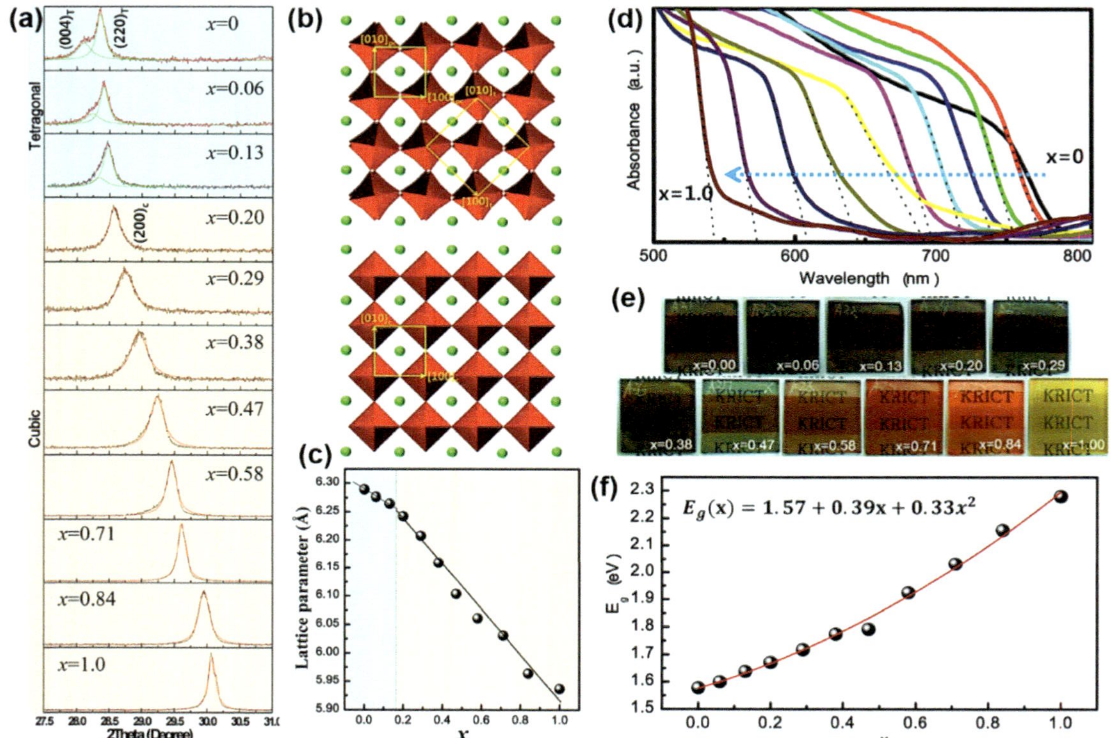

Figure 8.7 (a) XRD patterns of $MAPb(I_{1-x}Br_x)_3$ for different Br doping contents (x), (b) crystal structures of tetragonal (top) and cubic (bottom) phases, (c) lattice parameters, (d) UV–Vis absorption spectra, (e) film photos, and (f) bandgaps of $MAPb(I_{1-x}Br_x)_3$ for different Br doping contents (x). Reprinted with permission from [22], Copyright 2013, American Chemical Society.

to fabricate perovskite top cells in tandem solar cells [39]. The corresponding two-terminal monolithic tandem solar cells showed 27% efficiency and long-term stability over 1000 h under 100-sun illumination. Other anions having a similar ionic radius to halide ions such as SCN^- (2.13 Å), BF_4^- (2.18 Å), and PF_6^- (2.55 Å) were also used to fabricate highly efficient and stable PSCs [40–42]. From these works, it is believed that appropriate X-site doping enables high-quality structural stability and adjustable wide-bandgap MHP films.

8.3 History of Perovskite Solar Cell

The organic–inorganic lead halide perovskite-based PVs was first studied by Miyasaka and co-workers in 2005 [43]. In 2009, the first reported $MAPbI_3$ and $MAPbBr_3$-based solar cells showed an efficiency of 3.81% despite being unstable in all devices due to the use of liquid electrolytes. The liquid electrolytes can damage quickly the perovskites. As shown in Fig. 8.8a (left), the perovskite nanocrystals can serve as sensitizers adsorbed on the TiO_2 surface and transport electrons well to TiO_2 under visible light. Note that the iodine- and bromine-based perovskite films show different absorption wavelengths, and especially $MAPbI_3$ has a wide wavelength from 300 nm to 800 nm (Fig. 8.8a, right), indicating that halide perovskites are highly suitable for PV applications. In 2011, Nam-Gyu Park et al. reported a 6.5% efficiency of perovskite-based DSSCs with a higher concentration of perovskite precursor [45].

The advent of solid-state PSCs is an important milestone to boost device efficiency improvement. As shown in Fig. 8.8b, Park, Grätzel and co-workers first developed a solid-state perovskite solar cell structure by using a solid HTM (Spiro-OMeTAD), resulting in an impressive efficiency of 9.7% in 2012 [2]. Almost simultaneously, Snaith, Miyasaka, and colleagues reported stable mixed halide $MAPbI_{3-x}Cl_x$ with Al_2O_3 as a scaffold-based meso-superstructure PSCs, which showed a remarkable efficiency of 10.9% with higher a V_{oc} [3]. Such results indicate a sign of the long electron–hole diffusion length of carriers in perovskite [46]. Thus, a new semiconductor star was born in PV technology, and it shone bigger and brighter in the shortest time in PV technology history. PSCs have demonstrated an incredibly rapid ascent in PCE, reaching a record efficiency of 25.8% (certified 25.5%, 2021) [47]. By materials innovation and device configuration design, OMHPs-based PSCs have great potential to achieve over 30% efficiency [48, 49]. There is no doubt that this new PV technology has a great opportunity to compete, conquer, and promote commercialization in the near future.

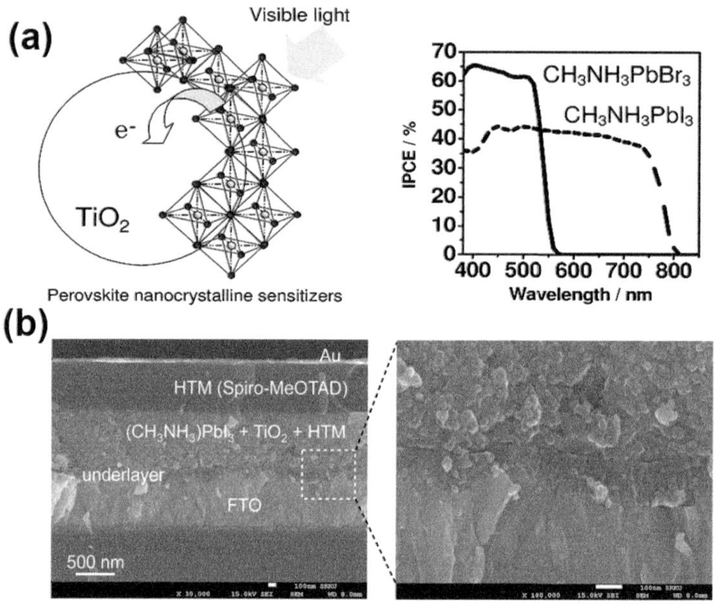

Figure 8.8 (a) Schematic of perovskite sensitized TiO_2 and electron-transport path, SEM cross-sectional image. Reprinted with permission from [1], Copyright 2009, American Chemical Society. (b) Pioneer solid-state perovskite solar cell. Reprinted with permission from [2], Copyright 2013, Springer Nature.

8.4 Basic Working Principles

The original structure of the PSC is derived from DSSC, but there are some differences in operating principles. Usually, DSSCs require a semiconductor oxide as a scaffold, but PSCs based on insulating mesoporous Al_2O_3- or planar heterojunction with compact TiO_2 also work well, suggesting that electrons and holes can be transported through the perovskite layer. As widely accepted today, the working principles of PSCs are more like solid-state p-n junction solar cells.

The presence of a continuous perovskite layer in charge-transport layers is important for enhancing light harvesting and conversion efficiency of photons to electrons. The operation principle of PSC is more consistent with those of n-i-p and p-i-n solar cells, where perovskites act as intrinsic absorbers sandwiched between two transport layers (n and p).

The working principles mainly include (1) the perovskite layer absorbs photons to generate electrons and holes (carrier generation process); (2) the electrons and holes are selectively collected and transported by the n-type ETL and p-type HTL (charge separation and transport process), respectively; (3) then current is produced by an external circuit connect (charge collection). Besides, the quasi-Fermi levels (QFLs) between transport layers and perovskite layer for n-i-p or p-i-n type PSCs show different influences on V_{oc}. Thus, a mismatch in energy band diagram levels severely affects the interface charge-transport efficiency, resulting in lower PCEs. In other words, the CB of n-type ETL and the VB of p-type HTL have to match the CB and VB of the perovskite layer, respectively.

In addition, OMHPs present a lot of distinctive photophysical properties, which indeed show excellent PV characteristics. Several theoretical studies have presented the potential role of intrinsic ferroelectricity or bulk crystal properties of OMHPs that affects the separation and transportation of charge carriers [50–52]. Some other studies have focused on the role of film surface/buried interface of the perovskite layer and their defects. It was found that the use of passivated layer on the top of the perovskite surface or under its buried interface can remarkably affect the properties of perovskite film and device interfaces [53, 54].

References

1. Kojima, A., Teshima, K., Shirai, Y., and Miyasaka, T. (2009) *J. Am. Soc.*, **131**, pp. 6050–6051.
2. Kim, H-S., Lee, C-R., Im, J-H., Lee, K-B., et al. (2012) *Sci. Rep.*, **2**, pp. 591.
3. Lee, M.-M., Teuscher, J., Miyasaka, T., Murakami, T. N., and Snaith, H. J. (2012) *Science*, **338**, pp. 643.
4. Jeon, N. J., Noh, J. H., Yang, W. S., Kim, Y. C., et al. (2015) *Nature*, **517**, pp. 476–480.

5. Smith, I. C., Hoke, E. T., Solis-Ibarra, D., McGehee, M. D., and Karunadasa, H. I. (2014) *Angew. Chem. Int. Ed.*, **53**, pp. 11232–11235.
6. Tsai, H. H., Nie, W. Y., Blancon, J. C., Toumpos, C. C. S., et al. (2016) *Nature*, **536**, pp. 312–316.
7. Quan, L. N., Yuan, M. J., Comin, R., Voznyy, O., et al. (2016) *J. Am. Chem. Soc.*, **138**, pp. 2649–2655.
8. Cao, D. H., Stoumpos, C. C., Farha, O. K., Hupp, J. T., and Kanatzidis, M. G. (2015) *J. Am. Chem. Soc.*, **137**, pp. 7843–7850.
9. Chen, Y. T., Liu, X. M., Wang, T. F., and Zhao, Y. X. (2021) *Acc. Chem. Res.*, **54**, pp. 3452–3461.
10. Deng, J. P., Li, J. L., Yang, Z., and Wang, M. Q. (2019) *J. Mater. Chem. C.*, **7**, pp. 12415.
11. Gratzel, M. (2017) *Acc. Chem. Res.*, **50**, pp. 487–491.
12. Saliba, M., Matsui, T., Domanski, K., Seo, J. Y., et al. (2016) *Science*, **354**, pp. 206–209.
13. Kim, M., Kim, G. H., Lee, T. K., Choi, I. W., et al. (2019) *Joule*, **3**, pp. 1–14.
14. Wang, Y., Mahmoudi, T., Rho, W-Y., Yang, H-Y., et al. (2017) *Nano Energy*, **40**, pp. 408–417.
15. Wang, Y., Mahmoudi, T., Yang, H.-Y., Bhat, K. S., et al. (2018) *Nano Energy*, **49**, pp. 59–66.
16. Wang, Y., Mahmoudi, T., and Hahn, Y.-B. (2020) *Adv. Energy Mater.*, **10**, pp. 2000967.
17. Wang, Y., Ju, H., Mahmoudi, T., Liu, C., et al. (2021) *Nano Energy*, **88**, pp. 106285.
18. Euvrard, J., Yan, Y., and Mitzi, D. B. (2021) *Nat. Rev. Mater.*, **6**, pp. 531–549.
19. Hao, F., Stoumpos, C. C., Chang, R. P. H., and Kanatzidis, M. G. (2014) *J. Am. Chem. Soc.*, **136**, pp. 8094–8099.
20. Li, C., Pan, Y., Hu, J., Qiu, S., et al. (2020) *ACS Energy Lett.*, **5**, pp. 1386–1395.
21. Xiao, K., Lin, R., Han, Q., Hou, Y., et al. (2020) *Nat. Energy*, **5**, pp. 870–880.
22. Noh, J., Im, H. S. H., Heo, J. H., Mandal, T. N., and Seok, S. I. (2013) *Nano Lett*, **13**, pp. 1764–1769.
23. Jeon, N. J., Noh, J. H., Yang, W. S., Kim, Y. C., et al. (2015) *Nature*, **517**, pp. 476–480.

24. Xie, F. X., Chen, C. C., Wu, Y. Z., Li, X., et al. (2017) *Energy Environ. Sci.*, **10**, pp. 1942–1949.
25. Binek, A., Hanusch, F. C., Docampo, P., and Bein, T. (2015) *J. Phys. Chem. Lett.*, **6**, pp. 1249–1253.
26. Leijtens, T., Bush, K., Cheacharoen, R., Beal, R., et al. (2017) *J. Mater. Chem. A*, **5**, pp. 11483–11500.
27. Son, D. Y., Kim, S. G., Seo, J. Y., Lee, S. H., et al. (2018) *J. Am. Chem. Soc.*, **140**, pp. 1358–1364.
28. Abdi-Jalebi, M., Andaji-Garmaroudi, Z., Cacovich, S., Stavrakas, C., et al. (2018) *Nature*, **555**, pp. 497–510.
29. Zhang, J., Gan, X., Sun, H., Yuan, H., et al. (2020) *Sol. RRL*, **4**, pp. 1900227.
30. Yang, Z., Wei, M., Voznyy, O., Todorovic, P., et al. (2019) *J. Am. Chem. Soc.*, **141**, pp. 8296–8305.
31. Zhou, Y., Chen, J., Bakr, O. M., and Sun, H. T. (2018) *Chem. Mater.*, **30**, pp. 6589–6613.
32. Eames, C., Frost, J. M., Barnes, P. R. F., O'Regan, B. C., et al. (2015) *Nat. Commun.*, **6**, pp. 7497.
33. Wang, L., Zhou, H., Hu, J., Huang, B., et al. (2019) *Science*, **363**, pp. 265–270.
34. Zhang, H., Wang, H., Williams, S. T., Xiong, D. H., et al. (2017) *Adv. Mater.*, **29**, pp. 1606608.
35. Li, M., Wang, Z. K., Zhuo, M. P., Hu, Y., et al. (2018) *Adv. Mater.*, **30**, pp. 1800258.
36. Xu, W. Z., Zheng, L. Y., Zhang, X. T., Cao, Y., et al. (2018) *Adv. Energy Mater.*, **8**, pp. 1703178.
37. Lin, Y. Z., Shao, Y. C., Dai, J., Li, T., et al. (2021) *Nat Commun.*, **12**, pp. 7.
38. Kim, M., Kim, G. H., Lee, T. K., Choi, I. W., et al. (2019) *Joule*, **3**, pp. 2179–2192.
39. Xu, J. X., Boyd, C. C., Yu, Z. S. J., Palmstrom, A. F., et al. (2020) *Science*, **367**, pp. 1097–1104.
40. Tai, Q., You, P., Sang, H., Liu, Z., et al. (2016) *Nat Commun.*, **7**, pp. 11105.
41. Chen, J., Rong, Y., Mei, A., Xiong, Y., et al. (2016) *Adv. Energy Mater.*, **6**, pp. 1502009.
42. Chen, J., Kim, S. G., and Park, N. G. (2018) *Adv. Energy Mater.*, **30**, pp. 1801948.
43. Jena, A. K., Kulkarni, A., and Miyasaka, T. (2019) *Chem. Rev.*, **119**, pp. 3036–3103.

44. Kojima, A., Teshima, K., Shirai, Y., and Miyasaka, T. (2007) *210th Meeting Abstr.*, MA2007-02, pp. 352.
45. Im, J-H., Lee, C. R., Lee, J. W., Park, S. W., and Park, N. G. (2011) *Nanoscale*, **3**, pp. 4088.
46. Stranks, S. D., Eperon, G. E., Grancini, G., Menelaou, C., et al. (2013) *Science*, **342**, pp. 341–344.
47. Min, H., Lee, D., Kim, J., Kim, G., et al. (2021) *Nature*, **598**, pp. 444–450.
48. Wang, Y., Arumugam, G. M., Mahmoudi, T., Mai, Y., and Hahn, Y.-B. (2021) *Nano Energy*, **87**, pp. 106141.
49. Ma, C. and Park, N.-G. (2020) *Chem*, **6**, pp. 1254–1264.
50. Liao, W. Q., Zhang, Y., Hu, C. L., Mao, J. G., et al. (2015) *Nat. Comm.*, **6**, pp. 7338.
51. Thompson, C. V. (1985) *J. Appl. Phys.*, **58**, pp. 763.
52. Xue, J. J., Wang, R., Wang, K. L., Wang, Z. K., et al. (2019) *J. Am. Chem. Soc.*, **141**, pp. 13948–13953.
53. Yang, X. Y., Luo, D. Y., Xiang, Y. R., Zhao, L. C., et al. (2021) *Adv. Mater.*, **33**, pp. 2006435.
54. Chen, B., Chen, H., Hou, Y., Xu, J., et al. (2021) *Adv. Mater.*, **33**, pp. 2103394.

Problems

8.1 Describe methods for composing various perovskites. Also discuss the advantages of different perovskite compounds by using A-site doping, B-site doping, and X-site doping strategy.

8.2 Why is the discovery of PSCs a milestone in the photovoltaic community?

8.3 What are the fabrication methods of PSCs? Describe their advantages.

8.4 What is current challenge of perovskite materials for solar cell applications?

8.5 What are the strategies to improve the stability of PSCs?

Chapter 9

Structures, Transport Materials, and Deposition Methods for Perovskite Solar Cells

The performance of PSCs is largely dependent on the device configurations, charge transport materials, and deposition methods for PSC fabrication. Therefore, in order to design an efficient PSC, it is important to understand the types of cell structure, transport materials for electrons and holes, and fabrication methods of PSCs. This chapter discusses the details of device structure, organic and inorganic transport materials, and deposition methods for device fabrication such as spin-coating, vapor-assisted deposition, and printing technique.

9.1 Perovskite Solar Cell Configurations

After decades of development for PSCs, there have been four types of configuration: traditional n-i-p and hole-conductor free (HCF) PSCs, and inverted p-i-n and HCF PSCs, as shown in Fig. 9.1. Currently, both traditional n-i-p and inverted p-i-n structures can achieve over 23% efficiency for lab-scale cells developed by many research groups. More importantly, a record efficiency of 25.8% (certified 25.5%) has been achieved in the traditional n-i-p structure. Undoubtedly, higher efficiencies of PSCs will continue roaring toward their theoretical efficiency limit. However, these n-i-p and p-i-n devices show poor

Next-Generation Solar Cells: Principles and Materials
Yoon-Bong Hahn, Tahmineh Mahmoudi, and Yousheng Wang

ISBN 978-981-4968-66-9 (Hardcover), 978-1-003-37238-7 (eBook)
www.jennystanford.com

operational stability attributed to material and device interface defects. Thus, the HCF-PSCs, especially traditional HCF-PSCs, become promising configurations to improve long-term stability. The main barrier to HCF-PSCs is that, despite much effort, efficiency still is less than 20%. In the following, we will discuss the structures, performances, and advantages in detail for those four types of solar cell configuration.

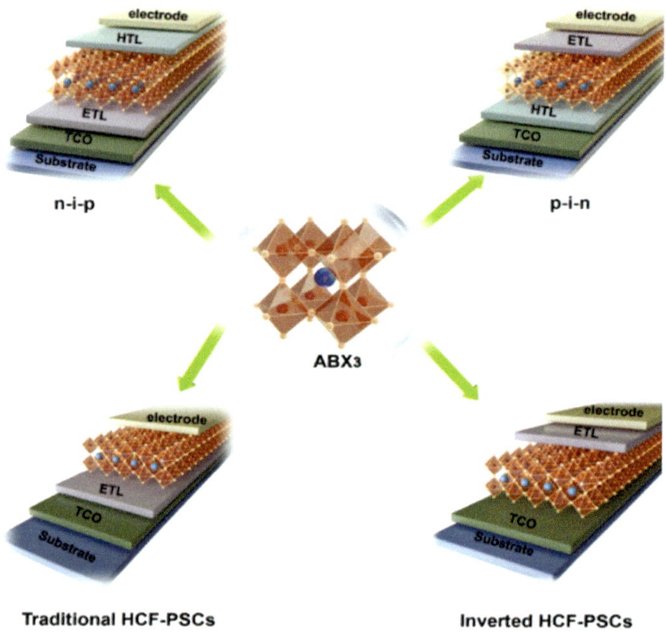

Figure 9.1 Typical perovskite solar cell configurations: traditional n-i-p structure, inverted p-i-n structure, traditional HCF structure, and inverted HCF structure solar cells.

9.1.1 n-i-p Structure

As discussed earlier, the first solid-state PSC was actually an n-i-p architecture, but there was no perovskite capping layer at that time. In fact, the traditional n-i-p configurations have two types of architecture, i.e., mesoporous- and planar-based structures. In general, mesoporous type PSC comprises an electron-transport layer (ETL) scaffold (such as TiO_2 and Al_2O_3) with a compact ETL, where perovskite precursor can infiltrate into the scaffold, forming a perovskite-metal oxide composite layer. Thus, the light absorber layers include both perovskite capping and composite layers, leading to high light-harvesting ability. The traditional plannar n-i-p type architecture does not have an ETL mesoporous scaffold or uses a full-organic charge-transport layer. Currently, most reported mesoporous PSCs can achieve over 22% efficiency by materials defect passivation and device innovation strategies [1–4].

The first n-i-p planar type PSCs were reported by Snaith et al., which only used compact TiO_2 (c-TiO_2) as the ETL and followed by deposition of the perovskite layer by a dual-source vapor method. Interestingly, compact SnO_2 ETL based planar PSCs have been considered the most promising solar cell architecture to achieve over 30% efficiency. For example, You et al. developed SnO_2ETL based planar solar cells with a surface-passivation strategy, achieving a certificated efficiency of 23.32% by NREL in 2019 [5]. Afterward, Seo et al. reported SnO_2 ETL based planar solar cells with a certified efficiency of up to 25.2% by improved carrier management [6]. Subsequently, Soek et al. achieved a PCE of 25.8% (certified 25.5% by NREL) SnO_2 ETL based planar solar cells by interface modification strategy between SnO_2 ETL and perovskite active layer [7]. Such results suggest that SnO_2 can be selected as one of the most effective ETLs in perovskite planar solar cells, which is mainly ascribed to the excellent optical and electronic properties of SnO_2 (see Section 9.2).

9.1.2 p-i-n Structure

Similarly, the inverted p-i-n structure also has two kinds of architecture: mesoporous and planar-based structures. Usually, NiO_x nanoparticles (NPs) can be used as scaffolds in inverted mesoporous PSCs. The perovskite precursor can infiltrate into the scaffold, forming the perovskite-NiO_x NPs composite layer. Wang et al. firstly reported NiO_x-based mesoscopic PSC in 2014, which resulted in 9.51% efficiency [8]. Afterward, efficiencies over 17% were achieved by rational design of mesoporous NiO_x layer and metal-doped NiO_x layer, respectively [9, 10]. However, the efficiency is limited (<20%) with such architecture, which may be attributed to the hole transportation limitation by the thicker mesoporous NiO_x layer. Thus, NiO_x-based planar PSCs are popular due to their simpler structure. The key to NiO_x-based planar PSCs is the fabrication of uniform and high-coverage NiO_x films. Due to high transparency, thermal stability, the well-aligned

VB edge, and high electron blocking properties, NiO_x become one of the most excellent inorganic hole-transport materials (HTMs). Until now, many efforts have been devoted to the optimization of surface morphology, defects, and optical and electronic properties of NiO_x, thus NiO_x-based planar PSCs have achieved a PCE of 22.9% [11].

Interestingly, the organic-based HTL [such as poly(triarylamine), PTAA] planar PSCs showed better efficiency than NiO_x-based planar PSCs. Jen et al., achieved a PCE of 23.37% (certified 22.75%, 2020) inverted PSCs by surface termination of perovskite films using a bifunctional molecule. Such surface defects passivation can release surface residual stress, resulting in less non-radiative recombination and V_{oc} deficit of only 0.33 V [12]. A higher PCE of 23.8% was achieved by using an organic bulk-heterojunction (BHJ) structure as ETL by the same group. Besides, they also fabricated highly efficient and mechanically durable flexible planar solar cells with a PCE of 21.73% [13]. However, compared to NiO_x-based inverted PSCs, organic HTL-based inverted PSCs show poor operational stability.

9.1.3 Hole-Conductor Free Perovskite Solar Cells

Early in 2012, Etgar et al. first developed simple mesoscopic $MAPbI_3$/TiO_2 heterojunction solar cells without a hole-transport layer (later called HCF-PSCs), where the device could work well and showed a PCE of 5.5% [14]. Such results suggest that the perovskite layer acts as both light absorber and hole conductor, which play a role in p-n junction type-based solar cells. In addition, most of the reported highly efficient PSCs have used quite expensive and unstable Spiro-OMeTAD or PTAA as a hole-transport material (HTM), which limits their application. Thus, HCF-PSCs have been widely studied to improve their efficiencies.

An important turning point for HCF-PSCs happened in 2014 when Han et al. developed printable mesoscopic HCF-PSCs-based mesoscopic triple-layer architecture and carbon counter electrode. The mp-TiO_2, -ZrO_2, and carbon layer were constructed on the as-prepared FTO/c-TiO_2, and $MAPbI_3$ precursor could infiltrate into three mesoporous layers to form $MAPbI_3$/TiO_2 heterojunction. Such HCF-PSCs achieved a certified PCE of 12.8% with superb stability over 1000 h in ambient air under continuous full sunlight.

Wu et al. developed inverted p-n junction HCF-PSCs by p-type molecular doping strategy. Firstly, they used p-type small-molecule

2,3,5,6-tetrafluoro-7,7,8,8-tetracyanoquinodimethane (F4TCNQ), as additives added to $MAPbI_3$ precursor. Then F4TCNQ molecular doping of halide perovskite films was formed after annealing. It was found that the F4TCNQ molecular doping not only could improve the conductivity of $MAPbI_3$ films but also modify the ITO/$MAPbI_3$ interface with favorable band bending. The better conductivity of $MAPbI_3$ films can reduce series resistance, and the favorable band bending facilitates the extraction of photo-excited holes from perovskite to the transparent conducting oxide (TCO) electrode. As a result, the inverted HCF-PSCs showed a higher PCE over 20% with almost no *J–V* hysteresis [17].

9.1.4 Flexible Perovskite Solar Cells

Due to the outstanding bending characteristics of OMHP films, there is an opportunity to fabricate lightweight, wearable, and flexible PSCs (F-PSCs). Besides, the F-PSCs using a plastic substrate can be the most competitive power per weight among the solar cells. More importantly, the F-PSCs also can be fabricated by using a solution-based roll-to-roll (R2R) process, which presents a great potential for large-scale perovskite solar module commercialization [18]. For example, Kim et al. reported a PCE of 12.2% F-PSCs by using the atomic layer deposition (ALD) method to fabricate TiO_2 ETL. They showed the excellent bending durability of OMHP films for the first time [19].

Unlike rigid PSCs, charge-transport layers such as metal oxides should be formed at a low temperature below 100°C in F-PSCs. However, a high annealing temperature is required to obtain high-quality metal oxides having excellent optical and electronic properties. In addition, OMHPs are polycrystalline materials, which are composed of crystalline grains and grain boundaries. Thus, the excess bending or crumpling will reduce the solar cell efficiency. Therefore, the main challenge is how to obtain high-efficiency and excellent bending durability of F-PSCs. Inspired by the robust crystallization and flexible structure of vertebrae, interestingly, Chen et al. [20] developed a conductive and glued polymer (poly(3,4-ethylenedioxythiophene):poly(ethylene-co-vinyl acetate), PEDOT:EVA) between TCO and perovskite layers. The adhesive EVA not only can act as the HTL but also facilitate the oriented growth of perovskite crystals with strong bending durability. The F-PSCs obtained a high

PCE of 19.87% and 17.55% for area of 1.01 cm^2 and 31.20 cm^2, respectively. Besides, the device can maintain over 85% of initial efficiency after 7000 narrow bending cycles.

9.2 Transport Materials for Perovskite Solar Cells

9.2.1 Electron-Transport Materials

9.2.1.1 Organic materials

As it is well known, the widely used organic materials such as electron-transport material (ETM) in PSCs are fullerene (C60 and C70) and its derivatives (such as phenyl-C61-butyric acid methyl ester and PCBM), and nonfullerenes such as perylene diimides. Seok et al. reported metal-oxide-free n-i-p type PSCs based on full-organic charge-transport layers, where solar cells presented a low efficiency of 15.3% [21]. Such relatively low efficiency was attributed to the low transparency of PCBM and partial light absorption by PCBM, which remarkably affects perovskite light harvesting. Thus, the most reported high-efficiency PSCs used metal oxides as ETLs. However, these fullerenes, nonfullerenes, and their derivatives are more appropriate for the fabrication of inverted p-i-n PSCs. Note that C60 and PCBM are one of the most efficient ETMs in p-i-n PSCs attributed to good band alignment with perovskite, good electronic properties, and passivation effects of the perovskite layer defects by interacting with under-coordinated halide or Pb-I anti-site defect [22].

9.2.1.2 Inorganic materials

The function of electron-transport materials is mainly to accept electrons and facilitate electron transportation. In general, the value of CB for the perovskite layer is around 3.9–4.5 eV, thus most n-type metal oxides such as TiO_2 (~ –4.2 eV), SnO_2 (~ –4.5 eV), and ZnO (~ –4.4 eV) can act as ETMs in PSCs. More importantly, their bandgaps with CB and VB can be controlled by different synthesis conditions that present great potential to break through efficiency limits in the n-i-p type PSCs. Through decades of development experience, SnO_2 and TiO_2 have been considered as one of the best ETMs.

It is worth noting that compared to TiO_2 and other ETMs, the SnO_2 presents several advantages: (1) a deeper conduction band that can form a good band alignment with the perovskite layer, resulting in better electron extraction and hole blocking; (2) high bulk electron mobility and conductivity that can facilitate electron transport and reduce the series resistance; (3) excellent chemical stability, UV-resistance properties, and less photocatalytic activity [23]. Due to above fabulous properties, SnO_2-based n-i-p type PSCs achieved a certified efficiency record for several times (23.3%, 25.2%, 25.5%, and 25.7%) [5–7]. Except for TiO_2 and SnO_2, metal oxides such as SrTiO3, BaSnO3 and Nb2O3, metal-doped metal oxides, and composite materials are also widely studied in PSCs.

9.2.2 Hole-Transport Materials

9.2.2.1 Organic materials

Ideal HTMs should have high hole mobility, good energy band alignment with perovskites, and long-term stability in outdoor conditions. According to the experience of DSSCs, the Spiro-OMeTAD has been used as an effective HTM for fabricating solid-state PSCs with better efficiency. So far, Spiro-OMeTAD is still one of the most efficient HTMs for high-performance PSCs, although it is expensive and less stable when using additives such as 4-tert-butylpyridine (TBP) and bis (trifluoromethane) sulfonimide lithium salt (Li-TFSI) [5–7]. The Li-TFSI and TBP are essential to improving electronic properties and the qualities of the Spiro-OMeTAD layer. Other HTMs, such as P3HT [3], poly[bis(4-phenyl)(2,4,6-trimethylphenyl)amine] (PTAA) [12, 13], and PEDOT:PSS [24] are also used for fabricating high-performance PSCs.

Furthermore, HTMs can be utilized to obtain highly efficient PSCs according to the bandgap of the perovskite compositional layer. Seo et al. developed a fluorene-terminated hole-transporting material M (N2, N2′, N7, N7′ -tetrakis(9,9-dimethyl-9H-fluoren-2-yl)-N2, N2′, N7, N7′ - tetrakis(4-methoxyphenyl)-9,9′-spirobi[fluorene]-2,2′,7,7′-tetraamine) (shorten to DM) with a fine-tuned energy level and a high glass transition temperature for highly efficient and stable PSCs [2]. Compared to the Spiro-OMeTAD-based PSCs, the optimized DM-based PSCs showed a higher PCE due to higher V_{oc}. Furthermore, the devices with DM present better stability than that of Spiro-OMeTAD-based devices.

9.2.2.2 Inorganic materials

Compared to expensive and environmentally unstable p-type organic materials, inorganic HTMs have been widely studied for highly efficient and stable PSCs. However, perovskite films present poor compatibility with many solvents, resulting in the deposition barrier of inorganic HTMs on the top of the perovskite layer by the solution process. Thus, inverted p-i-n type PSCs are preferred because most solution-processed inorganic HTMs should be deposited first on the TCO followed by high-temperature annealing. Several inorganic HTMs have been used to fabricate inverted p-i-n PSCs, such as nickel oxide (NiO), copper (I) thiocyanate (CuSCN), copper iodide (CuI), copper oxides (CuO_x), etc.

In particular, NiO or its doped materials and wide bandgap (~3.6 eV) p-type semiconductors with good transparency, thermal and chemical stability, and relatively high carrier mobility have been successfully used for highly efficient and stable inverted PSCs. For example, Chen et al. [25] reported highly efficient, stable, and large-area PSCs with doped NiO (i.e., NiMgLiO) as an HTL. The device presents remarkable photo-stability over 1000 h under ambient conditions.

The modified CB of NiMgLiO can effectively block electrons, thus facilitating carrier extraction and transportation. Encouragingly, numerous methods have been explored to synthesize novel nanostructures of NiO. Mali et al. [26] achieved efficient and air-stable p-i-n inverted PSCs by employing nanoporous NiO flakes. Interestingly, perovskite precursors effectively penetrated vertically aligned NiO walls/flakes and formed the perovskite-NiO composite layer, which was followed by the formation of an extra capping layer. These solar cells exhibit improved PCEs by over 19% due to excellent charge extraction and reduced interface barriers.

9.3 Deposition Methods for Perovskite Solar Cells Fabrication

For the fabrication of a perovskite solar cell, the main procedures include:

(1) Chemical or laser etching of TCO to obtain desired patterns.
(2) Deposition of a compact layer (thickness ~100 nm) on the etched TCO to reduce charge recombination by avoiding the direct contact between the perovskite layer and TCO, or HTL and TCO.
(3) Deposition of a perovskite active layer (thickness ~1000 nm) on the compact layer for planar structure. Before the active layer deposition, a mesoporous layer (mp-TiO_2 or mp-Al_2O_3, thickness ~500 nm) should be coated on the top of the compact layer.
(4) Deposition of an HTL in n-i-p or an ETL in p-i-n configuration by spin coating or printing method.
(5) Deposition of a thin layer (~100 nm) of a metal electrode (Au, Ag, Al, etc.) by thermal evaporation method on the top of HTL or ETL.

Note that the thickness of HTL is also a key factor that remarkably affects solar cell efficiency. For example, if the HTL is too thick, it will reduce the efficiency of hole extraction and transportation; or if the HTL is too thin, the perovskite layer probably directly contacts TCO or metal electrode for p-i-n or n-i-p configurations, respectively. The direct contact will increase charge recombination, leading to lower PCEs.

To achieve high-efficiency perovskite PVs, it is important to achieve high-quality perovskite thin films. Due to the low activation energy of perovskites, an inhomogeneous, island-type, and low-coverage perovskite layer can be observed by a simple solution-processed deposition, which is mainly attributed to the rapid crystal growth during the annealing process [27]. This severely limits the applicability of the solution process and efficiency improvement. In the following, deposition methods for obtaining high quality, homogeneous, and large-size perovskite crystalline grains are presented.

9.3.1 Spin-Coating Deposition

Organic–inorganic lead halide perovskites, such as $MAPbI_3$, can be synthesized by reacting methylammonium iodide (MAI) with lead iodide (PbI_2). Besides, a solution-based method can be used to synthesize complicated mixed-cation and -halide perovskites, such as $FAMACsPbX_3$, by using different cations- and halides-based precursors (MAI, MACl, FAI, FACl, CsI, etc.) to react with all kinds of lead halides (PbI_2, $PbCl_2$, and $PbBr_2$). For the case of $MAPbI_3$, the same molar ratio of MAI and PbI_2 is dissolved in mixed solvent dimethylformamide (DMF) and dimethyl sulfoxide (DMSO).

One-step coating. The $MAPbI_3$ film is formed by one-step spin coating and followed by annealing to vapor solvents, as shown in Fig. 9.2a. However, DMF and DMSO solvents have relatively low vapor pressures at room temperature and high boiling points. Thus, slow evaporation of the solvent can impede the nucleation rate in the one-step deposition method, resulting in low nucleation density and fast crystal growth. Therefore, poor-quality perovskite films with many pin holes and grain boundaries can be observed. To obtain homogenous high-quality perovskite films, solvent and/or anti-solvent engineering strategy is used to form the intermediate phase film after spin-coating perovskite precursor and then followed by heating [27]. This is because the solvent or anti-solvent engineering strategy can obtain a high degree of supersaturation and more crystal nuclei. In addition, an interesting strategy related to the vacuum-flashed assisted solution process (VASP) was utilized for the fast removal of solvents [28]. The main steps involve an as-coated substrate with perovskite precursor transfer into a vacuum chamber to rapidly vaporize solvents. Such a method also can obtain a high degree of supersaturation for fast crystallization.

Two-step coating. Except for one-step perovskite deposition, a two-step deposition method was also developed to fabricate high-quality perovskite films [29]. For example, $MAPbI_3$ formation: PbI_2 solution in DMF solvent is firstly coated on the as-prepared ETL layer-based substrate, followed by a lower temperature annealing treatment (70°C); MAI solution in 2-propanol (IPA) is then coated on the PbI_2 layer and then heated at 100°C to form black color perovskite films. Such methods also can obtain uniform and high-coverage perovskite thin films, resulting in good reproducibility of solar cell performance. The key to two-step deposition is that PbI_2 film should fully convert to $MAPbI_3$ film after deposition of MAI solution. However, if the residence time of the MAI solution in the PbI_2 film is short, it is difficult to convert the PbI_2 close to ETL into $MAPbI_3$.

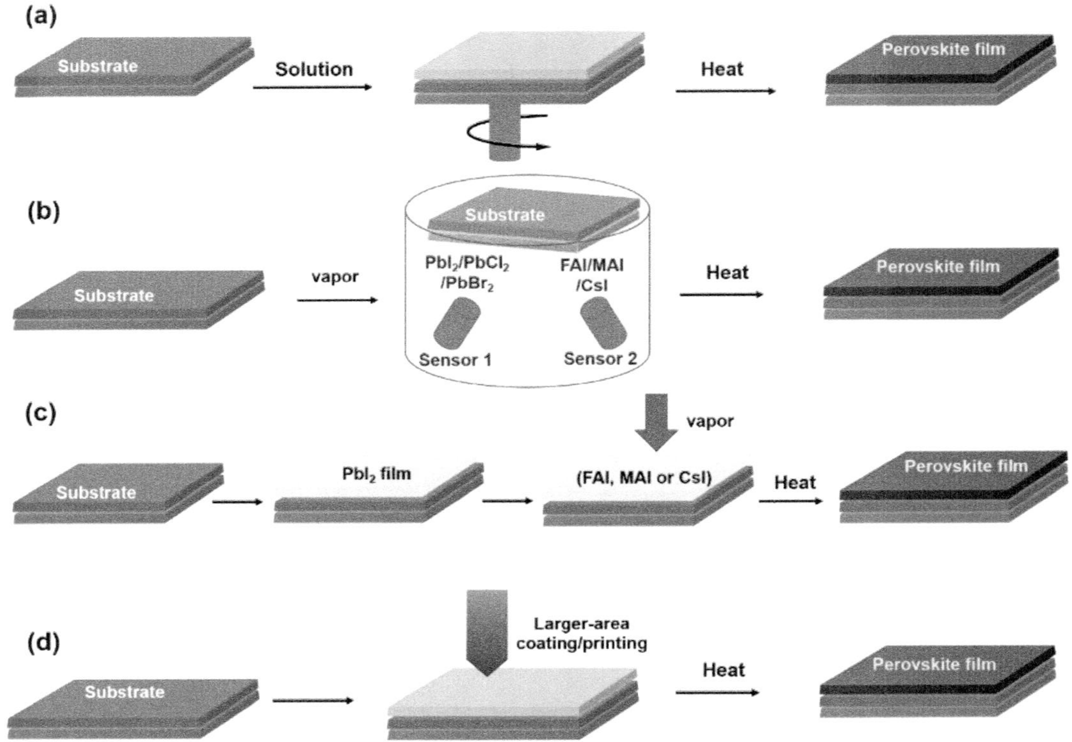

Figure 9.2 All kinds of deposition methods for perovskite thin films (a) Typical spin coating, (b) all-vapor-processed deposition, (c) vapor-assisted deposition, and (d) large-area printing or coating methods.

Note that it is also not a good idea to keep the MAI solution on the PbI_2 film for a long time because excess MAI solids are formed on the top of the $MAPbI_3$ film to reduce charge transportation. Therefore, careful control of the residence time of the MAI solution on the PbI_2 film is vitally important to obtain pure phase $MAPbI_3$ films.

9.3.2 Vapor-Assisted Deposition

As it is well known, vapor deposition is a solvent-free process that has been widely used to fabricate all kinds of thin film–based solar cells. In 2013, Snaith et al. pioneered the use of dual-source vapor deposition (i.e., MAI and $PbCl_2$) to obtain pin-hole free and large grain size perovskite crystal films [30]. Figure 9.2b shows the basic setup and vapor process. Compared to perovskite films fabricated by the spin-coating method, the co-evaporation method can achieve an efficient planar heterojunction PSC with an efficiency of over 15%. Subsequently, a higher PCE of over 20% was obtained by such a co-evaporation method, where the HTL and ETL were also formed by evaporation [31]. Since then, vapor-based deposition has been widely studied for the fabrication of high-quality and large-area perovskite films. Compared to the solution-processed perovskite films, vacuum deposition is good for the fabrication of high-efficiency tandem cells due to no solvent damage factor.

However, achieving a stoichiometric perovskite composition film remains a challenge, as the deposition rate and chemical sources must be carefully controlled. Thus, a single source was proposed for better control of the vacuum deposition parameters. Figure 9.2c presents the perovskite film formation process based on both solutions- and vapor-assisted deposition methods. For example, Chen et al. firstly deposited PbI_2 film by solution process and followed by a lower annealing temperature (70°C); then, organic MAI was vapored on the PbI_2 film and further heating for the formation of high-quality and -surface coverage $MAPbI_3$ [32].

9.3.3 Printing Deposition

Obtaining large-area perovskite films is a significant key toward perovskite PVs commercialization. There have been several scalable solution deposition methods for large-scale perovskite growth, such as blade-coating, slot-die coating, spray coating, inkjet printing, meniscus, etc. [33–36]. The main printing process can be briefly summarized in the flow chart shown in Fig. 9.2d. Here, cost-effective and widely applied printing methods are presented.

Blade-coating. Blade-coating is a widely recognized deposition method for obtaining solution-processed large-area thin films and has been widely used for various types of thin-film solar cells. In general, the role of a blade is to spread precursor solution on the preheated substrate and form a wet thin film, followed by further annealing treatment. The quality and thickness of films are generally determined by solution concentration, the distance between the blade and substrate, the blading speed, and the temperature of the preheated substrate. In the field of blade-coating perovskite films, Huang et al. firstly reported doctor-blade coating of $MAPbI_3$ films on a preheated substrate in 2015, which resulted in large grain size $MAPbI_3$ films and a PCE of over 15% [37]. A higher PCE of 22% PSCs with a 0.08 cm^2 active area was achieved using the blade-coating method by the same group in 2020 [38]. Subsequently, the above group fabricated a certified PCE of 19.25% and 19.15% solar cell modules for different module dimensions of 18.1 cm^2 and 50.0 cm^2, respectively [39]. Interestingly, blade-coating perovskite films with a vacuum-assisted method were also reported [40, 41]. Firstly, the PbI_2 solution was blade-coated on as-prepared substrates, followed by a vacuum process to remove the excess solvent. Then, as-synthesized MAI solution was blade-coated on the PbI_2 wet film and followed by heat treatment [40]. More importantly, high-quality 2D–3D heterostructure Sn-Pb mixed perovskite films can be obtained via one-step blade-coating with a vacuum process [41]. It is worth noting that preheated substrate in the vacuum-assisted blade-coating deposition method is unnecessary.

Slot-die coating. Slot-die coating was invented by A. E. Beguin in 1954, which is a pre-metered process for the deposition of uniform thin films [48]. The structure and working mechanism can be found in the literature [33, 34]. In 2015, Hwang et al. firstly reported the application of the slot-die coating to fabricate large-area PSCs by using an air knife to dry the wet film [43]. Several factors significantly determine the quality of slot-die coating perovskite films, such as solution concentration, ink flow rate, substrate moving speed, substrate temperature and air knife, etc. The perovskite solar cells fabricated by slot-die coating method showed high efficiency of 20.42% (certified 19.3%) and 19.54% for an active area of 17.1 and 65.0 cm^2, respectively [44]. Except for the perovskite layer, stacked layers including ETL and HTL should be also deposited by a large-area coating process, which will be beneficial for the high-throughput, scalable production of PSCs.

Roll-to-roll (R2R) process. For the purpose of low-cost and high-throughput, the R2R process would be a breakthrough in realizing the commercialization of PSCs by various printing techniques. For example, a complete printing of all stacked layers for fabrication of flexible PSCs was reported by Seo et al. [45]. First, the patterned ITO was deposited by rotary screen printing; second, the SnO_2 and perovskite layers were printed on ITO by gravure printing in the queue, where the perovskite wet film was treated by anti-solvent bath; third, HTL P3HT solution was gravure printed on the top of perovskite layer; finally, Au top electrode was deposited by the evaporation method. Such deposition methods enable large-area and cost-effective perovskite solar modules' future commercialization.

References

1. Yang, W. S., Park, B.-W., Jung, E. H., Jeon, N. J., et al. (2017) *Science*, **356**, pp. 1376–1379.
2. Jeon, N. J., Na, H., Jung, E. H., Yang, T. Y., et al. (2018) *Nat. Energy*, **3**, pp. 682–689.
3. Jung, E. H., Jeon, N. J., Park, E. Y., Moon, C. S., et al. (2019) *Nature*, **576**, pp. 511–515.
4. Kim, M., Kim, G. H., Lee, T. K., Choi, I. W., et al. (2019) *Joule*, **3**, pp. 2179–2192.
5. Jiang, Q., Zhao, Y., Zhang, X. W., Yang, X. L., et al. (2019) *Nat. Photon.*, **13**, pp. 460–466.
6. Yoo, J. J., Seo, G., Chua, M. R., Park, T. G., et al. (2021) *Nature*, **590**, pp. 587–593.

7. Min, H., Lee, D., Kim, J., Kim, G., et al. (2021) *Nature*, **598**, pp. 444–450.
8. Wang, K.-C., Jeng, J.-Y., Shen, P.-S., Chang, Y.-C., et al. (2014) *Sci. Rep.*, **4**, pp. 4756.
9. Zhang, H., Cheng, J. Q., Lin, F., He, H. X., et al. (2016) *ACS Nano*, **10**, pp. 1503–1511.
10. Yao, K., Li, F., He, Q. Q., Wang, X. F., et al. (2017) *Nano Energy*, **40**, pp. 155–162.
11. Chen, B., Chen, H., Hou, Y., Xu, J., et al. (2021) *Adv. Mater.*, **33**, pp. 2103394.
12. Li, F. Z., Deng, X., Qi, F., Li, Z., et al. (2020) *J. Am. Chem. Soc.*, **142**, pp. 20134–20142.
13. Wu, S. F., Li, Z., Zhang, J., Wu, X., et al. (2021) *Adv. Mater.*, **27**, pp. 2105539.
14. Etgar, L., Gao, P., Xue, Z., Peng, Q., et al. (2012) *J. Am. Chem. Soc.*, **134**, pp. 17396–17399.
15. Mei, A., Li, X., Liu, L., Ku, Z., et al. (2014) *Science*, **345**, pp. 295–298.
16. Wang, Y. S., Rho, W. Y., Yang, H. Y., Mahmoudi, T., et al. (2016). *Nano Energy*, **27**, pp. 535–544.
17. Wu, W. Q., Wang, Q., Fang, Y. J., Shao, Y. C., et al. (2018) *Nat. Comm.*, **9**, pp. 1625.
18. Jung, H. S., Han, G. S., Park, N.-G., and Ko, M. J. (2019) *Joule*, **3**, pp. 1850–1880.
19. Kim, B. J., Kim, D. H., Lee, Y. Y., Shin, H. W., et al. (2015). *Energy Environ. Sci.*, **8**, pp. 916–921.
20. Meng, X. C., Cai, Z. R., Zhang, Y. Y., Hu, X. T., et al. (2020). *Nat. Comm.*, **11**, pp. 3016.
21. Ryu, S., Seo, J., Shin, S. S., Kim, Y. C., et al. (2015). *J. Mater. Chem. A*, **3**, pp. 3271.
22. Said, A. A., Xie, J., and Zhang, Q. C. (2019) *Small*, **15**, pp. 1900854.
23. Jiang, Q., Zhang, X. W., and You, J. B. (2018) *Small*, **14**, pp. 1801154.
24. Wang, Z., Lu, Y., Xu, Z., Hu, J., et al. (2021) *Adv. Sci.*, **8**, pp. 2101856.
25. Chen, W., Wu, Y., Yue, Y., Liu, J., et al. (2015) *Science*, **350**, pp. 944–948.
26. Mali, S. S., Kim, H., Kim, H. H., Shim, S. E., and Hong, C. K. (2018) *Mater. Today*, **21**, pp. 483–500.

27. Heo, J. H., Im, S. H., Noh, J. H., Mandal, T. N., et al. (2013) *Nat. Photon.*, **7**, pp. 486–491.
28. Li, X., Bi, D. Q., Yi, C. Y., Decoppet, J. D., et al. (2016) *Science*, **353**, pp. 58–62.
29. Burschka, J., Pellet, N., Moon, S. J., Humphry-Baker, R., et al. (2013) *Nature*, **499**, pp. 316–319.
30. Liu, M., Johnston, M. B., and Snaith, H. J. (2013) *Nature*, **501**, pp. 395–398.
31. Momblona, C., Gil-Escrig, L., Bandiello, E., Hutter, E. M., et al. (2016) *Energy Environ. Sic.*, **9**, pp. 3456–3463.
32. Chen, Q., Zhou, H., Hong, Z., Luo, S., et al. (2014) *J. Am. Chem. Soc.*, **136**, pp. 622–625.
33. Li, Z., Klein, T. R., Kim, D. H., Yang, M. J., et al. (2018) *Nat. Rev. Mater.*, **3**, pp. 18017.
34. Park, N.-G. and Zhu, K. (2020) *Nat. Rev. Mater.*, **5**, pp. 333–350.
35. Kim, Y. Y., Yang, T. Y., Suhonen, R., Kemppainen, A., et al. (2020) *Nat. Comm.*, **11**, pp. 5146.
36. Wang, F. F., Cao, Y. Z., Chen, C., Chen, Q., et al. (2018) *Adv. Funct. Mater.*, **28**, pp. 1803753.
37. Deng, Y. H., Peng, E., Shao, Y. C., Xiao, Z. G., et al. (2015) *Energy Environ. Sci.*, **8**, pp.1544–1550.
38. Wu, W. Q., Rudd, P. N., Wang, Q., Yang, Z. B., and Huang, J. S. (2020) *Adv. Mater.*, **32**, pp. 2000995.
39. Chen, S., Dai, X., Xu, S., Jiao, H., et al. (2021) *Science*, **373**, pp. 902–907.
40. Guo, F., He, W. X., Qiu, S. D., Wang, C., et al. (2019) *Adv. Func. Mater.*, **29**, pp. 1900964.
41. Zeng, L. X., Chen, Z. G., Qiu, S. D., Hu, J. L., et al. (2019) *Nano Energy*, **66**, pp. 104099.
42. Beguin, A. E. (1954) *US Patent*, pp. 2681294.
43. Hwang, K., Jung, Y. S., Heo, Y. J., Scholes, F. H., et al. (2015) *Adv. Mater.*, **27**, pp. 1241–1247.
44. Bu, T. L., Li, J., Li, H. Y., Tian, C. C., et al. (2021) *Science*, **372**, pp. 1327–1332.
45. Kim, Y. Y., Yang, T. Y., Suhonen, R., Kemppainen, A., et al. (2020) *Nat. Comm.*, **11**, pp. 5146.

Problems

9.1 How many types of perovskite solar cell configurations are there and what are their advantages?

9.2 What is the design principles of charge-transport materials for perovskite solar cells?

9.3 Describe the fabrication processes of perovskite solar cells.

Chapter 10

Defects and Ions Migration in Perovskite Solar Cells

The performance of perovskite solar cells is affected by the quality of interfaces and crystallographic defects such as point defects, dislocations, grain boundaries, etc. Therefore, it is important to understand the defect physics so that appropriate methods can be developed to suppress the formation of harmful defects. In addition, ions migration in the perovskite layer is a major cause of anomalous device behaviors, including light-soaking effect, current–voltage hysteresis, and slow decay of open-circuit voltage. Under the electric field, charged ions could migrate or diffuse along the migrating channels transitioning from the internal bulk to the electrodes, thus increasing defect state density and causing interfacial polarization, resulting in device performance degradation. This chapter discusses the details of crystallographic defects and ions migration, which are generally occurring in perovskite solar cells.

10.1 Nature of Defects

In inorganic crystalline semiconductors, defects can be classified into three categories: point defects, impurity defects, and planar defects. Figure 10.1 shows schematic illustration of defects.

Next-Generation Solar Cells: Principles and Materials
Yoon-Bong Hahn, Tahmineh Mahmoudi, and Yousheng Wang

ISBN 978-981-4968-66-9 (Hardcover), 978-1-003-37238-7 (eBook)
www.jennystanford.com

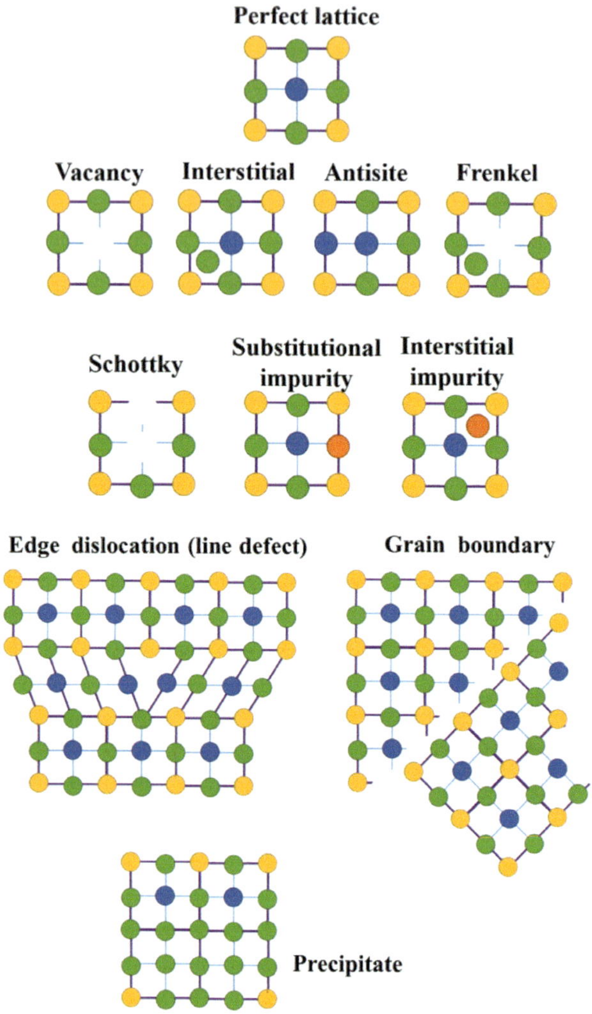

Figure 10.1 Schematic illustration of defects in perovskite crystal lattice (blue, yellow, and green dots represent the A-, B-, and X-site ions in ABX_3, respectively).

(1) Point defects

Point defects are formed by interruptions of perfect patterns in the crystalline lattice, also known as 'crystal defect'. The perfect crystal has a periodic structure in which the positions of atoms/molecules are at a fixed distance, which are called unit cell parameters. However, in crystal point defects these parameters are imperfect. Point defects occur when the crystallization process occurs at a fast rate, and they are mainly due to deviations in the arrangement of the constituent atoms. There are three types of point defects: Stoichiometric defect, Frenkel defect, and Schottky defect.

Stoichiometric defect. There are three types of stoichiometric defect, in which the ratio and electrical neutrality of the positive and negative ions (stoichiometry) of the crystal are not disturbed:

- ***Vacancy defect*:** When an atom is missing from crystal lattice sites, then that lattice site is vacant, generating a vacancy defect.
- ***Interstitial defect*:** When an atom occupies the intermolecular spaces in crystal, an interstitial defect is formed.
- ***Antisite defect*:** When atoms of different type exchange positions in an ordered alloy or compound, an antisite defect occurs.

Frenkel defect. This type of defect arises when the smaller ion (usually cation) is dislocated its original site to the interstitial site, this is also known as 'dislocation defect'.

Schottky defect. Schottky defect is caused by a vacant position that is generated in a crystal lattice due to the atoms or ions moving out from the interior to the surface of the crystal. In this type of defect, anion and cation vacancies are formed at the same time.

(2) Impurity defects

Impurity defect is caused by distortion of crystal lattice when an impurity (i.e., foreign atom or ion) occupies an interstitial site in the lattice or replaces the parent atom or ion in the regular sites [1–3]. There are two of types:

- ***Substitution impurity defect*:** This defect occurs when a cationic impurity is substituted in the place of a regular cation.
- ***Interstitial impurity defect*:** This defect arises when a cationic impurity is present in the interstitial positions of a regular lattice and changes the properties of an original crystalline solid.

(3) Planar defects

A planar (or line) defect is a discontinuity of the perfect crystal structure across a plane, such as dislocations, grain boundaries, antiphase, and precipitates.

In organic–inorganic halide perovskite, the most studied defects are point defects. For example, $MAPbI_3$ has twelve-point defects including vacancy defect of MA, Pb, and I ions, interstitials of MA, Pb, and I, and anti-site between MA-Pb, MA-I, Pb-MA, I-MA, and I-Pb. To improve the performance of solar cells it is better to know the qualitative tendency of defect formation and to design an appropriate fabrication technique. To do that, there are theoretical calculations to study defect formation energies in stable and pure conditions and equilibrium between perovskite $MAPbI_3$ and I_2, Pb, and MA. In perovskite materials, these calculated quantities are projected by density functional theory (DFT).

10.2 Defects and Charge-Transport Processes

In a perovskite solar cell, the photo-absorber layer should be thick enough to enhance absorption and thin enough to transfer photo-generated charges without quenching and recombining. Thus, the efficiency of the active layer depends on the recombination and transport of photo-generated charges to the electrodes. However, defects substantially affect the charge-transport and recombination processes. In perovskite crystals, photo-generated electrons can flow through the crystals under the applied electric field until they encounter scattered phonons, defects and impurities, and then change the direction of the accelerated carriers. Charge transport in crystalline perovskite is typically affected by the scattering of charge carriers near a point defect. This defect is limited only at low temperatures because, at high temperatures, the high thermal velocity of electrons causes faster carriers with higher energy, which can overcome such weak deflection [4].

The potential barrier at grain boundaries is a type of defect with an energy level widened bandgap. These barriers trap charge carriers and reduce conductivity.

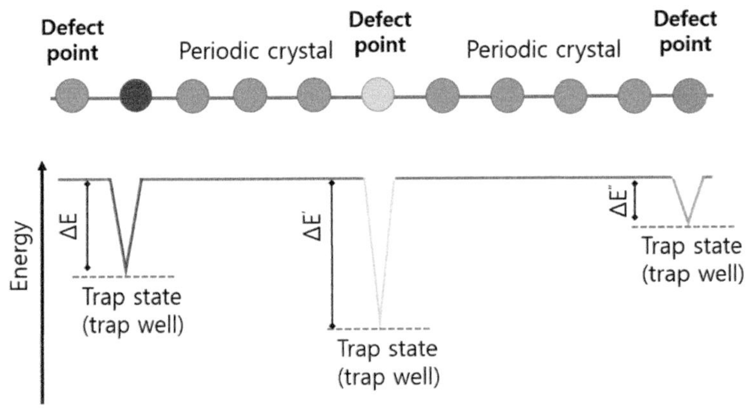

Figure 10.2 Schematic illustration of non-periodic disorder and resulting defect in a crystal lattice, which forms new energy levels, causing charge trapping and slower charge-carrier transport.

The trap density directly affects the charge transport until the traps are fully filled, and then the remaining free carriers flow through the active layer. A very high ratio of charge density/trap density may have little effect on charge transport. However, if the ratio is low, most carriers are trapped and charge transport is controlled by the hopping or trapping–detrapping process. For an average trap density, charge transport is dependent on the applied bias voltage; the carrier density is low in a short-circuit condition but high in an open-circuit condition.

In the ideal structure of crystals, each atom should be located in its designed location, but the collapse of this state will cause a crystal defect. When a dislocation causes a disorder in the lattice network, it will affect the moving carriers and their properties. Figure 10.2 illustrates that crystal imperfections can generate new electron states with different energy levels at which photo-generated carriers can be trapped. Compared to immobilized moving charges in the conduction band, the trapped charges have different energy states with restricted mobility in the crystal network. After a charge is trapped at a defect site, it can escape (detrapping) and return to the transport levels by absorbing extra energy [5]. The additional energy can be provided *either by* optical excitation or by absorbing thermal energy. It is worth noting that not all the defects contribute to charge trapping. Those defects with energy levels located between VB and

CB can create trap states, while those with energy levels below VB and above CB considered harmless.

Defect traps can be classified into two types: shallow traps or deep traps. Shallow traps have relatively low energy differences from their CB or VB edges, and detrapping can happen by energy absorption. However, deep traps have a deeper energy level (in the middle of the bandgap), which restricts the detrapping and can cause non-radiative recombination (Fig. 10.3). The main role of shallow and deep traps is to decrease the majority of mobile charge carriers. In semiconductor materials, deep traps are electronic states that lie deep in the middle of the energy gap due to disorder or impurity that acts as recombination centers. In contrast, shallow traps lie near the band edges. Charges are captured for short periods of time in shallow traps, could be released to nearby bands by thermal activation. Contrary to this, deep traps retain charges for a much longer period of time after capturing them, immobilizing mobile charges and preventing them from conduction.

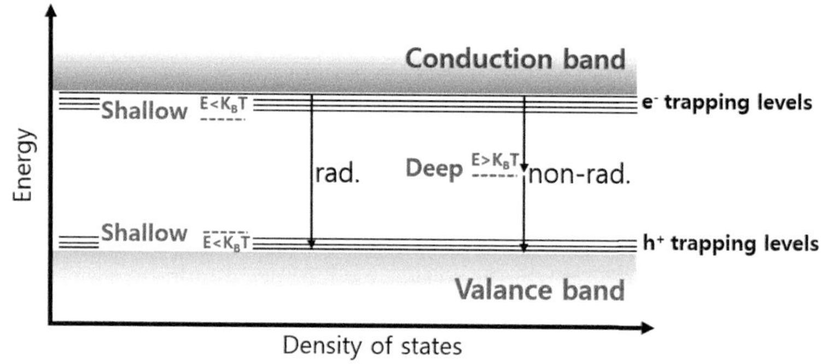

Figure 10.3 Density of states in a defected semiconductor, with band-to-band radiative (rad.) decay and non-radiative (non-rad.) decay recombination.

Carrier trapping disrupts the charge flow and trapped charges cannot participate in net charge transport. Captured carriers in shallow traps can be thermally excited and return to the conduction band above the mobility edge and can contribute to net charge transport. Thus, the carrier mobility of a semiconductor highly depends on carrier concentration, types of trap and trap densities, thermal energies, and so on.

At room temperature, in the presence of shallow traps, the model of multiple trapping and release (MTR) is commonly used to explain the mobility properties of crystalline semiconductors at ambient temperature. Effective mobility is directly related to trap-free mobility and is reduced by temporal trappings. Charge trapping can be suppressed by illumination intensity. High illumination intensities enhance the concentration of photo-excited carriers which can fill the traps and reduce their influence on the charge transport. However, deeply trapped carriers can reduce the power-conversion efficiency through non-radiative recombination. This phenomenon occurs when the trapped energy states are located in the middle of the bandgap (i.e., $\Delta E \gg k_B T$), and the required energy for de-trapping is very high and the transition is impossible.

In solar cells, non-radiative recombination is an unfavorable process because it restricts the carrier diffusion length (L_D) which must be higher than the thickness of perovskite film for the device to operate efficiently. The diffusion length is highly proportional to the mobility and lifetime of charge: $L_D=[\mu k_B T/e]^{1/2}$ [6, 7].

For an efficient $MAPbI_3$ perovskite solar cells, the diffusion length needs to be in the range of hundreds of nanometers to micrometers. Generally, shallow traps have a minor effect on carrier mobility and only delay the charge delivery, but deep traps can cause non-radiative recombination which alters the charge extraction process and can limit the maximum current density of the device. Perovskite structures are naturally disorderly and complicated. Even high-quality perovskites inherently have some level of disorder, because their crystal structure is easy to disrupt and its complex lattice vibrations with complicated ions migration generate non-radiative recombination centers.

DFT is the most recognized technique for studying the origin of point defects. DFT calculations propose the energy states of intrinsic point defects which are located near electronic states of CB and VB. The point defects are unlikely to form deep mid-gap defect states at room temperature. In perovskite materials, the defect states arise from anti-bonding and bonding orbitals at max-VB and min-CB, in contrast to the conventional III–V semiconductors band structure.

10.3 Formation of Intrinsic Defects

The defects formation through perovskite growth depends on crystallization dynamics and kinetics. Defect formation is a multi-parameter process highly related to growth conditions such as material stoichiometry, temperature, humidity, and pressure. It can be suppressed by optimizing thermal treatments and the molar ratio of the precursor agents (Pb source and MAI). For example, heat treatment can remove extra MAI and turn p-type perovskite to n-type perovskite. Besides, ions migration can be observed under operating conditions or substantially high temperatures. The perovskite films exposed to small amounts of oxygen (or dry air) can lead to iodine interstitial and lower deep level trap density. Compositional alloying can help to suppress traps and to lessen the defect formation and charge trapping. The introduction of Rb^+ or K^+ can cause interstitial defects in $MAPbI_3$ and create MA vacancies and improve the electrical properties by preventing iodide defect formation. Introducing K^+ cations can also reduce the hysteresis effect and could improve crystallinity and reduce surface potential [8]. Similarly, halide doping with bromide and chloride can effectively deactivate iodine hole traps, as discussed in Chapter 8.

10.4 Light Soaking and Trap Filling

The light-soaking effect occurs while a solar cell is exposed to light, thus it affects the performance of solar cells. As an example, the PCE of amorphous silicon (a-Si) decreases after hundred hours of illumination due to Si–Si bond breaking by thermalization. For thin-film solar cells such as CdTe- and CIS/CIGS-based SCs, continuous illumination increases the device productivity by filling up trap states. For perovskite solar cells, material properties such as light-induced ions migration and compositional relocation mechanisms highly affect the trap density and create a complex combination of parameters, which are still in dispute when the environment (O_2, moisture, etc.) has an effect on the process. Some studies have shown that light soaking can help reduce trap density through the extinction of iodide vacancy defect and iodide ion migration defect

[9]. Moreover, light-activated degradation of the PSCs was reversed by a self-healing process when the device was left in the dark.

10.5 Extrinsic Defects

Extrinsic defects result from impurities derived from processing. In general, solution processing can easily cause impurity formation during crystallization because of stoichiometry deviations. Accidental impurities in the solution can also introduce foreign elements into the lattice, which can lead to doping and defects such as lattice vacancies, substitutions, or interstitial defects, thereby introducing traps into the band structure. In perovskite solar cells, the extrinsic defects occur in grain boundaries or surface morphology of the thin film. For example, a single-crystalline $MAPbI_3$ has $\sim 10^{10}$ cm^{-3} defect density while a $MAPbI_3$ thin film has $\sim 10^{16}$ cm^{-3}, attributed to the difference in grain size and surface area due to external defects. Polycrystalline perovskite shows much lower charge mobility and carrier diffusion length than single-crystalline ones. Generally, in solution-processed thin films, there is a variety of mismatch bonding at grain boundaries and surfaces, which increases the concentration of deep traps and causes non-radiative recombination losses, thereby reducing cell efficiency [8–10].

10.6 Techniques to the Probe Defect States

Most methods for experimental probing of defect states are designed based on properties, which are directly or indirectly affected by charge trapping. For instance, electrical properties are not dependent on the type of trap (i.e., shallow trap or nonradiative deep trap), implicating that such traps cannot be detected by electrical properties. The high rate of charge injection raises the trap filling rate and reduces the trap effect on charge dynamics of solar cell devices. Hence, even highly defected materials can still perform well with the high injection rate and trap filling. Trap concentration can be estimated by PL decay vs. excitation power density. However, this method requires modeling, knowledge of doping concentration, and recombination mechanisms which are not practically available. Obviously, each technique for trap detection has its sensitivity limits.

A large change in PL intensity determines different concentrations of defects. Also, fluctuation and flickering in PL intensity at a different site in a film show the presence of nano- and micro-crystals. PL blinking is a result of chemical reactions, ion migration, and self-healing properties which could cause the reversible formation of metastable non-radiative sites [10]. In lead halide perovskite, the PL and its intensity fluctuations are an effective way to identify non-radiative centers presented in small crystals. PL analysis can only detect metastable defects, while the permanent traps can be seen by PL lifetime studies.

10.7 Ions Migration

The effect of mobile charged ions (i.e., ions migration) in MA and FA-based perovskite is an important topic. This charged defect occurs by the migration of ionic atoms with low activation energies as an inherent plausible mechanism inside perovskites. The ions migration is responsible for slow photoconductivity response, hysteresis, and switchable photocurrent in perovskite solar cells [11]. Iodine vacancies and interstitials are highly influenced by the externally applied voltage and can easily diffuse through the perovskite crystals by receiving low activation energy of ~0.1 eV. Under the applied energy field, iodine defects migration becomes an origin of network redistribution in perovskite and its density of states as well as electrical properties. The migration of MA and Pb ions requires higher activation energies of ~0.5 and 0.8 eV, respectively, and could cause a slow photoconductivity response. The migration time of iodine defect is calculated to be less than 1 μs, and for MA and Pb ions, it is calculated to be tens of microseconds to minutes. Ion migration does not directly contribute to charge trapping, but it causes a soft solid structure that is prone to complications, forms defects, and affects carriers inside the crystal.

Figure 10.4 illustrates the migration routes for ions and defects. For iodide vacancy defect (V_I), the vacancy in an equatorial position can migrate toward an axial site or equatorial site since both migrations are basically isoenergetic (within the difference < 0.01 eV) [12]. For bromide vacancy defect (V_{Br}) the axial site with an activation energy of 0.07 eV is favorable (Fig. 10.4a). Regarding methyl ammonium vacancy defect (V_{MA}), hopping off the defect between neighboring voids of an inorganic network is suggested. V_{MA} is expected to have in-plane migration through a Pb_4I_4 framework (Fig. 10.4b). The same behavior for Pb vacancy defect (V_{Pb}) is expected with in-plane migration, where V_{Pb} migrates along the side of the square formed by four Pb and four I atoms, (Fig. 10.4c). Another less likely migration path for V_{Pb} could occur along the diagonal of the Pb_4I_4 framework with a large activation energy of 0.80 eV [13].

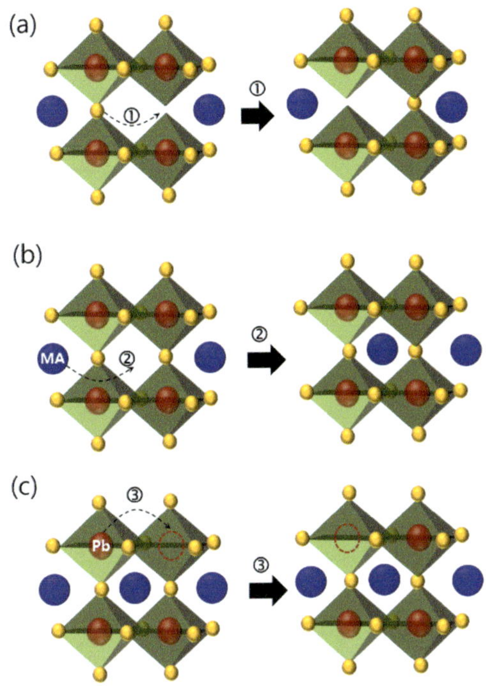

Figure 10.4 Migration path for the (a) V_I, (b) V_{MA}, and (c) V_{Pb} defects. Dashed circles refer to vacancies. Interstitial defects are shown in red atoms. Migrations are shown by solid lines and trails of the vacancy are exhibited by dashed lines. Blue = MA molecules, yellow = I atoms, and red = Pb atoms.

10.8 Ions Migration in Operating Solar Cell

Among the ion and vacancy migrations, the V_I and V_{MA} are the most moving components in bulk perovskite. Figure 10.5a shows an operating device under a photo-generated field along with the direction from ETL to HTL. Such a built-in potential field will

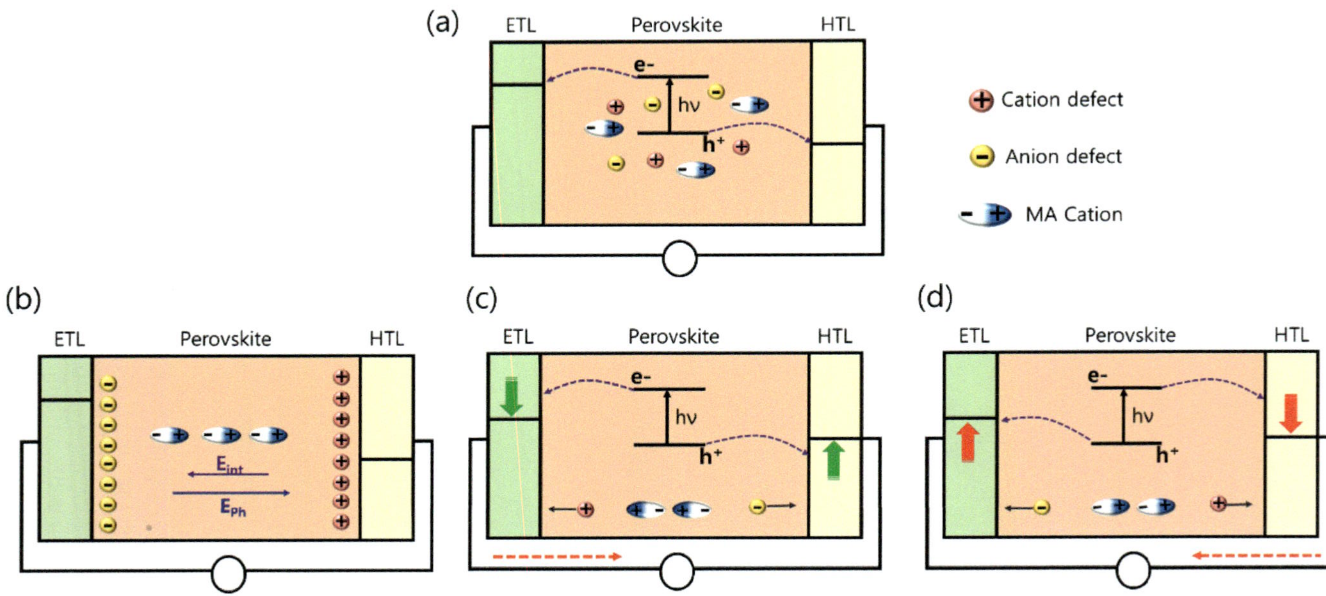

Figure 10.5 Schematic illustration of defect migration impact on the perovskite device operation. (a) Separation of the carriers by the photo-generated potential (E_{Ph}), with an even distribution of defects and the random orientation of the MA^+; (b) Migration of the defects and the reorientation of the MA^+ under photo-generated field (E_{Ph}), and formation of the internal potential (E_{int}); (c) and (d) Switchable current as a function of the external field.

initiate migration of vacancy defects caused by absence of MA (V_{MA}) and I (V_I) toward the HTL and ETL. Figure 10.5b exhibits that for some time, the concentration of V_I (or V_{MA}) defects near the HTL (or ETL) increases to create an internal electrostatic field over the perovskite film, which is the opposite of the internal field that is applied for charge separation in the solar cell. This internal field intrudes the charge transport and/or charge collection at contacts and ultimately affects the device performance. This microscopic mechanism can explain the dependency of the hysteresis J–V curve in a PSC on the scanning rate and direction. The slow scan rate causes defect migration, to partially help with the photo-generated electric field. On the other hand, biasing the PSC at V_{oc} would impede ions migration, and the device performance is little affected. In other words, in a fast voltage sweep, the measurement is less sensitive to the direction of the scan, since defects have less time to migrate. The effect of ion migration on the switchable photovoltaic effect can be explained as V_I will be collected at the anode, positively charged defects form shallow states near the CB of the perovskite, inducing n-doping of the perovskite/anode interface (Fig. 10.5c). In turn, negatively charged ions form the same electronic levels above the VB, resulting in p-doping of the perovskite/cathode interface. Under positive doping, the carriers follow the gradient field (Fig. 10.5c). For negative doping, iodine vacancies move in the opposite direction and invert the sense of the current (Fig. 10.5d) [14].

References

1. Stranks, S. D., Eperon, G. E., Grancini, G., Menelaou, C., et al. (2013) *Science*, **342**, pp. 341–344.
2. Wehrenfennig, C., Eperon, G. E., Johnston, M. B., Snaith, H. J., and Herz, L. M. (2014) *Adv. Mater.*, **26**, pp. 1584–1589.
3. Yang, B., Dyck, O., Poplawsky, J., Keum, J., et al. (2015) *J. Am. Chem. Soc.*, **137**, pp. 9210 –9213.
4. Ohmann, R., Ono, L. K., Kim, H.-S., Lin, H., et al. (2015) *J. Am. Chem. Soc.*, **137**, pp. 16049–16054.
5. Vorpahl, S. M., Stranks, S. D., Nagaoka, H., Eperon, G. E., et al. (2015) *Science*, **348**, pp. 683–686.
6. Ball, J. M. and Petrozza, A. (2016) *Nat. Energy*, **1**, pp. 16149.

7. Ueki, T., Itsumi, M., and Takeda, T. (1997) *Appl. Phys. Lett.*, **70**, pp. 1248–1250.
8. Xiao, Z., Yuan, Y., Shao, Y., Wang, Q., et al. (2014) *Nat. Mater.*, **14**, pp. 193–198
9. Chen, Q., Zhou, H., Song, T.-B., Luo, S., et al. (2014) *Nano Lett.*, **14**, pp. 4158–4163.
10. Xiao, Z., Dong, Q., Bi, C., Shao, Y., et al. (2014) *Adv. Mater.*, **26**, pp. 6503–6509.
11. Nie, W., Tsai, H., Asadpour, R., Blancon, J.-C., et al. (2015) *Science*, 347, pp. 522–525.
12. Ran, C., Xu, J., Gao, W., Huang, C., and Dou, S. (2018) *Chem. Soc. Rev.*, **47**, pp. 4581–4610.
13. Ayres, J. (1993) *J. Appl. Phys.*, **74**, pp. 1787–1792.
14. Azpiroz, J. M., Mosconi, E., Bisquert, J., and De Angelis, F. (2015) *Energy Environ. Sci.*, **8**(7), pp. 2118–2127.

Problems

10.1 What is the nature of defects in crystalline semiconductors?

10.2 In perovskite crystals, photo-generated electrons can flow through the crystals under the applied electric field until they encounter traps and potential barriers. Explain the cause of these barriers.

10.3 What is ions migration in perovskite solar cells?

10.4 What are shallow and deep traps? How do they affect solar cell efficiency?

Chapter 11

Quantum Dots, Tandem, and Lead-Free Perovskite Solar Cells

Metal halide perovskites have great potential in the optoelectronic fields such as solar cells, light-emitting diodes, and photodetectors. The ability to design the composition of ABX_3 perovskites facilitates fine-tuning of optoelectronic properties and bandgap energy of perovskites for different applications. In the past few years, fundamental studies on compositional and interfacial engineering as well as optimization of device configuration have shown remarkable progress in single-junction PSCs with the highest power-conversion efficiency of over 25%, which is comparable to commercial PV devices such as c-Si solar cells. However, PSCs suffer from efficiency degradation and device instability due to the deterioration of perovskite molecules when operating for a long time under harsh conditions such as high temperature, humidity, and strong illuminance.

To solve such problems, quantum dots photovoltaics and tandem PSCs have been developed. The tandem solar cell is a kind of multi-junction solar cell, made of different materials with different bandgaps and absorption spectra, resulting in absorption of a broad

Next-Generation Solar Cells: Principles and Materials
Yoon-Bong Hahn, Tahmineh Mahmoudi, and Yousheng Wang

ISBN 978-981-4968-66-9 (Hardcover), 978-1-003-37238-7 (eBook)
www.jennystanford.com

range of wavelengths and an increase in the efficiency of converting more photons into electrical energy. Another problem with PSCs is the toxicity of lead (Pb), a limiting factor that, in addition to operational instability, prevents commercial use of perovskite solar cells. To replace lead with non-toxic material, the Pb-free absorbers should have low toxicity and proper stability.

This chapter deals with quantum dots, tandem, and Pb-free PSCs in detail.

11.1 Perovskite Quantum Dots Solar Cells

11.1.1 From Perovskite Thin Films to Quantum Dots

To overcome the instability problems of hybrid perovskite solar cells, all-inorganic solar cells were introduced by replacing organic MA^+ and FA^+ cations with Cs^+. In the ABX_3 structure, using inorganic Cs^+ cations in the A-site is believed to improve the thermal stability of PSCs. However, the small size of Cs^+ cations could cause a low Goldschmidt tolerance factor which leads to poor phase stability at room temperature. The efficiency of all-inorganic PSCs with β-$CsPbI_3$ perovskite has exceeded 18% but the desired α-$CsPbI_3$ perovskite that is highly appropriate for solar cell applications only is stable at high temperatures (over 200 °C).

All-inorganic perovskite quantum dots or nanocrystals exhibit improved phase stability due to the less surface strain caused by reducing the size of perovskite grains. Properties of perovskites quantum dots (PQDs) can be controlled by the halide composition, size, and shape of PQDs as well as multiple generation effect [1, 2]. The multiple exciton generation effect enables the PQDs to absorb high-energy photons and overcome the Shockley–Queisser efficiency limitation for single-junction PSCs solar cells [3]. The PCE of perovskite quantum dot solar cells (PQDSCs) has shown outstanding improvement in a short time and potential in becoming high-performance solar cells [4–8].

However, PQDs face the same challenge as other QDs, that is, there are a large number of surface defects that can be centers of non-radiative recombination of photo-generated carriers, limiting the carrier extraction and leading to energy loss in PQDSC devices. Surface passivation and controlling interfacial properties are critical to reducing interfacial recombination and thus enhancing device photovoltaic performance [9–11].

11.1.2 Crystal Structure and Properties of Perovskite QDs

The perovskite lattice has ABX_3, where the A-site is occupied by inorganic and or organic cations such as Cs^+, MA^+, and FA^+, and the B-site is filled with metallic cations such as Pb^{+2} and Sn^{+2} and the X-site has a single or mixed halogen anion (Cl^-, Br^-, or I^-). In 2015, for the first time, the hot-injection method was used to fabricate inorganic PQDs. These PQDs were color-tunable by composition engineering of halide ions [14]. Among all, $CsPbX_3$ PQDs with high surface energy showed phase stability at room temperature [15]. $CsPbX_3$ PQDs showed bright and narrow PL emission and fast radiative lifetime compared to the bulk counterparts. The bandgap of PQDs can be tuned by QD size. As the PQD size increases, the light absorption edge shows a red shift in the spectrum. In PQDs, the surface defects, which act as trap states, are caused by under-coordinated atoms and form sub-bandgap non-radiative recombination centers. These surface defects can be passivated by surface ligand formulation or surface chemistry. Moreover, low-energy point defects can be formed during the purification process within the bandgap or near the edge of the electronic band. In PQDs, due to multiple exciton generation (MEG), QDs can produce multiple excitons by receiving high-energy photons and producing a high photocurrent in PQDSCs.

11.1.3 Emerging of Perovskite QDSCs

Highly efficient quantum dot solar cells (QDSCs) have been known by common chalcogenide colloidal quantum dots, such as lead sulfide (PbS) QDs. PbS-QDSCs have made a significant progress, with increasing the efficiency from 2.9% in 2010 to more than 13% recently.

Perovskite QDs are a class of solution-processable semiconductors with tunable and strong light absorption spectrum and high quantum yields. Figure 11.1a shows an exceptional increase in the number of publications on PQDSCs, signifying a good trend in the PQDSCs development. Figure 11.1b shows the certified efficiencies of QD-based SCs by the National Renewable Energy Laboratory (NREL). PQDSC with an efficiency of 10.77% was first reported in 2016, with all-inorganic $CsPbI_3$ PQDs material. In PQDs, due to negative surface energy and size-induced lattice strain the $CsPbI_3$ QD films showed improved stability than bulk $CsPbI_3$ perovskite films [12]. With fundamental improvement in compositional engineering and surface chemistry of PQDs, the performance of PQDSCs was highly improved to 16.6% [13], which is higher than that of PbS-based QDSCs, showing great potential for next-generation QD-based SCs.

In a PQDs solar cell, an efficient charge extraction can be obtained by reducing charge recombination in main layers and interfaces via various methods, such as surface chemistry and compositional engineering, and optimization of device architecture. Though the performance of PQDSCs has been highly improved in the past few years, it is still behind that of single-junction bulk perovskite solar cells. One of the main concerns about PQDSCs is PQD surface chemistry and its determining effect on photovoltaic performance and stability of devices.

The surface chemistry of PQDs is highly dependent on surface ligands, which affects the charge mobility in PQD films. The carrier mobility can be improvised via ligand replacements or post-treatment to further promote device efficiency. However, numerous defects can be formed on the surface of PQD during the ligand exchange and become recombination centers in the PQD films. In PQDSCs, the PQDs film is sandwiched between ETL and HTL charge-transport materials, therefore high mobility, adjusted energy levels, and PQD/ETL and PQD/HTL interfaces are important for obtaining a high-performance device.

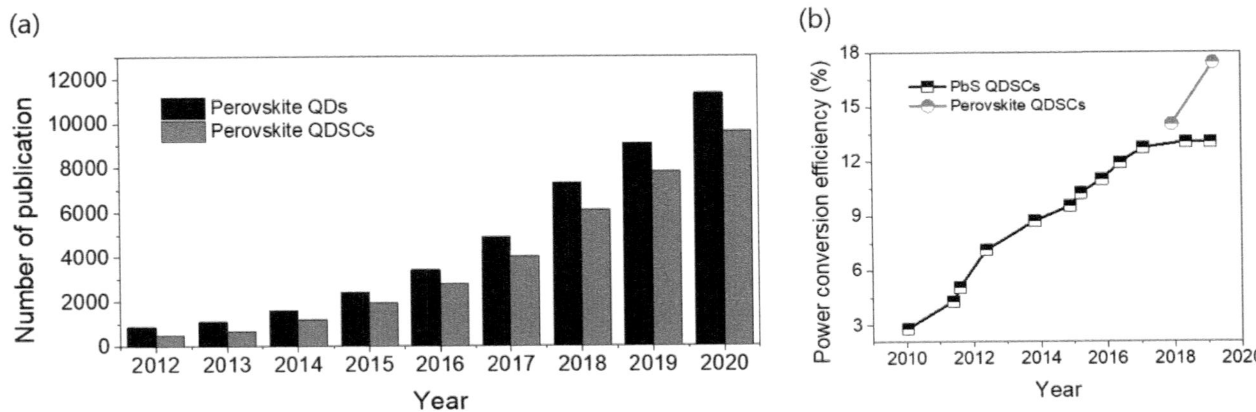

Figure 11.1 (a) The number of annual publications of perovskite QDs (search keyword: perovskite quantum dot) and perovskite QDSCs (searched keyword: perovskite quantum dots). The data were retrieved from google scholar. (b) Improvement of the certified efficiency of QDSCs by the NREL [92].

11.1.4 Methods of Enhancing the Device Performance of PQDSCs

Ligand-exchange processing resulted in long electron diffusion length, in the device with the configuration of FTO/TiO_2/$CsPbI_3$ QD/Spiro-OMeTAD/MoO_3/Al, achieving a large V_{oc} of 1.23 V and power-conversion efficiency of 10.77%, similar to thin-film perovskite cells [25].

Surface modifications are significantly effective due to higher dynamic surface and low material formation energy. By replacing oleate ligands with acetate, cationic oleylammonium ligands were detected to remain in the QD films and further improved the mobility of $CsPbI_3$ QD films. Using smaller size salts, such as formamidinium iodide (FAI), on the surface instead of longer ligands could induce larger grain size in the films with improved the electronic coupling between QDs, which enhanced the film mobility, resulting in a high photocurrent density of 14.37 mA/cm^2 and PCE of 13.43% [18]. $CsPbI_3$ QD PVs have delivered high V_{oc} but lattice and surface defects after ligand exchange in $CsPbI_3$ QD limit the solar cell performance. Surface passivation using cesium salts (e.g., cesium acetate) was reported to not only fill up the surface vacancy sites but also improve electronic coupling in $CsPbI_3$ film, resulting in efficiency of 14.1% and improved moisture stability [19]. Cs^+ treatment was also utilized to modify the mesoporous TiO_2 (mp-TiO_2) ETL, improve the interface chemistry of $CsPbI_3$ QDs and mp-TiO_2, and enhance electron extraction and the best PCE of 14.32% was recorded [20].

The use of micron-size graphene (μGR) and QDs composite based film enhances carrier transport and conductivity with better thermal and moisture stability. μGR/$CsPbI_3$ solar cell showed a high PCE of 12%, with improved V_{oc} of 1.18 V, J_{sc} of 13.59 mAcm^{-2}, and FF of 72.6%. Moreover, instability caused by the thermal/moisture-induced QD agglomeration was suppressed by the μGR incorporation [21].

The use of anti-solvent effectively removed the insulating ligands on $FAPbI_3$ QDs, with preserving perovskite crystal. It resulted in improved electrical coupling and a PCE of 8.38% was achieved for $FAPbI_3$ CQD solar cells [22]. An effective method to enhance charge separation in $FAPbI_3$ QD was introduced by using conjugated small molecules and the resulting PQDSCs exhibited a PCE of 13% [23].

The cation-exchange strategy was proposed for PQDSCs PV with $Cs_{1-x}FA_xPbI_3$ photo active QDs, which exhibited a lower V_{oc} loss than the thin-film perovskite SCs with similar compositions. This study stepped beyond simple $CsPbI_3$ or $FAPbI_3$ QDSCs devices, and the latest certified PCE of 16.6% has been recorded using $Cs_{0.5}FA_{0.5}PbI_3$ composition [24]. A summary of reported PQDSC with different PQDs is listed in Table 11.1.

Table 11.1 Summary of reported PQDSC with different PQDs

PQDs layer	V_{OC} (V)	J_{SC} (mA/cm^2)	FF	PCE (%)	Ref.
$CsPbI_3$	1.23	13.47	0.65	10.77	25
	1.20	14.37	0.78	13.40	26
	1.27	12.39	0.80	12.55	27
	1.25	14.96	0.76	14.10	28
	1.24	14.14	0.74	13.12	29
	1.18	15.21	0.74	13.30	30
	1.06	17.77	0.76	14.32	31
	1.22	15.10	0.74	13.8	32
$CsPbBr_3$	1.42	7.01	0.53	5.60	33
	1.50	5.60	0.62	5.40	34
$CsPbBrI_2$	1.30	13.13	0.70	12.02	33
$CsPbBr_{0.6}I_{2.4}$	1.30	5.30	0.77	5.34	35
$FAPbI_3$	1.10	15.40	0.75	12.70	23
	1.12	16.70	0.71	13.20	32
$Cs_{0.75}FA_{0.25}PbI_3$	1.18	14.36	0.68	11.14	36
$CsPbI_3$ + $FAPbI_3$	1.22	17.26	0.74	15.60	37
$Cs_{0.5}FA_{0.5}PbI_3$	1.17	18.30	0.78	16.60	30
$Cs_{0.25}FA_{0.75}PbI_3$ + $CsPbI_3$	1.20	18.91	0.76	17.39	38

Unlike conventional perovskite solutions, QDs can be suspended in nonpolar solvents, create structures that are not easily formed in thin-film perovskite, and engineer deposition of various layers. PQD cells can be fabricated with an internal heterojunction of $CsPbI_3$ QDs/$Cs_{1-x}FA_xPbI_3$ QDs layers, leading to PCE of 17% with a negligible hysteresis due to unbalanced interfacial charge transport [37]. Replacing the ETL and HTL with SnO_2 and PTAA, respectively, balances the interfacial charge transport. Using the concept, a solar cell with $CsPbI_3$ QDs/$CsPbI_2Br$ thin films improved exciton lifetime, carrier mobility, and well-matched energy levels [38]. More interestingly, the combination of perovskite thin films and PQDs leads to simultaneous bulk and surface passivation, yields a PCE of 21.5% with $MAPbI_3$ thin films, and $CsPbBrCl_2$ QDs [39].

11.2 Perovskite Tandem Solar Cells

The tandem solar cell (TSC) is a kind of multi-junction solar cell, made of different materials with different bandgaps and absorption spectra. Different materials are layered on top of each other to absorb more incident light than a single material. Each p-n junction generates a current in response to various wavelengths of light. Multiple semiconductors with different bandgaps receive a broader range of wavelengths and increase the efficiency of converting photons into electrical energy. The theoretical efficiency of a single-junction cell was estimated to be 33.16%, while that of a multi-junction solar cell with multiple junctions was estimated to be 86.8% under highly concentrated sunlight [41–43]. Currently, the commercial case of tandem cells has PCE of ~30% under one-sun illumination and increases up to ~40% under concentrated sunlight. However, due to its complexity and high price-to-performance ratio, its application has been limited to aerospace industry which requires a high power-to-weight ratio. On the planet, these solar cells are developing in concentrator PV industries, and the number of installations is increasing.

Perovskite is the perfect material for tandem solar cells for their tunable bandgaps, high light conversion efficiency, and facile low-temperature processing. With low-temperature fabrication and high efficiency, perovskite can be easily combined with various types of solar cells such as silicon (Si), CIGS, DSSC, OPV, and QDSCs. Perovskite tandem cells show great potential in both research laboratories and industrial applications. In this section, design of tandem cells, measurement and characterization techniques, and PCE improvement strategies for tandem solar cells will be discussed.

11.2.1 Working Principles of Tandem Solar Cells

Tandem cells are stacks of p-n junctions formed of semiconductors with different bandgap energies. Each junction responds to a different section of the solar spectrum. Multiple single-junction cells are stacked on top of each other with the highest bandgap on the top and the lowest at the bottom. Therefore, photons having a wide range of wavelengths could be absorbed, thereby reducing thermalization loss and yielding a high PCE [Fig. 11.2(a,b)].

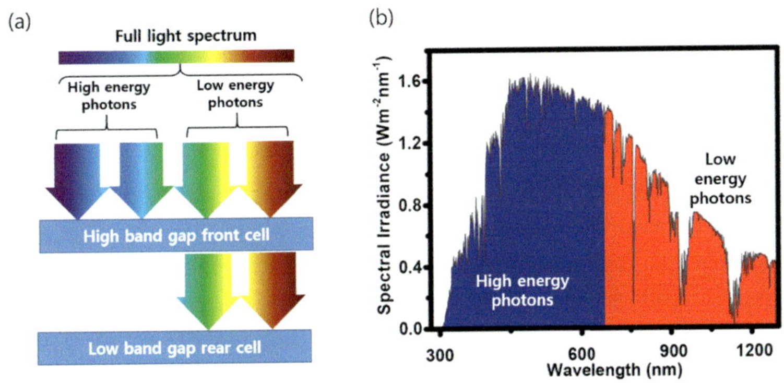

Figure 11.2 (a) Working principle of tandem cell. (b) Absorption spectra for tandem cells.

Tandem solar cells have two major intrinsic losses: the radiative recombination and the instability of a cell junction to match the solar spectrum. As the PCE is highly dependent on bandgap, it is important to select absorbers with matching bandgaps for both sub-cells. Under standard 1 sunlight intensity, the highest theoretical PCEs for four terminal (4T) and two-terminal (2T) TSCs are ~45.7% and ~86.8%, respectively, for an ideal infinite multi-junction solar cell under concentrated sunlight. Figure 11.3 shows that for tandem cells there are two configurations of mechanically integrated 4T and monolithically integrated 2T.

In the 4T configuration, the cells can be fabricated separately and connected later. The sub-cells, with their active layer, can operate individually up to their highest efficiency. In this configuration, the optical losses are high due to the presence of TCO layers at interfaces which cause optical losses. Despite the simple process 4T tandem devices are not commercially developed due to their complex manufacturing process. Furthermore, the operating individual sub-cells require additional units, such as inverters, wirings, and connections from individual cells, which greatly increase the system costs [44].

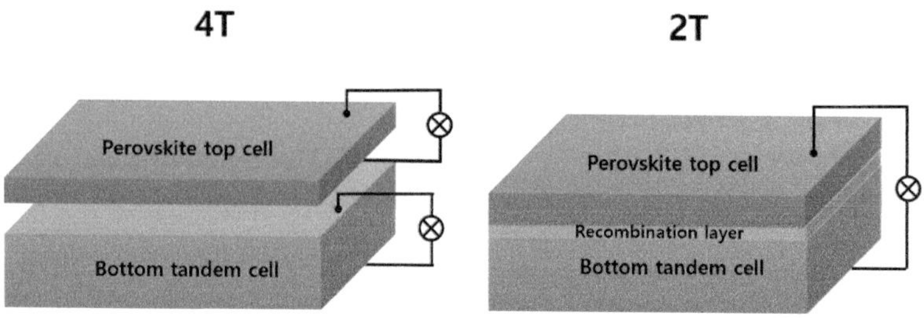

Figure 11.3 Schematic diagrams of 4T and 2T TSCs.

In the 2T tandem SCs, sub-cells are fabricated on the same substrate and the whole device has two electrodes. The 2T cells require an interconnecting layer that can either be a highly doped tunnel junction or a TCO layer to provide optical and electrical connection of sub-cells. The 2T tandem cells contain fewer electronic circuits with lighter weight and lower cost. However, the 2T cells have several challenges because they require careful optimization of current and lattice matching, compatibility of sub-cells in the fabrication process, and proper recombination and tunneling junctions.

11.2.2 Perovskite Tandem Solar Cells

Metal halide perovskites are noble light-absorbing semiconductors suitable for both top and bottom cells in tandem SCs because they have tunable bandgaps from 1.17 eV to 3.1 eV by compositional engineering. In general, perovskite solar cells are used as a top cell in tandem SCs because the highly efficient PSCs have a bandgap of ~1.7eV which is more suitable for a top cell. Si (with a bandgap of 1.12 eV) and CIGS (bandgap ~1.08–1.15 eV) are the most promising materials for bottom cells. PSCs with the third-generation PVs such as DSSCs, OPVs, and QDSCs are also very efficient for fabricating tandem cells [45]. Third-generation solar cells not only have improved PCE, but also their fabrication process is compatible with the PSCs with low-cost materials and roll-to-toll production.

A tunable bandgap is a significant advantage of perovskites for tandem devices. Perovskite also has a tolerance factor in structure, which can reduce the limitation of lattice mismatch of sub-cells. The Goldschmidt tolerance factor (t) is calculated by [46]:

$$t = \frac{r_A + r_B}{\sqrt{2}\left(r_A + r_X\right)} \tag{11.1}$$

where r_A, r_B, and r_X are the ionic radii of the A, B, and X ions, respectively. 3D (three-dimensional) perovskite with cubic structure has the t value in the range of 0.81–1.00.

A value of t close to 1.0 indicates a perfect fit, but a smaller tolerance factor may be attributed to tilting of the BX_6 octahedral and lower symmetry. A value of t less than 0.81 means that the cation is too small and leads to an alternative structure, and if t >1.00, the hexagonal perovskite structure is formed.

The tolerance factor is very successful in describing perovskite stability, for instance, in ABX_3 compounds where the X-site is filled with a high electronegative anion, the high electronegativity of the anions leads to a large charge pulling in the bonding, and forms a hard-sphere model.

Aside from the structure tolerance, PSCs also benefit from their electrical tolerance suitable for tandem SCs which can efficiently reduce the deep defect centers caused by lattice mismatch between sub-cells. Perovskite is a material with good defect tolerance, high absorption coefficient (1.5×10^4 cm^{-1}), high carrier mobility (1.60–170 $cm^2V^{-1}s^{-1}$), large charge diffusion length (>1 μm) and long carrier lifetime (>1 μs). Intrinsic defects are only in shallow levels within the bandgap. Shallow defect states are due to the strong Pb *s* orbital and I *p* orbital anti-bonding coupling and the high $MAPbI_3$ ionicity. The excellent characteristics of perovskite mentioned above are promising to develop perovskite tandem SC in combination with any kind of solar cell.

11.2.2.1 Bandgap dependency of perovskite

Wide-bandgap perovskite. Bandgap tuning of perovskite from 1.5 to 2.3 eV is important for tandem cell applications. However, the progress on wide-bandgap PSCs has been slow due to the intrinsic photo-instability of the materials. Wide-bandgap (E_g)

PSCs (E_g = 1.7–1.9 eV) are ideal for tandem top cells with Si, CIGS, and low-E_g PSCs as tandem bottom cells. A higher bandgap can be achieved by substituting I with Br in iodide-based perovskites, where greater content of Br leads to a wider bandgap. Br content of ~15% is found to be optimal for high-quality perovskite films and efficient PSCs. Devices with $MAPbBr_xI_{3-x}$ having a bandgap of 1.7–1.75 eV yield PCE > 16%. Unlike low-bandgap (E_g < 1.6 eV) perovskite cells, wide-bandgap (E_g > 1.7 eV) perovskite solar cells have a great challenge to achieve high V_{oc}.

The loss in V_{oc} is due to energy-level mismatch between wide-bandgap perovskite and ETL/HTL, photo-induced phase segregation, and deep defects in the bulk and grain boundaries of perovskite films. Another efficient wide-bandgap perovskite is based on FA/Cs mixed cations, which show more structural stability than the MA-cation under thermal and photo stresses. Devices with $FA_{1-y}Cs_yPb(I_{1-x}Br_x)_3$ having E_g of ~1.7 eV yield high PCE (~ 17%) and V_{oc} (~1.2 V), becoming one of the most popular wide-bandgap perovskites.

Enhancement in electronic properties, PCE, and stability of wide bandgap PSCs can be obtained by processing strategies such as interface modification, additive incorporation, co-solvent, etc. Ionic defects in the surfaces and grain boundaries, which could cause recombination and limit photovoltage of solar cells, can be suppressed by surface modification and defect passivation. Surface modifications help bypass the intermediate phase formation and reduce extrinsic defects.

Low-bandgap perovskite. Low-bandgap perovskite can be obtained by replacing a certain amount of lead (Pb) with other elements. Sn-Pb mixed perovskites show a smaller bandgap (~ 1.17 eV) and harvest more infrared photons. Br-incorporated perovskite (i.e., $MAPb_{1-x}Sn_x(I_{1-y}Br_y)_3$) gives an ideal bandgap of 1.35 eV for single-junction solar cells with PCE > 17%. Low-bandgap Sn-Pb perovskites with mixed organic cations (MA/FA) also show high quantum efficiency and decent storage stability because of the low bandgap of ~1.25 eV and long carrier lifetime over hundreds of ns. Among various compositions, $(FASnI_3)_{1-x}(MAPbI_3)_x$ showed good optoelectronic properties with the highest performance for low-bandgap stable perovskite solar cells [47].

Halide (Cl and Br) incorporation in Pb-Sn-based perovskite helps to passivate grain boundary via improved electronic properties and reduced recombination, leading to high PCEs. Figure 11.4 shows the efficiency improvement of single-junction and tandem PSCs in recent years [48]. Advances in the wide and low-bandgap PSCs provide an opportunity to develop all types of perovskite tandem solar cells over the past few years. 2-T and 4-T perovskite/Si and 4-T perovskite/CIGS tandem cells showed higher PCEs than the world record of single-junction PSCs. With increasing PCE, the efficiency of PSCs is approaching the S-Q limit for single-junction solar cells, and their tandem solar cells have become very promising in the future.

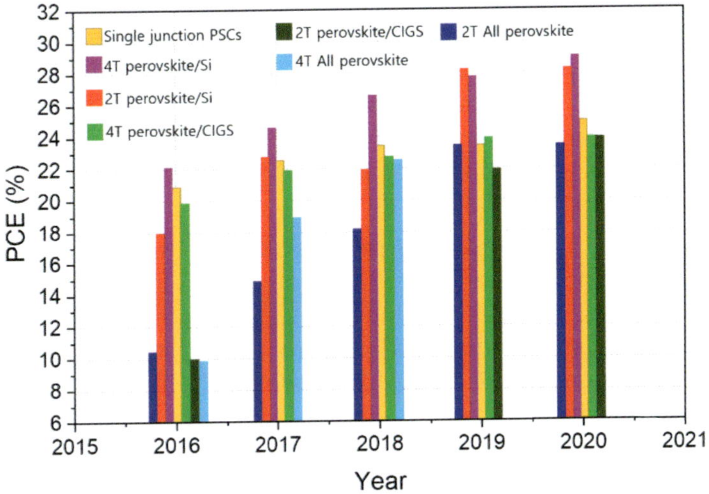

Figure 11.4 Efficiency evolution of single-junction and tandem PSCs in recent years [48].

11.2.2.2 Perovskite/Si tandem solar cells

Four-terminal perovskite/Si tandem solar cells. The c-Si-based PVs with high-power output, field durability, and reduced processing costs are ideal choices to make tandem devices with wide-bandgap perovskites. Tandem devices of c-Si (Eg ~1.1 eV) with wide-bandgap perovskite (1.6 to 1.75 eV) achieved high PCEs. The first development of perovskite/Si tandem cells was reported in 2014 with 4-T $MAPbI_3$/c-Si and 4-T $MAPbI_3$/m-Si tandem cells, yielding efficiencies of 13.4% and 17%, respectively [50]. For $MAPbI_3$/m-Si tandem cells, the top and bottom cells showed PCEs of only 12.7% and 4.3%, resulting in a total PCE of 17.0%. Combining a wide-bandgap PSC with a Si heterojunction (SHJ) cell for a 4-T perovskite/Si tandem device with efficiencies of 15.1% and 7.3% for the top and filtered bottom cells, respectively, resulted in a tandem device with an efficiency of 22.4%, attributed to the photocurrent improvement for both top and bottom cells [51].

The FA/Cs-based perovskite compositions have become the common choice for the top cell of perovskite tandem devices. 4-T perovskite/Si tandem cells with 1.63 eV quadruple-cation $RbFAMACsPb(IBr)_3$ perovskite top cell and Si-bottom cell also could yield a PCE of 26.4% [52]. The ion-implanted polycrystalline silicon (IBC-Si) as bottom cells resulted in more absorption of near-IR wavelengths, thereby inducing a high J_{sc} of 18.7 mA cm^{-2}. A high PCE of 27.1% was reported for perovskite/IBC-Si 4-T tandem solar cells. With the combination of wide-bandgap $Cs_{0.15}FA_{0.85}Pb(I_{0.71}Br_{0.29})_3$ with IBC-Si as the bottom subcell, a high V_{oc} of 1.22 V was obtained for a tandem device [53].

Two-terminal perovskite/Si tandem solar cells. The first 2-T monolithic perovskite/Si-heterojunction (SHJ) tandem solar cell was reported in 2015. The tandem device was made of $MAPbI_3$ (top cell) and Si (bottom cell) connected by an n^{++}/p^{++} Si tunnel recombination junction (Fig. 11.5a), enabling the high V_{oc} of 1.65 V. Afterward, monolithic perovskite/Si tandem cells were developed with an FA-MA mixed cation and mixed halide perovskite top cell, a SnO_2/ITO interface layer, and a silicon heterojunction bottom cell, resulting in a PCE of 19.9% with high V_{oc} of 1.785 V and FF of 79.5% [54]. In late 2015, tandem devices with PCEs of 21.2% was demonstrated by semitransparent $MAPbI_3$ perovskite/double-side polished Si combination (Fig. 11.5b) [55]. Later, by texturing Si wafers and a c-Si recombination layer, PCE of 22.8% was obtained with stable FA/Cs-based perovskite ($FA_{0.83}Cs_{0.17}Pb(I_{0.83}Br_{0.17})_3$)/Si tandem cells with high V_{oc} of 1.75 V and FF of 77%. Using single-sided textured SHJ as the bottom cell with $FA_{0.83}Cs_{0.17}Pb(I_{0.83}Br_{0.17})_3$ perovskite improved the PCE up to 23.6% with V_{oc} of 1.65 V and FF of 79.0%. In this tandem cell, light enters from the perovskite top subcell with LiF/PCBM/SnO_2/(Zn, Sn)O_2 as ETL [56].

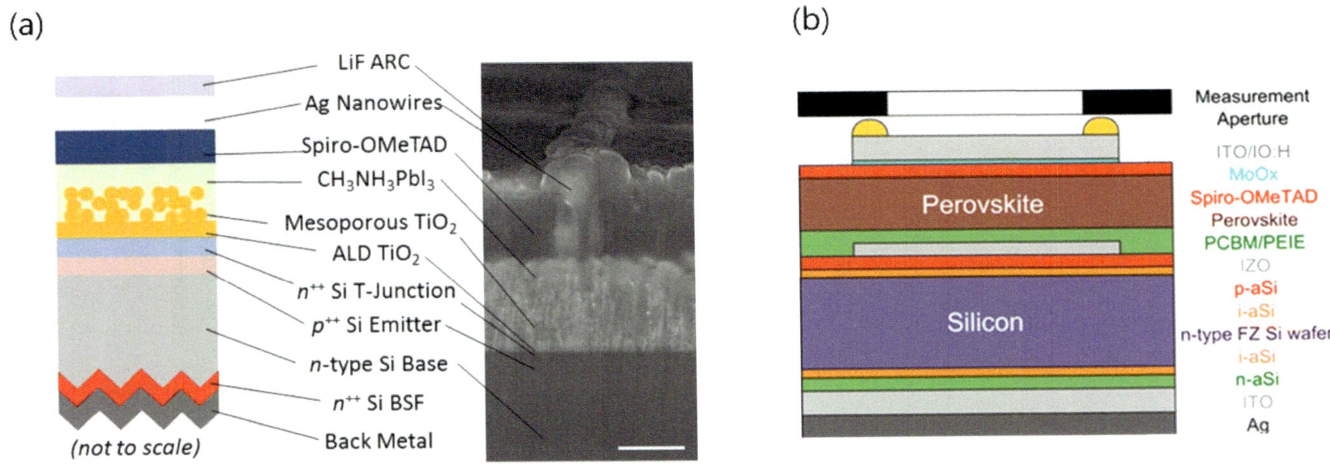

Figure 11.5 Schemes of 2-T monolithic perovskite/Si tandem solar cells, with (a) MAPbI3/c-Si, Reprinted with permission from [164], Copyright 2021, AIP publishing, and (b) MAPbI3/SHJ (double-side polished), Reprinted with permission from [55], Copyright 2021, American Chemical Society.

Most efficient 2T monolithic perovskite/Si tandem solar cells are designed based on Si-bottom cells with a polished front surface. Compared to single-side polished Si wafers, double-side textured c-Si wafers could yield a certified PCE of 25.2% with a high J_{sc} of 19.5 mAcm^{-2}. The optical design shows promising results that can reduce optical losses in 2-T monolithic perovskite/Si tandem cells. By combining a thin top transparent electrode, metal fingers, and anti-reflection foil, a tandem cell could deliver a PCE of 25.1% with V_{oc} of 1.75 V, J_{sc} of 18.4 mA cm^{-2}, and FF of 78% [57]. With the use of a light-management foil on the transparent side, a high PCE of 25.5% with *a* V_{oc} of 1.26 V was demonstrated with $Cs_{0.05}(MA_{0.17}FA_{0.83})Pb_{1.1}(I_{0.83}Br_{0.17})_3$ perovskite/Si tandem cell [58].

Besides the SHJ-based c-Si solar cells, the homojunction c-Si cells with a high market share of ~93% are ideal candidates for the commercialization of perovskite/Si tandem solar cells. Back-side-textured Si cells in a 2-T monolithic tandem device combined with $(FAPbI_3)_{0.83}(MAPbBr_3)_{0.17}$ perovskite verified a steady-state efficiency of 21.8% on a large area of 16 cm^2 [59]. This progress provides a way for commercially inexpensive perovskite/homojunction Si tandem cell technology.

11.2.2.3 Perovskite/chalcogenide thin-film tandem solar cells

Four-terminal perovskite/chalcogenide tandem solar cells The first tandem cell of CIGS/Perovskite was reported in 2014, showing PCE of 18.6% with $MAPbI_3$ top cell and CIGS bottom cell [62]. Thereafter, when the 4-T polycrystalline thin-film tandem cells were used, PCE of 19.5% was obtained, i.e., PCE of 12.1% for $MAPbI_3$ perovskite top cell and 7.4% for CIGS as bottom cell [63]. Semitransparent $MAPbI_3$ n-i-p perovskite with a configuration of FTO/ZnO/PCBM/perovskite/Spiro-OMeTAD/MoO_3/In_2O_3:H in the 4-T tandem device with CIGS resulted in PCE of 20.5% [64]. With the structure of glass/In_2O_3:H/PTAA/$MAPbI_3$/PCBM/ZnO/ZnO:Al/metal grid, the efficiency of perovskite/CIGS and perovskite/CIS tandem cells increased by 22.1% and 20.9%, respectively [60]. By replacing the medium bandgap $MAPbI_3$ with a wide-bandgap perovskite, the infrared wavelengths enter the bottom CIGS cell. For example, the semitransparent PSC with $Cs_{0.05}Rb_{0.05}FA_{0.765}MA_{0.135}Pb(I_{1-x}Br_x)_3$ absorber having E_g of 1.62 (x = 0.15) and 1.75 eV (x = 0.4) showed PCEs of 18.1% and 16.0%, respectively, and the 4-T tandem cells combined with the CIGS (PCE = 16.5%) bottom cell showed efficiency of 23.9% [61].

Two-terminal perovskite/chalcogenide tandem solar cells. The first fabrication of a monolithic tandem device of perovskite with $Cu_2ZnSn(S,Se)_4$ (CZTSSe) and CIGS as bottom cells resulted in PCEs of 4.6% and 10.9%, respectively, much inferior to the PCEs of single-junction cells [65]. There are only a few research reports on the fabrication of 2-T perovskite/CIGS tandem due to difficulty in solution processing when interconnecting the perovskite top cell on the CIGS thin-film bottom cell. Monolithically interconnected solution-processed perovskite ($MAPbI_3$)/chalcogenide [$CuIn(Se,S)_2$] tandem solar cells, yielded a PCE of 8.55% [66]. However, a breakthrough has been made recently by chemically and mechanically polishing the top contact of CIGS cells with heavily doping PTAA HTL. The 2-T perovskite/doped-CIGS tandem cell with sub-cells of 1.59 eV perovskite ($Cs_{0.09}FA_{0.77}MA_{0.14}Pb(I_{0.86}Br_{0.14})_3$) and 1.0 eV CIGS, showed a PCE of 22.4% with V_{oc} of 1.774 V, J_{sc} of 17.3 mA/cm^2, and FF of 73.1% [67].

11.2.2.4 All-perovskite tandem solar cells

In perovskite tandem technology, all-perovskite tandem cells are promising devices for their exceptional properties such as low-temperature processes for the whole device, flexibility and lightweight, and potentially low manufacturing costs. However, the lack of highly efficient low-bandgap perovskite is an issue to fabricate effective all-perovskite tandem cells. The recent development has boosted the PCEs of 4-T and 2-T all-perovskite tandem cells to 23.1% [81] and 21% [82], respectively, reaching the PCEs of perovskite/Si and perovskite/CIGS tandem cells.

Four-terminal all-perovskite tandem solar cells. The progress of all-perovskite tandem solar cells is slower than other perovskite tandem cells due to the difficulty in producing efficient and stable low-bandgap PSCs. The addition of a thin absorber layer of medium bandgap $MAPbI_3$ could help sufficient photons absorption in the bottom cell. In 2016, 4-T all-perovskite tandem solar cells with $MAPbBr_3/MAPbI_3$ as top/bottom cells showed a PCE of ~9.5% [68]. Furthermore, the 4-T all-perovskite tandem solar cell combining $MAPbI_3$ as low-bandgap top cell and $MA_{0.5}FA_{0.5}Pb_{0.75}Sn_{0.25}I_3$ as wide-bandgap bottom cell achieved a high PCE of 19.1%. $MAPbI_3$ with bandgap of 1.55 eV and $MA_{0.5}FA_{0.5}Pb_{0.75}Sn_{0.25}I_3$ with bandgap of 1.33 eV showed PCEs of 13.5% and 5.6% for the top and filtered bottom cells, respectively [69].

Using mixed cation and mixed anion in both the top and bottom cells demonstrated 4-T tandem cells with an optically balanced absorber. The 4-T tandem cell made of $FA_{0.83}Cs_{0.17}Pb\ (I_{0.83}Br_{0.17})_3$ top cell with a wide bandgap of 1.6 eV and $FA_{0.75}Cs_{0.25}Sn_{0.5}Pb_{0.5}I_3$ bottom cell with a low bandgap of 1.2 eV resulted in a PCE of 20.3% [70]. By combining wide-bandgap $FA_{0.3}MA_{0.7}PbI_3$ (E_g = 1.58 eV) and low-bandgap $(FASnI_3)_{0.6}(MAPbI_3)_{0.4}$ (E_g = 1.25 eV), the efficiency of the 4-T tandem cell was further increased to 21.2% [72], and by replacing top cell with $FA_{0.8}Cs_{0.2}Pb(I_{0.7}Br_{0.3})_3$ the PCE was further improved to 23.1% [73].

Two-terminal all-perovskite tandem solar cells. In a 2-T all-perovskite tandem solar cell, the most important layer is the interconnecting layer (IL) which connects the top and bottom cells and protects the bottom subcell from any damage caused by processing the top cell. The monolithic 2-T all-perovskite tandem solar cell was first reported in 2015 by pairing two $MAPbI_3$ cells connected by organic ILs yielded a low PCE of 7%, but a high V_{oc} up to 1.89 V. $MAPbBr_3$-$MAPbI_3$ 2-T tandem solar cell was fabricated by attaching two separated sub-cells with a PCE of 10.8% and a high V_{oc} of 2.25 V. These devices showed a low J_{sc} of 8.3 $mAcm^{-2}$ attributed to the large bandgap of a bottom cell ($MAPbI_3$, 1.55 eV) and the physically attached IL [74].

2-T monolithic all-perovskite tandem cell using low-bandgap Sn-Pb PSCs was first reported by pairing $FA_{0.83}Cs_{0.17}Pb(I_{0.5}Br_{0.5})_3$ (E_g = 1.8 eV) as top cell and $FA_{0.75}Cs_{0.25}Sn_{0.5}Pb_{0.5}I_3$ (E_g = 1.2 eV) as bottom cell and ZTO/SnO_2/ITO ILs [75]. In this design, the sputtered ITO layer forms an optical and electrical connection between two cells to protect the bottom cell from damages that occur during the processing of the second subcell in solution. The best device showed a PCE of 16.9%. The 2-T perovskite tandem cell fabricated by pairing a photo-stable $FA_{0.83}Cs_{0.17}Pb(I_{0.7}Br_{0.3})_3$ top cell and $FA_{0.75}Cs_{0.25}Sn_{0.5}Pb_{0.5}I_3$ bottom cell showed a PCE of 19.1% with high V_{oc} of 1.81 V and J_{sc} of 14.8 mA/cm^2 [75].

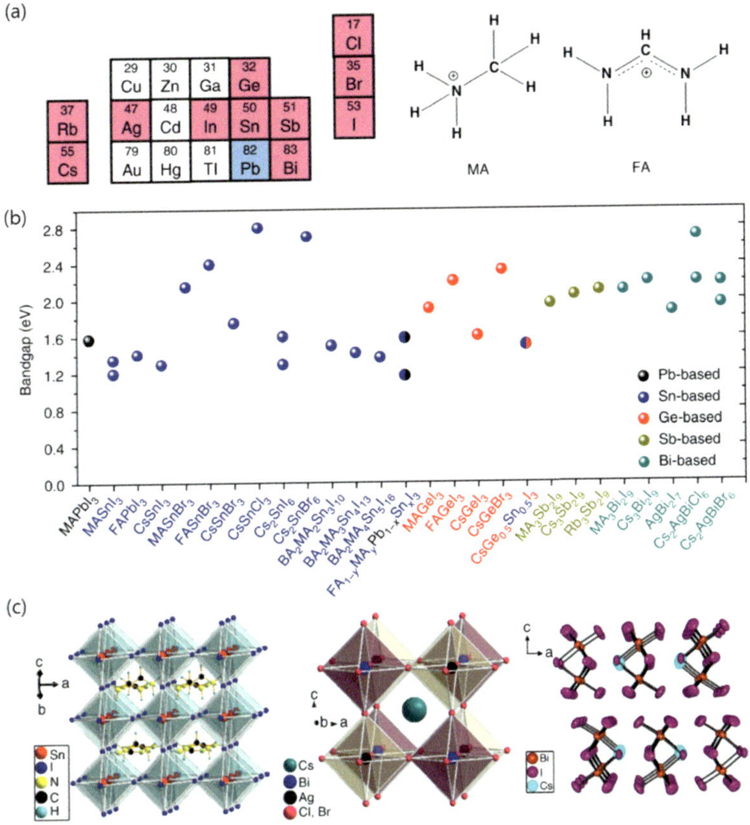

Figure 11.6 Potential perovskite materials for lead-free solar cell absorbers. (a) Potential A-site cations in ABX3 perovskite. (A=organic MA and FA or inorganic Cs and Rb), B-site metals, and X-site halides (I, Br, Cl) for perovskite structure. (b) Bandgaps of suitable materials for perovskite solar cells. (c) Crystal structures of $FASnI_3$ (left), $Cs_2AgBi(ClBr)_6$ (middle), and $Cs_3Bi_2I_9$ (right). Reprinted with permission from [192], Copyright 2021, Nature.

11.3 Lead-Free Perovskite Solar Cells

11.3.1 Limitations of Pb-Based Perovskite Materials

Lead-based perovskites are an excellent material for single-junction photovoltaic devices for their simple fabrication processes, outstanding optoelectronic properties, and high-power conversion efficiency (PCE > 25%). However, lead toxicity, except for the operational instability currently being studied, is a limiting factor that prevents commercial use of perovskite solar cells. In the past few years, considerable research has been conducted on alternative metals to replace lead. If performance does not deteriorate, the development of perovskites that are less toxic or lead-free is always preferred in the solar cell market. Pb-free absorbers should have low toxicity, direct low bandgap, high optical-absorption coefficient and mobility, low exciton binding energy, and proper stability. There are several candidates to replace lead but maintain similar properties to lead-based perovskite structures, such as Sn-, Ge-, Sb-, Bi-based perovskites, and double perovskite-like structures (Fig. 11.6) [76]. This section provides an overview of lead-free perovskite absorbers and their photovoltaic properties.

11.3.2 Tin-Based Perovskites

A typical strategy for Pb-free perovskite involves the replacement of Pb(II) with Sn(II). The Sn-based lead-free perovskite photovoltaics produced the highest PCE of 13.24% [80]. Sn^{2+} having a similar ionic radius to that of Pb^{2+} forms $ASnX_3$ perovskites (monovalent cation A and halide anion X) with a three-dimensional (3D) structure similar to that of Pb-based perovskites. $MASnI_3$, $FASnI_3$, and $CsSnI_3$ have been the most studied Sn-based perovskites. These absorbers have similar structural and optoelectronic properties to Pb-based counterparts with a small direct gap of ~1.3 eV, ideal for single-junction PV devices.

The Sn-based PSCs showed a short-circuit current as high as >20 $mAcm^{-2}$ but a low open-circuit voltage (<0.6 V), resulting in low PCEs (<7%) [77]. It has been found that the intrinsic property of Sn^{2+} oxidizing to Sn^{4+} creates defect density and morphological deficiencies in perovskite structure with a large carrier background due to *p*-type doping. [78]. Suppression of Sn^{2+}/Sn^{4+} oxidation with the use of reducing additives improves the device stability by increasing PCE up to ~10% [79]. A PCE of up to 13.24% (the highest to date) was presented through defect passivation and reduced lattice strain (Fig. 11.7) [80].

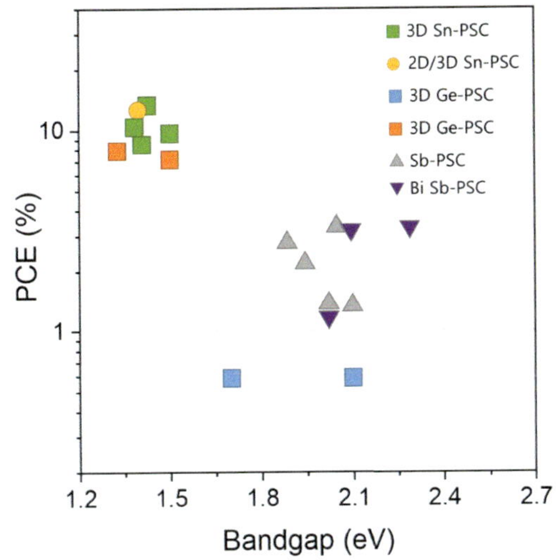

Figure 11.7 Single-junction PCE of lead-free perovskite absorbers [93].

To overcome the problems of $ASnX_3$ photovoltaics, several alternative Sn-based approaches have been examined based on the engineering of perovskite structure, including tin-based vacancy-ordered double perovskites, which delivered high stability but rather low PCE. Another approach is to use hollow perovskites to create structural voids within a 3D structure using medium-sized cations. This additive-free $ASnX_3$ has shown a promising PCE of ~7% with improved stability [81]. Dimensional manipulation achieved higher PCE with 2D/3D perovskites, as high as 9.41% in a planar p-i-n device [82]. A high voltage of 0.94 V and PCE up to 12.4% for tin-based PSCs are obtained by replacing commonly used PCBM electron-transport material with indene-C_{60} with a higher-energy level for $PEA_xFA_{1-x}SnI_3$ perovskite solar cells [84]. It has been identified that large cations in the layered phase help improve stability, resulting in PCEs of 9 ~ 12.4%.

Despite the instability issues, the non-toxic Sn PSCs have a great progress prospect for future development. According to structural and device performance analysis, there are several ways to promote the stability of Sn PSCs:

- Use highly pure Sn-halide precursor. Preparation and improvement of pure precursor materials, especially in source, suppress oxidation during film processing.
- Optimize the open-circuit voltage by reducing the non-radiative recombination. The low value of V_{oc} is a major obstacle in the development of Sn-perovskites, which is induced by high recombination and mismatched energy levels. Hence, controlling the film formation process and crystallization of perovskite reduces the trap state in tin-based perovskites.
- Reduce oxygen contact. Inherently, on the surface and grain boundaries of perovskite film, Sn^{2+} ions tend to be oxidized to a more stable Sn^{4+} state. Deposition of an antioxidant protective layer on the surface is an effective way of avoiding oxygen contact and improving the efficiency and stability of Sn PSCs.
- Develop compact inorganic ETL and HTL materials. The development of an all-inorganic Sn-PSC with inorganic ETL and HTL can further isolate oxygen and moisture and prevent the oxidation of Sn^{2+} [85]. Aside from high crystallinity and tunable energy levels, metal oxides are chemically compatible with perovskite due to the intrinsic passivation effect.
- Develop inverted p-i-n Sn PSCs. An inverted p-i-n Sn-PSC has high and stable performance, and tin oxidation is less in inverted devices. Fabrication of inverted structure combining new HTL and ETL helps to develop efficient and stable tin-based PSCs.

The low V_{oc} of Sn PSCs is attributed to a high defect density that causes non-radiative recombination. Thus, defect passivation is needed, and it should be a combination of compositional and morphological optimization of the Sn-based absorber layer (by A-site mixed cation and incorporation of additives), high stability against Sn oxidation, and careful optimization of device layers to minimize interfacial recombination. 3D perovskites with mixed cations and mixed 2D/3D showed promising performance with low defect density and improved stability. The use of additive composites also helps with better film formation and enhanced crystallinity.

Another parameter affecting the V_{oc} loss is energetic disorder. The energetic disorder has been rarely investigated in the Sn-based perovskites. The energetic disorder can be studied with Urbach energy which is used to quantify energetic disorder in the band edges of a semiconductor. As it is often used to describe electron transport in structurally disordered materials, the Urbach energy is highly dependent on structural disorder, stoichiometric imperfection, and surface passivation. Lower Urbach energy indicates the reduced defect concentration. An optimized Sn PSCs having a PCE of 12.4% and a record high V_{oc} of 0.94 V have a low Urbach energy of 65 meV [86]. Much lower Urbach energy values (16 meV – 32 meV) were observed from $CsSnI_{3-x}Br_x$.

Recently, Hahn et al. reported the use of a photoactive composite of tin halide perovskite with graphene-tin quantum dots (rGO-Sn QDs) in the active layer to suppress Sn^{2+}/Sn^{4+} oxidation and improve perovskite film morphology in the mesoporous Sn-PSCs [47]. The tin halide perovskite/rGO-Sn QDs composites in the active layer not only reduce structural defects but also enhance light-harvesting and carrier transport. With the optimization of rGO-Sn QDs content in the photoactive composite layer, significant improvements in device performance and stability were achieved with a substantial reduction in trap state density. Compared to a reference cell without rGO-Sn QDs, the composites-based Sn-PSC devices maintained 42% of their initial PCE over 30 days (Fig. 11.8a). The composite-based cell showed even better stability than the reference cell retaining 85% of initial PCE in the Ar atmosphere over 30 days (Fig. 11.8b). It was thanks to an effective role of rGO-Sn QDs in the active layer, i.e., suppression of Sn^{+2}/Sn^{+4} oxidation. Under the built-in energy field, Sn QDs oxidize to Sn^{2+}, leading to suppression of Sn^{+2}/Sn^{+4} oxidation by supplying more electrons and generating more Sn^{2+}.

A suppresion mechanism of Sn^{+2}/Sn^{+4} oxidation by rGO-Sn QDs in the active layer under built-in energy field is illustrated in Fig. 11.8c. The photoactive composite chemical reactions in the active layer are likely to occur [47]:

$$Sn^0 \rightarrow Sn^{+2} + 2e \tag{11.1}$$

$$Sn^{+4} + 2e \rightarrow Sn^{+2} \tag{11.2}$$

$$Sn^{+4} + Sn^0 \rightarrow 2Sn^{+2} \tag{11.3}$$

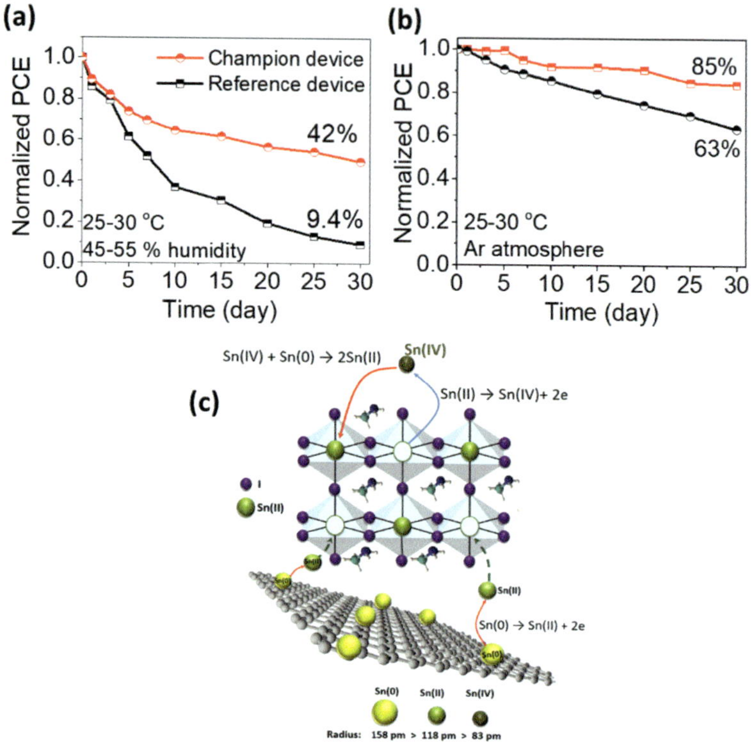

Figure 11.8 Stability test of Sn perovskite devices. The normalized power-conversion efficiency of Sn PSCs. The reference device is without and champion device with rGO-Sn QDs stored in (a) ambient atmosphere and (b) in dry Ar atmosphere. (c) Schematic illustration of Sn QD oxidation and its effect on Sn^{+2}/Sn^{+4} oxidation. Reprinted with permission from [47], Copyright 2021, Elsevier.

The Sn QDs under the built-in energy field play three important roles: oxidation to Sn^{+2} (Eq. 11.1), stimulating the reduction of Sn^{+4} to Sn^{+2} (Eq. 11.2), and generating more Sn^{+2} (Eq. 11.3). In light-activated solar cells, Sn QDs have a great tendency to lose electrons and oxidize to Sn^{2+}. This oxidation of Sn^0/Sn^{2+} in rGO-Sn is favorable because it not only compensates lack of Sn^{2+} but also decelerates the Sn^{2+}/Sn^{4+} oxidation in perovskite by supplying more electrons in the redox reaction (Eq. 11.2). It is also worth noting that the role of conductive graphene sheets in the active layer is probably to provide fast conduction paths of electrons so that Sn^{+4} is readily reduced to Sn^{+2}. Thus, the chemical reactions above strongly support the suppression mechanism of Sn^{+2}/Sn^{+4} oxidation with the aid of Sn QDs in the photoactive composite active layer. They also support the lower recombination centers, lower trap density, and longer exciton lifetime, resulting in enhancement of the performance of the composite-based device.

11.3.3 Germanium-Based Perovskites

Germanium (Ge) having similar electronic properties to Pb^{2+} enables the formation of 3D $AGeX_3$ perovskites. $CsGeI_3$ with a direct bandgap of 1.63 eV is the most promising candidate for single-junction photovoltaics. Despite their high theoretical efficiency in the Shockley–Queisser limit (~30% for $CsGeI_3$), few studies have been reported on Ge-based PSCs due to the tendency of Ge^{2+} to be oxidized to Ge^{4+}. $AGeX_3$ perovskites yield the lowest PCE among all the lead-free perovskites due to large losses in both V_{oc} (~0.5 V) and J_{sc} (~2 mA/cm^2). An alternative Ge-based perovskite is alloyed with tin, i.e., $ASn_{1-x}Ge_xX_3$. With compositional engineering, Sn–Ge mixed perovskites have achieved much higher efficiency than the Ge-only devices ($AGeX_3$). Recently, an increase in the efficiency of Sn–Ge perovskite solar cells with $CsSn_{0.5}Ge_{0.5}I_3$ (PCE ~7.1%) and $FA_{0.75}MA_{0.25}Sn_{1-x}Ge_xI_3$ (PCE ~7.9%) was reported, attributed to trap passivation and stabilization [88].

11.3.4 Antimony- and Bismuth-Based Perovskites

Antimony (Sb) and bismuth (Bi) are also other candidates for Pb-free perovskite due to their low toxicity and similar electronic properties to Pb^{2+}. $A_3B_2X_9$ absorbers (B^{3+} = Sb^{3+} or Bi^{3+}) have two phases: a dimer phase (0.5-dimensional, 0.5D) with isolated face-sharing metal halide bi-octahedra and a layered phase (2.5-dimensional, 2.5D) with planes of corner-sharing metal halide octahedra. 0.5D phased $A_3B_2I_9$ (with A = Cs^+, MA^+) having a wide bandgap of 2.1–2.4 eV is a potential candidate for tandem photovoltaics. Low dimension 0.5D absorbers showed a low PCE (< 1%), but the use of additives, optimized transport layers, and keen deposition process led to performance improvement, reaching PCE of 2.8% for 0.5D $MA_3Sb_2I_9$ and 3.2% for 0.5D $A_3Bi_2I_9$ (A = Cs^+, MA^+).

The higher dimensionality of 2.5D Sb- and Bi-based $A_3B_2X_9$ absorbers have a narrower bandgap and low exciton energy and effective mass. Despite keen processing protocols and composition engineering PCEs of 1.4%, 2.2% and 3.34% have been achieved with 2.5D $Rb_3Sb_2I_9$, $Cs_3Sb_2Cl_xI_{9-x}$ and $MA_3Sb_2Cl_xI_{9-x}$, respectively [88]. Improving the efficiency of 2.5D $A_3B_2X_9$ systems can be explored further through developing processing strategies. Moreover, Urbach energy values of 60 meV or larger have been reported for some Sb- and Bi-based perovskites, which shows high chances to reduce the V_{oc} loss by improving device structures, composition, and processing conditions. It is worth noting that some Sb- and Bi-based perovskites are reported to have more than 60 meV of Urbach energy, which is likely to reduce V_{oc} loss through improving device structure, composition, and processing conditions.

11.3.5 Halide Double Perovskites

Halide double perovskites have a general formula $A_2BB'X_6$, exhibiting a 3D structure with wider bandgaps of around 2 eV and higher stability in air. In $A_2BB'X_6$, B and B′ are monovalent cations and trivalent cations alternating at the octahedra centers of a perovskite lattice, respectively. Halide double perovskites have a broad range of materials, more than 350 compounds synthesized to date [89]. These absorbers are not ideal for single-junction solar cells, but their low toxicity, 3D structure, long carrier lifetimes (≈600 ns), and air stability have promoted research on the solar cell application. The highest PCE attained in double perovskite single-junction devices is ~2.8%, where the highest performance was achieved with an organic interlayer which improved overall photon absorption beyond the onset of the absorber material [90]. Cs_2SnX_6 compounds with a direct bandgap of 1.3 eV–1.6 eV achieved a PCE of 2.0% with promising stability in the air [91]. Another attractive absorber is titanium-based Cs_2TiBr_6 with a direct bandgap of 1.8 eV and a carrier diffusion length >100 nm delivered a PCE of 3.3% in single-junction devices with good environmental stability (non-encapsulated, retain 94% of initial PCE after two weeks at 70°C in the air under illumination) [45].

References

1. Protesescu, L., Yakunin, S., Bodnarchuk, M. I., Krieg, F. et al. (2015) *Nano Lett.*, **15**, pp. 3692–3696.
2. Sun, S., Yuan D., Xu, Y., Wang, A., and Deng, Z. (2016) *ACS Nano*, **10**, pp. 3648–3657.
3. Weerd, C., Gomez, L., Capretti, A., Lebrun, D. M., et al. (2018) *Nat. Commun.*, **9**, pp. 4199.
4. Hao, M., Bai, Y., Zeiske, S., Ren, L., et al. (2020) *Nat. Energy*, **5**, pp. 79–88.
5. Swarnkar, A., Marshall, A. R., Sanehira, E. M., Chernomordik, B. D., et al. (2016) *Science*, 354, pp. 92–95.
6. Pan, A., He, B., Fan, X., Liu, Z., et al. (2016) *ACS Nano*, **10**, pp. 7943–7954.
7. Sanehira, E. M., Marshall, A. R., Christians, J. A., Harvey, S. P., et al. (2017) *Sci. Adv.*, **3**, pp. eaao4204.
8. Ling, X., Yuan, J., Zhang, X., Qian, Y., et al. (2020) *Adv. Mater.*, **32**, pp. 2001906.
9. Ling, X., Zhou, S., Yuan, J., Shi, J., et al. (2019) *Adv. Energy Mater.*, **9**, pp. 1900721.
10. Xue, J., Wang, R., Chen, L., Nuryyeva, S., et al. (2019) *Adv. Mater.*, **31**, pp. 1900111.
11. Yuan, J., Ling, X., Yang, D., Li, F., et al. (2018) *Joule*, **2**, pp. 2450–2463.
12. Zhao, Q., Hazarika, A., Schelhas, L. T., Liu, J., et al. (2019) *ACS Energy Lett.*, **5**, pp. 238–247.
13. Hao, M., Bai, Y., Zeiske, S., Ren, L., et al. (2020) *Nat. Energy*, **5**, pp. 79–88.
14. Protesescu, L., Yakunin, S., Bodnarchuk, M. I., Krieg, F., et al. (2015) *Nano Lett.*, **15**, pp. 3692–3696.
15. Zhao, Q., Hazarika, A., Schelhas, L. T., Liu, J., et al. (2019) *ACS Energy Lett.*, **5**, pp. 238–247.
16. Chen, J., Jia, D., Johansson, E. M., Hagfeldt, A., and Zhang, X. (2021) *Energy Environ. Sci.*, **14**, pp. 224–261.
17. Zolfaghari, Z., Hassanabadi, E., Pitarch-Tena, D., Yoon, S. J., et al. (2019) *ACS Energy Lett.*, **4**, pp. 251–258.
18. Wheeler, L. M., Sanehira, E. M., Marshall, A. R., Schulz, P., et al. (2018) *J. Am. Chem. Soc.*, **140**, pp. 10504–10513.
19. Ling, X., Zhou, S., Yuan, J., Shi, J., et al. (2019) *Adv. Energy Mater.*, **9**, pp. 1900721.

20. Chen, K., Jin, W., Zhang, Y., Yang, T., et al. (2020) *J. Am. Chem. Soc.*, **142**, pp. 3775–3783.

21. Wang, Q., Jin, Z., Chen, D., Bai, D., et al. (2018) *Adv. Energy Mater.*, 8, pp. 1800007.

22. Xue, J., Lee, J., Dai, Z., Wang, R., et al. (2018) *Joule*, **2**, pp. 1866–1878.

23. Xue, J., Wang, R., Chen, L., Nuryyeva, S., et al. (2019) *Adv. Mater.*, **31**, pp. e1900111.

24. Hao, M., Bai, Y., Zeiske, S., Ren, L., et al. (2020) *Nat. Energy*, **5**, pp. 79–88.

25. Swarnkar, A., Marshall, A. R., Sanehira, E. M., Chernomordik, B. D., et al. (2016) *Science*, **354**, pp. 92–95.

26. Sanehira, E. M., Marshall, A. R., Christians, J. A., Harvey, S. P., et al. (2017) *Sci. Adv.*, **3**, pp. eaao4204.

27. Yuan, J., Ling, X., Yang, D., Li, F., et al. (2018) *Joule*, **2**, pp. 2450–2463.

28. Ling, X., Zhou, S., Yuan, J., Shi, J., et al. (2019) *Adv. Energy Mater.*, **9**, pp. 1900721.

29. Shi, J., Li, F., Yuan, J., Ling, X., et al. (2019). *J. Mater. Chem. A Mater. Energy Sustain.*, **7**, pp. 20936–20944.

30. Kim, J., Koo, B., Kim, W. H., Choi, J., et al. (2019) *Nano Energy*, **66**, pp. 104130.

31. Chen, K., Jin, W., Zhang, Y., Yang, T., et al. (2020) *J. Am. Chem. Soc.*, **142**, pp. 3775–3783.

32. Ji, K., Yuan, J., Li, F., Shi, Y., et al. (2020) *J. Mater. Chem. A Mater. Energy Sustain.*, **8**, pp. 8104–8112.

33. Hoffman, J. B., Zaiats, G., Wappes, I., and Kamat, P. V. (2017) *Chem. Mater.*, **29**, pp. 9767–9774.

34. Akkerman, Q. A., Gandini, M., Di Stasio, F., Rastogi, P., et al. (2017) *Nat. Energy*, **2**, pp. 16194.

35. Christodoulou, S., Di Stasio, F., Pradhan, S., Stavrinadis, A., and Konstantatos, G. (2018) *J. Phys. Chem. C*, **122**, pp. 7621–7626.

36. Hazarika, A., Zhao, Q., Gaulding, E. A., Christians, J. A., et al. (2018) *ACS Nano*, **12**, pp. 10327–10337.

37. Zhao, Q., Hazarika, A., Chen, X., Harvey, S. P., et al. (2019) *Nat. Commun.*, **10**, pp. 2842.

38. Bian, H., Bai, D., Jin, Z., Wang, K., et al. (2018) *Joule*, **2**, pp. 1500–1510.

39. Zheng, X., Troughton, J., Gasparini, N., Lin, Y., et al. (2019) *Joule*, **3**, pp. 1963–1976.

40. Duan, J., Wei, J., Tang, Q., and Li, Q. (2020) *Sol. Energy*, **195**, pp. 644–650.

41. Multi-Junction Solar Cell. https://en.wikipedia.org/wiki/Multi-junction_solar_cell

42. King, R. R., Boca, A., Hong, W., Liu, X.-Q., et al. (2009) In: *24th European Photovoltaic Solar Energy Conference*. EU: Hamburg, Germany, pp. 1–33.

43. Essig, S., Allebé, C., Remo, T., Geisz, J. F., et al. (2017) *Nat. Energy*, **2**, pp. 17144.

44. Geisz, J. F., Steiner, M. A., Jain, N., Schulte, K. L., et al. (2018) *IEEE J. Photovolt.*, **8**, pp. 626–632.

45. Shen, H. P., Walter, D., Wu, Y. L., Fong, K. C., et al. (2020) *Adv. Energy Mater.*, **10**, pp. 1902840.

46. Mailoa, J. P., Bailie, C. D., Johlin, E. C., Hoke, E. T., et al. (2015) *Appl. Phys. Lett.*, **106**(12), pp. 121105.

47. Mahmoudi, T., Rho, W. Y., Kohan, M., Yeon Ho Im, et al. (2021) *Nano Energy*, **90**, pp. 106495.

48. Jiang, Q., Zhao, Y., Zhang, X. W., Yang, X. L., et al. (2019) *Nat. Photon.*, **13**, pp. 460–466.

49. Jiang, Q., Zhao, Y., Zhang, X. W., Yang, X. L., et al. (2019) *Nat. Photon.*, **13**, pp. 460–466.

50. Werner, J., Niesen, B., and Ballif, C. (2018) *Adv. Mater. Interfaces*, **5**(1), pp. 1700731.

51. Ren, Z., Zhou, J., Zhang, Y., Ng, A., et al. (2018). *Sol. Energy Mater. Sol. Cells*, **179**, pp. 36–44.

52. Duong, T., Wu, Y., Shen, H., Peng, J., et al. (2017) *Adv. Energy Mater.*, **7**(14), pp. 1700228.

53. Hamada, K., Yonezawa, K., Yamamoto, K., Taima, T., et al. (2019) *Jpn. J. Appl. Phys.*, **58**(SB), pp. SBBF06.

54. Albrecht, S., Saliba, M., Baena, J. P. C., Lang, F., et al. (2016) *Energy Environ. Sci.*, **9**(1), pp. 81–88.

55. Werner, J., Weng, C. H., Walter, A., Fesquet, L., et al. (2016) *J. Phys. Chem. Lett.*, **7**(1), pp. 161–166.

56. Bush, K. A., Palmstrom, A. F., Yu, Z. J., Boccard, M., et al. (2017) *Nat. Energy*, **2**(4), pp. 1–7.

57. Wang, Y., Gu, S., Liu, G., Zhang, L., et al. (2021) *Sci. China Chem.*, **64**(11), pp. 2025–2034.

58. Jošt, M., Köhnen, E., Morales-Vilches, A. B., Lipovšek, B., et al. (2018) *Energy Environ. Sci.*, **11**(12), pp. 3511–3523.
59. Zheng, J., Mehrvarz, H., Ma, F. J., Lau, C. F. J., et al. (2018) *ACS Energy Lett.*, **3**(9), pp. 2299–2300.
60. Fu, F., Feurer, T., Weiss, T. P., Pisoni, S., et al. (2016) *Nat. Energy*, **2**(1), pp. 1–9.
61. Shen, H., Peng, J., Jacobs, D., Wu, N., et al. (2018) *Energy Environ. Sci.*, **11**(2), pp. 394–406.
62. Bailie, C. D., Christoforo, M. G., Mailoa, J. P., Bowring, A. R., et al. (2015) *Energy Environ. Sci.*, **8**, pp. 956–963.
63. Kranz, L., Abate, A., Feurer, T., Fu, F., et al. (2015) *J. Phys. Chem. Lett.*, **6**, pp. 2676–2681.
64. Fu, F., Feurer, T., Jager, T., Avancini, E., et al. (2015) *Nat. Commun.*, **6**, pp. 8932.
65. Todorov, T., Gershon, T., Gunawan, O., Lee, Y. S., et al. (2015) *Adv. Energy Mater.*, **5**(23), pp. 500799.
66. Uhl, A. R., Rajagopal, A., Clark, J. A., Murray, A., et al. (2018) *Adv. Energy Mater.*, **8**(27), pp. 1801254.
67. Han, Q., Hsieh, Y. T., Meng, L., Wu, J. L., et al. (2018) *Science*, **361**(6405), pp. 904–908.
68. Li, Z., Boix, P. P., Xing, G., Fu, K., et al. (2016) *Nanoscale*, **8,** pp. 6352–6360.
69. Yang, Z., Rajagopal, A., Chueh, C.-C., Jo, S. B., et al. (2016) *Adv. Mater.*, **28,** pp. 1602696.
70. Eperon, G. E., Leijtens, T., Bush, K. A., Prasanna, R., et al. (2016) *Science*, **354,** pp. 861–865.
71. Zhao, D., Yu, Y., Wang, C., Liao, W., et al. (2017) *Nat. Energy*, **2,** pp. 17018.
72. Zhao, D., Wang, C., Song, Z., Yu, Y., et al. (2018) *ACS Energy Lett.*, **3,** pp. 305–306.
73. Leijtens, T., Prasanna, R., Bush, K. A., Eperon, G. E., et al. (2018) *Sustain. Energy Fuels*, **2**, pp. 2450–2459.
74. Heo, J. H. and Im, S. H. (2016) *Adv. Mater.*, **28**, pp. 5121–5125.
75. Eperon, G. E., Leijtens, T., Bush, K. A., Prasanna, R., et al. (2016) *Science*, **354**, pp. 861–865.

76. Ke, W. and Kanatzidis, M. G. (2019) *Nat. Commun.*, **10**(1), pp. 1–4.

77. Ke, W., Stoumpos, C. C., and Kanatzidis, M. G. (2019) *Adv. Mater.*, **31**, pp. 1803230.

78. Noel, N. K., Stranks, S. D., Abate, A., Wehrenfennig, C., et al. (2014) *Energy Environ. Sci.*, **7**, pp. 3061.

79. Wang, T., Tai, Q., Guo, X., Cao, J., et al. (2020) *ACS Energy Lett.*, **5**, pp. 1741.

80. Nishimura, K., Kamarudin, M. A., Hirotani, D., Hamada, K., et al. (2020) *Nano Energy*, **74**, pp. 104858.

81. Ke, W., Stoumpos, C. C., Zhu, M., Mao, L., et al. (2017) *Sci. Adv.*, **3**, pp. e1701293.

82. Ryu, S., Seo, J., Shin, S. S., Kim, Y. C., et al. (2015). *J. Mater. Chem. A*, **3**, pp. 3271.

83. Wang, F., Jiang, X., Chen, H., Shang, Y., et al. (2018) *Joule*, **2**, pp. 2732.

84. Jiang, X., Wang, F., Wei, Q., Li, H., et al. (2020) *Nat. Commun.*, **11**, pp. 1245.

85. Zheng, D., Wang, G., Huang, W., Wang, B., et al. (2019) *Adv. Funct. Mater.*, **29**, pp. 1900265.

86. Jiang, X., Wang, F., Wei, Q., Li, H., et al. (2020) *Nat. Commun.*, **11**, pp. 1245.

87. Ng, C. H., Nishimura, K., Ito, N., Hamada, K., et al. (2019) *Nano Energy*, **58**, pp. 130–137.

88. Pecunia, V., Occhipinti, L. G., Chakraborty, A., Pan, Y., and Peng, Y. (2020) *APL Mater.*, 8(10), pp. 100901.

89. Igbari, F., Wang, Z. K., and Liao, L. S. (2019) *Adv. Energy Mater.*, **9**, pp. 1803150.

90. Yang, X., Chen, Y., Liu, P., Xiang, H., et al. (2020) *Adv. Funct. Mater.*, **30**, pp. 2001557.

91. Lee, B., Krenselewski, A., Baik, S. I., Seidman, D. N., and Chang, R. P. H. (2017) *Sustain. Energy Fuels*, **1**, pp. 710.

92. Best-research-cell-efficiencies. NREL, https://www.nrel.gov/pv/cell-efficiency.html.

93. Pecunia, V., Occhipinti, L. G., Chakraborty, A., Pan, Y., and Peng, Y. (2020) *APL Mater.*, **8**, pp. 100901.

Problems

11.1 What is a tandem solar cell?

11.2 Tandem cells are fabricated in two configurations as mechanically integrated four-terminal (4T) and monolithically integrated two-terminal (2T). Briefly introduce these structures.

11.3 What are wide-bandgap and low-bandgap perovskites and what is their application?

11.4 Pb-based perovskite materials have a great potential for high-performance single-junction photovoltaic devices. However, there is a main concern to replace them with lead-free perovskites. Explain why Pb-free perovskites are considered as the most promising candidates for solar cell applications.

11.5 Sn-based perovskite suffers from self-oxidation of Sn(II) to Sn(IV) during operation, which suppresses the stability of the device. What are the ways to promote the stability of Sn PSCs?

Chapter 12

Composites-Based Efficient and Stable Perovskite Solar Cells with Interface Engineering

A composite material is a combination of two or more materials having different chemical and physical properties. When other materials are combined, they produce a composite with different properties from individual materials. Compared to traditional materials, composites can improve the properties of base materials and can be applied in many situations. Composite materials have advantages such as design flexibility, specialized chemical, and physical properties, resistance to a wide range of chemicals, etc. Therefore, they may give benefit to solve critical issues related to the efficiency and operational stability of perovskite solar cells.

Factors detrimental to both efficiency and long-term stability of perovskite solar cells (PSCs) are the ionic nature and polycrystalline properties of organic–inorganic hybrid perovskites (OIHPs) [1], interface defects [2, 3], ions migration [4, 5], and thermal- and photo-instability [6, 7]. As reported, the bulk and interface trap density of polycrystalline perovskite film is much higher than that of single-crystal perovskite film. It is also known that interface trap density is higher than the bulk defect density in inverted (p-i-n) PSCs [8].

Next-Generation Solar Cells: Principles and Materials
Yoon-Bong Hahn, Tahmineh Mahmoudi, and Yousheng Wang

ISBN 978-981-4968-66-9 (Hardcover), 978-1-003-37238-7 (eBook)
www.jennystanford.com

In addition, the under-coordinated sites (such as lead and halide ions), lead clusters, and vacancy defects are the preponderant trap sources on the surfaces and interfaces (such as grain boundaries) of polycrystalline perovskite films, resulting in non-radiative recombination losses [9, 10].

The photo-induced holes and electrons have to transport across the interfaces in a device, resulting in problems of charge loss, charge recombination, and interfacial degradation due to possible interfacial defects. Thus, the interface engineering and surface-passivation processes are essential for reducing interface trap densities in polycrystalline perovskite thin films.

In this chapter, a perovskite-based composite that combines perovskite with one or more materials such as metal, metal oxide, and graphene will be discussed. In addition to composite engineering, interface engineering will be introduced to solve interfacial instability attributed to the interface contacts between perovskites and charge-transport materials.

12.1 Organic Materials–Based Perovskite Composites

12.1.1 Small-Molecule-Based Perovskite Composites

To restrain ion migration, the development of perovskite composites with fullerene derivatives has been proposed [11–13]. For example, $MAPbI_{3-x}Cl_x$-PCBM composites-based PSC showed highly efficient and thermally stable performance [11]. Thermal stability is attributed to the passivation of mobile halides by PCBM and strong chemical interaction between PCBM and $MAPbI_{3-x}Cl_x$ molecules. It is also noted that fullerene derivatives have moisture resistance. The perovskite composites with PCBB-OEG ([6,6]-phenyl-C61-butyric acid-(3,4,5-tris(2-(2-(2-methoxyethoxy)ethoxy)ethoxy)phenyl) methanol ester) fullerene derivative-based devices showed better performance and stronger resistance to moisture [13]. The improved moisture stability is attributed to the hydrophobic properties of PCBB-OEG molecules. Besides, a strong chemical interaction between PCBB-OEG and $MAPbI_3$ by a hydrogen bond between the –O of the OEG group and the $–NH_3$ of $MAPbI_3$ also enhances the stability of perovskite crystal lattice.

To suppress defects and ion migration at the grain boundaries (GBs), small-molecule organic materials have been incorporated into the GBs of perovskite films to form bulk heterojunction perovskite composites [14, 15]. For instance, Chen et al. reported F4TCNQ-$MAPbI_3$ composites-based solar cells [F4TCNQ: (2,3,5,6-tetra-fluoro-2,5-cyclohexadiene-1,4-diylidene)dimalononitrile], which possessed better air stability than the pristine $MAPbI_3$-based devices because the F4TCNQ molecules were filled into GBs to form bulk heterojunction films of F4TCNQ-$MAPbI_3$. The bulk heterojunction film effectively improves charge-transfer efficiency, decreases trap-assisted non-radiative recombination losses, and hinders moisture from infiltrating the perovskite layer. Another interesting work based on $MAPbI_3$-F4TCNQ composites used for the fabrication of stable HCF-PSCs with a high 20.2% efficiency was reported by Huang et al. [15]. The p-type F4TCNQ can be concentrated in the GB regions to improve electrical conductivity, perovskite morphology, and charge dissociation as well as to suppress trap density and ion migration in the composite film. Moreover, the F4TCNQ can further improve the interfacial contact of ITO/perovskite and particularly turn the $MAPbI_3$-F4TCNQ composites into more p-type properties.

12.1.2 Polymer-Based Perovskite Composites

To resist degradation of perovskite materials after exposure to moisture, Wu et al. developed self-assembled $MAPbI_3$-PMMA composite single crystals which showed well-ordered lattice structure and fewer internal defects, where PMMA denotes polymethyl methacrylate) [16]. The $MAPbI_3$-PMMA composite-based PSCs exhibited robust air stability under high humidity. The PMMA at GBs effectively passivate iodide vacancy defects and reduce trap densities. Moreover, the hydrophobic PMMA polymer chain prevents water molecules from contacting perovskite grains through grafting the PMMA on the surfaces of perovskite crystals and GBs. The presence of strong chemical interaction and bonding between PMMA and $MAPbI_3$ molecules stabilize perovskite phases.

Another interesting example is the perovskite composites with polyethylene glycol (PEG) (i.e., $MAPbI_3$-PEG) [17]. From Fig. 12.1a, it is observed that the PEG has a self-healing function for the structural recovery of $MAPbI_3$ after degradation. The pure $MAPbI_3$ film without

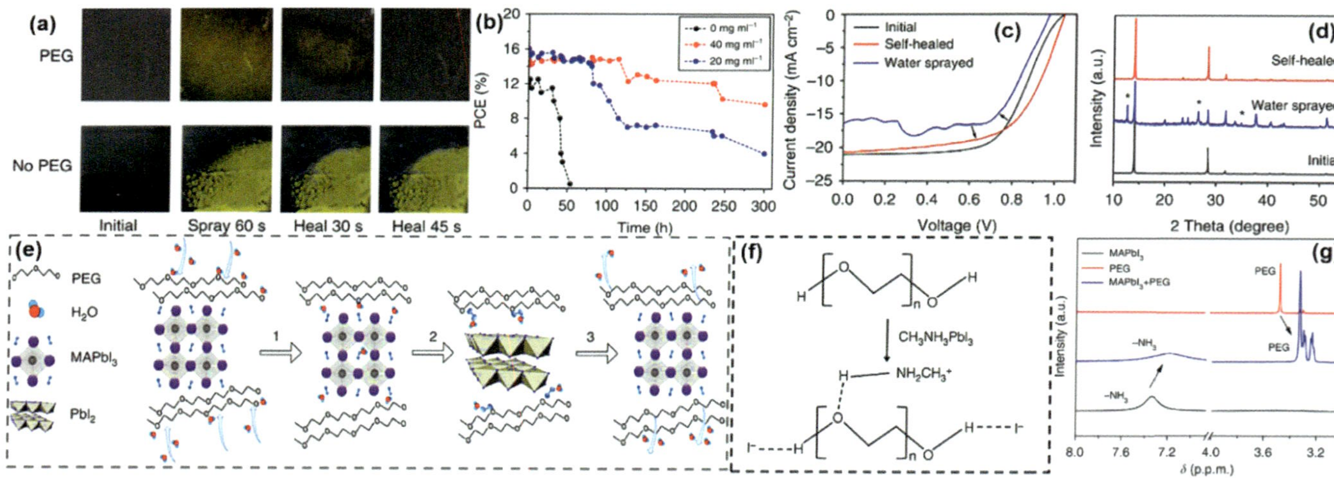

Figure 12.1 (a) Photographs of moisture self-healing of perovskite films with and without PEG. (b) Moisture stability of PSCs with 20 and 40 mg ml^{-1} PEG and without PEG scaffold exposed in a high humid (70% relative humidity) dark environment. (c) *J*–*V* curves and (d) XRD evolution of PEG-$MAPbI_3$ based PSCs before and after water spray. (e) Schematic diagram to show mechanisms for the self-healing properties in PSCs. (f) Illustration of hydrogen bonding formation between PEG molecules and $MAPbI_3$ and (g) NMR spectra of $MAPbI_3$, PEG, and $MAPbI_3$-PEG composites. Reproduced with permission from [17], Copyright 2016, Springer Nature.

PEG is degraded to yellow PbI_2. By contrast, the $MAPbI_3$-PEG composite film is initially yellow and then restored to black $MAPbI_3$ after 45 s. It was found that the moisture-resistant ability of the device is substantially affected by the amount of PEG (Fig. 12.1b). More interestingly, the $MAPbI_3$–PEG composites-based devices treated with water almost retain their primitive *J–V* curve behavior after self-healing as shown in Fig. 12.1c. Such results are in good agreement with XRD analysis. That is, from the absence of PbI_2 peak (original film) to the formation of PbI_2 peak (water sprayed film), and then to the elimination of PbI_2 peak (self-healed film) for the $MAPbI_3$–PEG composite film (Fig. 12.1d). From Fig. 12.1(d–g), it can be identified that the moisture barrier of PEG and PCBM molecules is located on the surfaces of $MAPbI_3$ grains through hydrogen bonding by MA with PCBM and PEG molecules. PEG molecules can effectively absorb water, protecting the crystal grains of $MAPbI_3$ from moisture penetration into the perovskite film. Hence, the PbI_2 in the film can react with MAI and rebuild the perovskite $MAPbI_3$ phase after water absorption by PEG.

12.1.3 Ammonium-Based Perovskite Composites

To chemically stabilize perovskite structures with crosslink neighboring molecules, a cross-linking agent is needed. Li et al. developed perovskite-based composites by adding small-molecular bifunctional alkyl phosphonic acid ω-ammonium chloride additives (4-ABPACl) into $MAPbI_3$ precursor [18]. The 4-ABPA not only acts as a crosslinker that crosslinks the adjacent perovskite grains and improves their cohesion and morphology but also effectively enhances both light-harvesting and charge transport in the as-formed composite film. The NMR analysis exhibits a strong chemical interaction and cross-linking between 4-ABPACl and $MAPbI_3$ by N-H...I and O-H...I hydrogen bonding interactions. Thus, the composite-based device shows robust air stability over 1000 h, thermal stability over 300 h, and photo-stability over 144 h. Similarly, some other interesting crosslinker-based perovskite composites, such as alkylamine ligands (AALs) [19], 1,3,7-trimethylxanthine (caffeine) [20], 2-aminoterephthalic acid [21], thioctic acid [22] and theophylline [23], were also developed and successfully implemented in the fabrication of highly efficient and stable PSCs.

It is also noteworthy that the presence of hydrophobic and hydrophilic groups in ammonium-based materials has successfully developed perovskite-based composites with these materials, increasing moisture stability. Bi et al. reported air-stable and highly effective PSCs by adding an aliphatic fluorinated amphiphilic additive such as 1,1,1-trifluoroethyl ammonium iodide (FEAI) to $MAPbI_3$ [24]. The enhanced moisture stability of the $MAPbI_3$-FEAI-based cell is mainly ascribed to both hydrophobic and hydrophilic sides in FEAI groups, which cover the surfaces of $MAPbI_3$ grains and effectively absorb moisture, thus avoiding the direct chemical reaction between $MAPbI_3$ and water molecules.

12.1.4 Low-Dimensional/Three-Dimensional Perovskite Composites

Low-dimensional perovskites have better stability against humidity, light exposure, and thermal stress than three-dimensional (3D) counterparts because of their hydrophobic properties generated in aliphatic- or aromatic-based organic spacers. The pioneering work with 2D perovskite-based PSCs obtained a lower efficiency of 4.73% [25]. Afterward, higher efficiency of up to 12.52% was reported based on a novel approach to producing thin films of near-single-crystalline quality having Ruddlesden–Popper phases (i.e., layered 2D perovskite films), which enhances charge transport [26, 27].

To further increase the efficiency of 2D perovskite-based solar cells, a feasible solution has been sought to build a 2D-3D perovskite complex by combining 2D perovskite with 3D perovskite. This strategy can effectively stabilize the perovskite phase and protect 3D perovskite from moisture by 2D perovskites. Mei et al. first reported 2D/3D perovskite composites fabricated through partial substitution of MA cations with 5-AVA cations (5-AVA: 5-ammoniumvaleric acid) to form $(5\text{-AVA})_x(MA)_{1-x}PbI_3$ composites [28]. These 2D/3D perovskite-based composites have lower defect concentration and higher exciton lifetime than pristine 3D $MAPbI_3$. More remarkably, the 2D/3D-based hole-conductor free (HCF) PSCs showed excellent stability. Furthermore, stable large-area (10×10 cm^2) and printable HCF-PSC modules based on 2D/3D composites were reported with PCE of 10.4% on an active area of 49 cm^2 [29]. Grancini et al. also used the same composition of 2D/3D $(5\text{-AVA})_2PbI_4/CH_3NH_3PbI_3$ by interface engineering strategy and obtained an efficient and stable large-area (10 ×10 cm^2) HCF-PSC module with 11.2% efficiency and improved stability over one year [30].

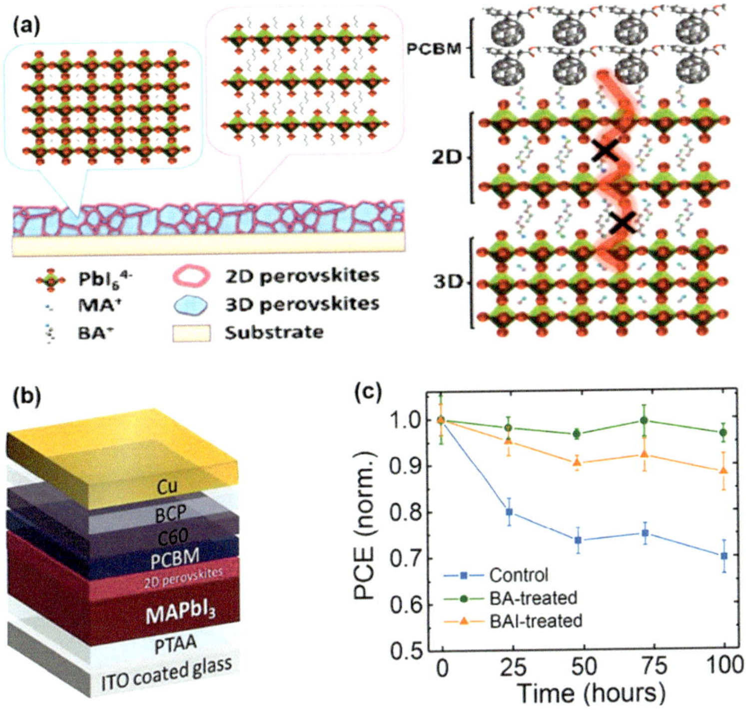

Figure 12.2 (a) Schematic figure of the perovskite film treated by BA to form a 2D/3D stacking structure. And proposed degradation mechanism in 2D/3D stacking device. (b) Device structure based on 2D/3D perovskite. (c) Thermal stability of three different devices. Reproduced with permission from [30], Copyright 2018, American Chemical Society.

The introduction of a thin 2D perovskite passivation layer on the top of the 3D perovskite layer to form 2D/3D heterostructures becomes another intriguing strategy to further enhance efficiency and stability. Wong et al. achieved thermal- and photo-stable PSCs by using 2D/3D perovskite composites [31]. From the formation of 2D BA_2PbI_4 (BA: n-butylammonium) on the top surface and at GBs of 3D perovskite films, it can be identified that the 2D BA_2PbI_4 perovskite effectively suppresses the defects and phase segregation of 3D perovskites. The BA-modified perovskite solar cells exhibit stability improvement, excellent photovoltaic performance, with PCE of 18.1% and an open-circuit voltage (V_{oc}) of 1.24 V. Subsequently,

Huang et al. reported thermally stable inverted PSCs based on the assembled 2D/3D stacking structures [32]. It was concluded that the 2D BA_2PbI_4 perovskites can effectively occupy the GBs of 3D $MAPbI_3$ grains [Fig. 12.2(a–c)] to protect 3D perovskite crystals from environmental moisture.

Interestingly, 1D/3D hybrid perovskite heterojunctions have been successfully employed in the fabrication of PSCs [33]. The 1D/3D perovskite-based devices exhibited a high *PCE* of 18.1% with excellent air- and thermal stability under an ambient environment of 50% humidity and heat treatment at 85°C. Furthermore, compared to the rapid degradation of pristine 3D perovskite-based devices, the 1D/3D perovskite-based cells showed thermodynamic self-healing capability. It can be noted that the 1D/3D halide perovskite exhibits a layer-by-layer arrangement (3D-1D-3D) along their preferred orientation giving rise to improved charge transport and reduced recombination [34]. Furthermore, the 1D/3D perovskite composite films dramatically passivate interface traps, suppress ions migration in the heterojunction region, and protect the 3D perovskite crystal grains from moisture.

12.2 Inorganic Material–Based Perovskite Composites

12.2.1 Metal Oxide–Based Perovskite Composites

Perovskite composites with inorganic materials have been widely explored to improve device stability. Hahn group first developed perovskite composites with p-type NiO nanoparticles (NPs) [35–37], which is a pioneering work to boost the development of composites-based stable and efficient PSCs that include photoactive functional composites made of perovskite molecules, metal oxide nanoparticles, metal quantum dots, and graphene and its derivatives. The perovskite-NiO composites have been successfully applied to hole-conductor free (HCF) and novel mesoscopic PSCs (Fig. 12.3). In the HCF device architecture, multiple heterojunctions are formed in the active layer and at the interface (i.e., $MAPbI_3$/NiO NPs, $MAPbI_3$/mp-TiO_2, and NiO NPs/mp-TiO_2) which could effectively inhibit recombination and enhance the charge-carrier conduction. P-type NiO has a wide bandgap of ~3.6 eV with CB of −1.8 eV and VB of

–5.3 eV, and its VB energy level is quite close to that of $MAPbI_3$ (right, Fig. 12.3). Therefore, the composite-based device has two conduction paths for electrons: direct transport to the c-TiO_2 and transport to mp-TiO_2 to c-TiO_2. Moreover, it also has two ways for hole conduction: direct transport to the Au electrode and transfer to the electrode through NiO NPs. Thus, it is believed that the NiO NPs in the composite act as multiple charge selective contacts in the active layer and provide efficient pathways of hole transport to the external circuit.

As-prepared $MAPbI_3$-NiO NPs composites can lead to 0.3 eV push-up of the valence band maximum of $MAPbI_3$, resulting in a lower bandgap relative to $MAPbI_3$ as indicated in Fig. 12.3 (right). Therefore, the $MAPbI_3$-NiO NPs composite-based HCF-PSC showed a high PCE of 12.14%, which is mainly due to a remarkable increase in photocurrent density, suggesting that the $MAPbI_3$-NiO composite enhances the charge-carrier generation via improving photo-absorption and fast hole conduction by NiO NPs.

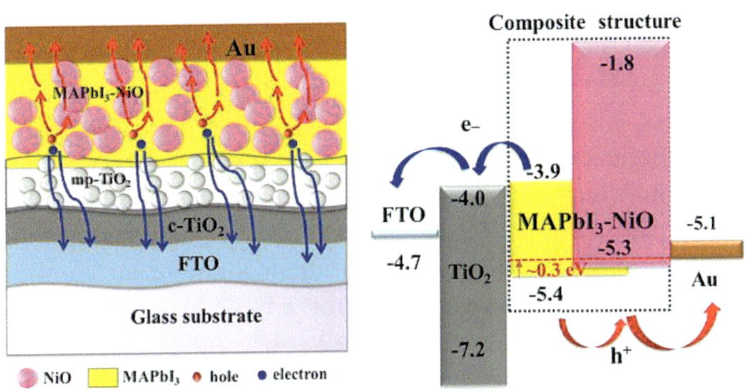

Figure 12.3 Schematic illustration (left) of band diagram (right) of a hole-conductor-free $MAPbI_3$–NiO composite-based PSC. Reprinted with permission from Ref. [35].

More importantly, the perovskite-NiO composites have strong chemical interaction and bonding between NiO and $MAPbI_3$ or $MAPbI_{3-x}Cl_x$ molecules, contributing to robust air stability. The composite-based HCF-PSCs without encapsulation showed remarkable air stability over 60 days under an ambient environment [35]. The formation of chemical bonds effectively stabilizes the structures of perovskite and thus suppresses ions migration of I^- and MA^+. Thereby the device with interface engineering of NiO thin layer had less hysteretic *J*–*V* behavior [36, 37]. Nie et al. and Li et al.

also found the chemical interactions at interfaces between NiO and perovskite-based on DFT calculations [38, 39]. Thus, the NiO layer as an HTL in p-i-n configuration-based PSCs has boosted a photo-stability [38, 40].

To further improve efficiency and stability, Hahn group developed $FA_xMA_{1-x}PbI_{3-y}Cl_y$-Cu:NiO composites and utilized them for the fabrication of highly stable and efficient PSCs through fully-ambient-air and anti-solvent-free processes [41]. The champion device based on the $FA_xMA_{1-x}PbI_{3-y}Cl_y$-Cu:NiO composite active layer yielded substantial enhancement of photovoltaic parameters and long-term stability, as shown in Fig. 12.4(a–c). The $FA_xMA_{1-x}PbI_{3-y}Cl_y$-Cu:NiO composite film shows highly crystallized dense morphology with large grains over 3 μm, fewer GBs, and a smoother surface without observable pinholes (Fig. 12.5d). The small quantity of Cu:NiO nanoparticles (size 10–15 nm) plays a significant role in perovskite mixed-solution to assist the formation of homogeneous perovskite nucleation and more orientated crystalline domains during crystallization, resulting in highly stable perovskite crystals attributed to the strong chemical interactions and bonding or cross-linking between Cu:NiO and perovskite molecules.

Besides, except for the presence of cross-linking and chemical bonding between Cu:NiO and perovskite molecules (Fig. 12.5h), a little amount of δ-$FAPbI_3$ only at GBs does not significantly affect solar cell performance [Fig. 12.5(e–g)], but it can effectively protect perovskite grains from degradation under ambient air conditions. Thus, these chemical bonding and/or cross-linking and δ-$FAPbI_3$ effectively suppress the soft nature of I^-, Pb^{2+}, and MA^+ or FA^+ ions in the perovskite phase with oxidizing to I^0 and reducing to metallic Pb^0 and MA (g) or FA (g) under device operation conditions, respectively. Due to the excellent properties of NiO in perovskite composites, Yun et al. also fabricated typical mesoporous PSCs and carbon-based HCF-PSCs based on $MAPbI_3$-NiO nanotubes (NTs) composite, resulting in a PCE of 19.3% and 12.82%, respectively [42]. This enhanced performance with $MAPbI_3$-NiO is accredited to improved perovskite grain size and the presence of NiO NTs at GBs, which effectively passivate the positive ion trap states and reduce charge recombination. Interestingly, Hsu group reported the presence of a chemical redox reaction at the interface between NiO/$MAPbI_3$ and the formation of hole-doping $MAPbI_{3-2\delta}O_\delta$ material, which facilitates hole transport and modifies the energy-level diagram, leading to the enhancement of photovoltaic performance [43].

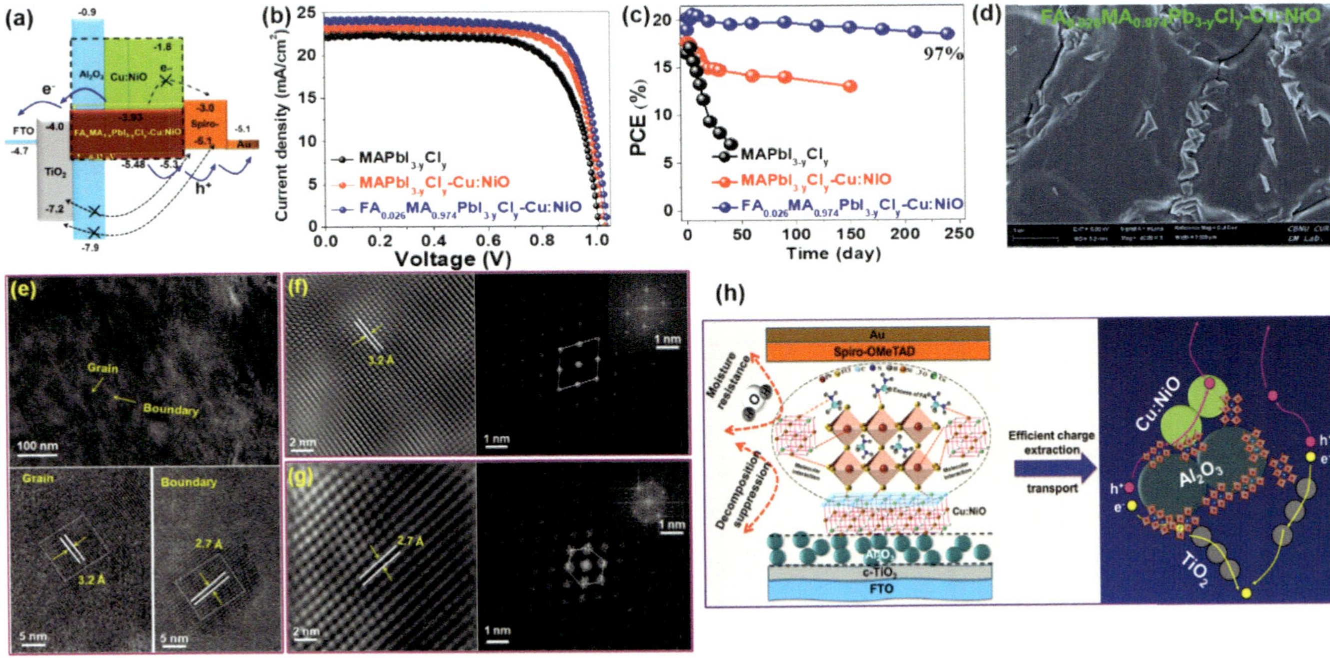

Figure 12.4 (a) Energy band diagram of $FA_xMA_{1-x}PbI_{3-y}Cl_y$-Cu: NiO-based devices and carriers blocking properties and working path. (b) *J–V* curves and (c) long-term stability of pristine $MAPbI_{3-y}Cl_y$, $MAPbI_{3-y}Cl_y$-Cu:NiO, and $FA_xMA_{1-x}PbI_{3-y}Cl_y$-Cu:NiO-based devices. (d) FE-SEM image of $FA_xMA_{1-x}PbI_{3-y}Cl_y$-Cu:NiO composite film. (e) HRTEM images of grain and boundary areas, (f) and (g) inverse FFT images of the selected regions of grain and boundary, respectively, and their corresponding SAED patterns (inset shows FFT phase images) from $FA_xMA_{1-x}PbI_{3-y}Cl_y$-Cu:NiO composite films. (h) Schematic illustration of chemical interaction, cross-linking and chemical bonding between Cu:NiO and perovskite molecules (Ni-I, Pb-O-Pb, Ni-N) (left), and illustration of highly efficient charge carriers extraction and transport behavior (right). Panels (a–h) reproduced with permission from [41], Copyright 2020, WILEY-VCH.

12.2.2 Carbon-Based Perovskite Composites

Perovskite composites with carbon-based materials such as graphene and its derivatives are attracting great attention for their low cost, eco-friendly, high efficiency, and stable PSC. The carbon-based derivatives have excellent electronic properties such as high conductivity, thermal stability, carrier mobility, and zero-bandgap. In particular, the graphene-based derivatives, such as graphene oxide (GO) and reduced graphene oxide (rGO), act as efficient charge-transport materials through controlled and modified surface chemistry. The graphene-based derivatives also have a hydrophilic property and its packed 2D network structure effectively facilitates device stability [44]. Chung et al. achieved effective and stable inverted PSCs by employing perovskite composites with GO (i.e., $MAPbI_3$-GO) [45]. The $MAPbI_3$-GO composites can balance electron–hole transport properties, which improve charge separation and reduce charge recombination. Furthermore, the $MAPbI_3$-GO hybrid composites-based device is more stable than pristine $MAPbI_3$-based devices. Yavari et al. developed perovskite composites with carbon nanoparticles (CNPs), which were successfully employed in PSCs [46]. The CNPs in composite films not only assist the growth of perovskite grain size but also tremendously improve thermal stability. Moreover, the perovskite-based composite film has more hydrophobic properties than the pristine perovskite film. In this case, the enhanced stability is mostly ascribed to strong chemical interaction between CNPs and perovskite molecules because of the presence of hydrogen bonds between FA fragments within perovskite and functional groups of CNPs.

As a noteworthy work, Hahn et al. developed perovskite/Ag-rGO composites as an active layer in PSCs, which not only suppress the migration of iodide ions but also enhance thermal- and photo-stability [47]. The improved stability is mainly attributed to two factors: (1) the graphene (i.e., rGO) has moisture and oxygen resistance; (2) thanks to a lattice constant (0.246 nm) smaller than the diameter (0.412 nm) of the iodide, the graphene sheet effectively suppresses the migration of iodide ions toward top electrodes under light or heat treatment (Fig. 12.5a). From the XPS spectra, it was observed that the iodide concentration on the Au electrode in pristine perovskite devices significantly increased with thermal

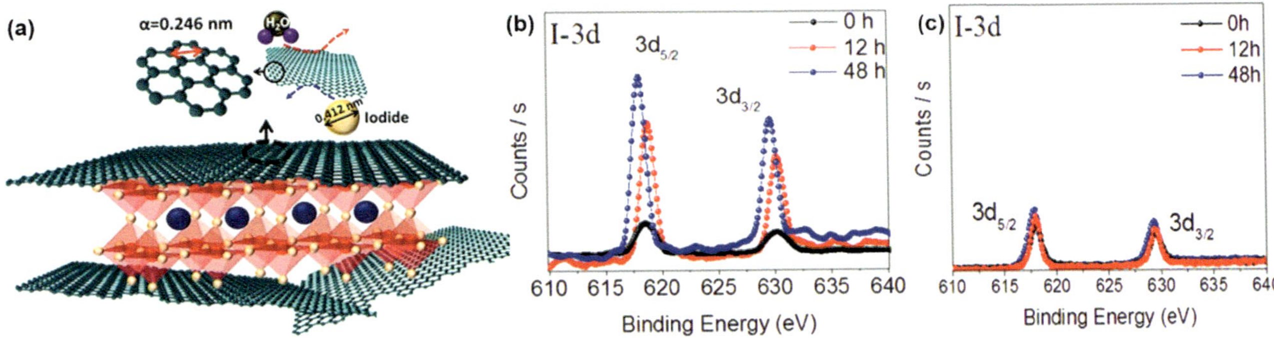

Figure 12.5 (a) The mechanism of suppression of I ions and H_2O molecules by rGO, (b) and (c) XPS spectra performance for pristine perovskite and perovskite/Ag-rGO after thermal treatment. Reproduced with permission from [46], Copyright 2019, American Chemical Society.

aging time and peaks shifted to lower binding energy (Fig. 12.5b), indicating the migration of more iodide ions to the upper electrode. By contrast, for the perovskite/Ag-rGO composite-based solar cells, the intensity of iodide peaks remained nearly unchanged, indicating that the ions migration is much suppressed in the active layer during thermal heating (Fig. 12.5c) [47].

12.2.3 Semitransparent PSCs with Metal Oxide–Based Composites

Semitransparent PSCs attract tremendous interest in developing tandem cells and building-integrated PVs (BIPVs). Generally, semitransparent solar cells are quantified in terms of average visible transmittance (AVT) versus PCE. Semitransparent PSCs are mainly dependent on film thickness and lower coverage with non-continuous islands-perovskite film. In 2014, Snaith et al. group pioneered neutral-color planar semitransparent PSCs with PCE of 6% at AVT of 28% by controlling the perovskite fractional surface to develop perovskite islands, which showed a microstructured array enough thick perovskite films to absorb the majority of light but the surface coverage on the compact TiO_2 layer was low to provide a maximum AVT [48].

In 2016, Wang et al. presented fully ambient-processed stable and mesoscopic semitransparent PSCs by non-continuous perovskite islands-structure-composites and interface engineering strategy [37]. As shown in Fig. 12.6 except for the islands-structure-$MAPbI_{3-x}Cl_x$-NiO capping layer, a uniform, thicker, and transparent $TiO_2/Al_2O_3/NiO/MAPbI_{3-x}Cl_x$ composite layer is formed, which can effectively reduce photocurrent density loss and suppress interface recombination. Two interfacial layers of NiO and Al_2O_3 eliminated the possibilities of direct contact between p-type Spiro-OMeTAD and n-type TiO_2, suggesting low recombination and high shunt resistance. Finally, the semitransparent device achieved a PCE of 10.06%, corresponding to AVT = 27%. In addition to mesoporous semitransparent PSCs, thickness-controlled planar semitransparent PSCs with fully organic charge-transport layers were also reported, for example, ITO/PEDOT:PSS/polyTPD/thermal evaporating $MAPbI_3$/PCBM60/Au (6 nm)/LiF (PCE of 7% with AVT of 33%) [49].

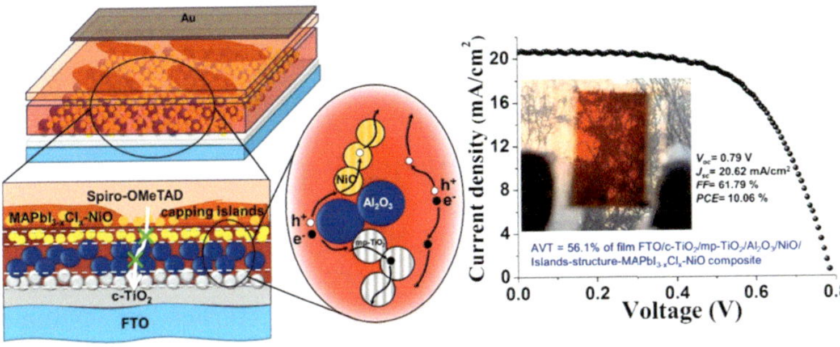

Figure 12.6 Conceptual illustration and performance of mesoscopic semitransparent PSCs by islands-structure-$MAPbI_{3-x}Cl_x$-NiO composite and Al_2O_3/NiO interface engineering. Reprinted with permission from reference [37].

12.2.4 Other Inorganic Halides–Based Perovskite Composites

Perovskite composites with a little amount of inorganic metal halide/thiocyanate materials have been studied and successfully employed in the fabrication of stable PSCs. More significantly, these perovskite composites-based devices (such as $MAPbI_{3-x}Cl_x$-Cu(Tu)I [50], $MAPbI_3$-$SrCl_2$ [51], $MAPbI_3$-$ZnCl_2$ [52], $(FAPbI_3)_{1-x}(MAPbBr_3)_x$-NaCl [53], $MAPbI_{3-x}Cl_x$-KCl [54], $MAPbI_3$-$CuBr_2$ [55], and AVAI/$MAPbI_3$-CuSCN [56]) exhibit enhanced *PCEs* and air stability. Interestingly, these metal halide materials not only achieve partial substitution of Pb with metal-based sources but also assist the growth of perovskite grains with fewer GBs and reduce trap density. For instance, Sr occupies Pb sites at crystal lattice to stabilize $MAPbI_3$ grains, which modifies the intrinsic electronic and optical properties of resultant perovskite composites. Owing to high-quality perovskite morphology with fewer defects, trap density passivation and stronger electrostatic attraction between Cl or Br of metal chloride/bromide and MA of perovskite, both efficiency and stability in PSCs have been improved.

12.3 Stability Enhancement with Interface Engineering

12.3.1 Why Is Interface Engineering Needed?

In addition to stabilizing and improving the structures of perovskite film itself through composite engineering, the interface contacts between perovskites and charge-transport materials are valuable for further studies toward commercialization, including interfacial energetics, charge transfer, recombination kinetics, hysteresis behaviors, and interfacial degradation. The photo-induced holes and electrons have to conduct across the interfaces in a device, resulting in occurrences of charge loss, charge recombination, and interfacial degradation. Particularly, the interfacial degradation between the perovskite layer and contact interface layers, attributed to interfacial defects, becomes a critical intrinsic factor inducing the poor stability of PSCs. The electron-transport layer (ETL)/perovskite and perovskite/HTL contacts induce an interfacial collapse of the perovskite structure. This is because perovskite atoms or groups at the interfaces can move disorderly under actual operation conditions. Therefore, interface engineering is considered very important in improving interface contact and passivation because interface engineering is useful in effectively reducing electron or hole-transmission barriers, non-radiative recombination, interface defects, and ions movement.

In this section, interface engineering strategies are systematically analyzed in determining device efficiency, stability, and *J*–*V* hysteresis of PSCs.

12.3.2 Interface Engineering at TCO/ETL Interface

The interface engineering strategy is to tailor energy band alignments toward high-performance PSCs [57]. Zhou et al. modified ITO electrode with polyethyleneimine ethoxylated (PEIE) to reduce the work function of ITO from 4.6 to 4.0 eV. Yttrium-doped TiO_2 (Y-TiO_2) was also used as the ETL to obtain minimum conduction band of 4.0 eV, which is similar to that of the PEIE-modified ITO (Fig. 12.7a). Thus, there is no electron-transport barrier, that is, the same conduction band of 4.0 eV in Y-TiO_2 and

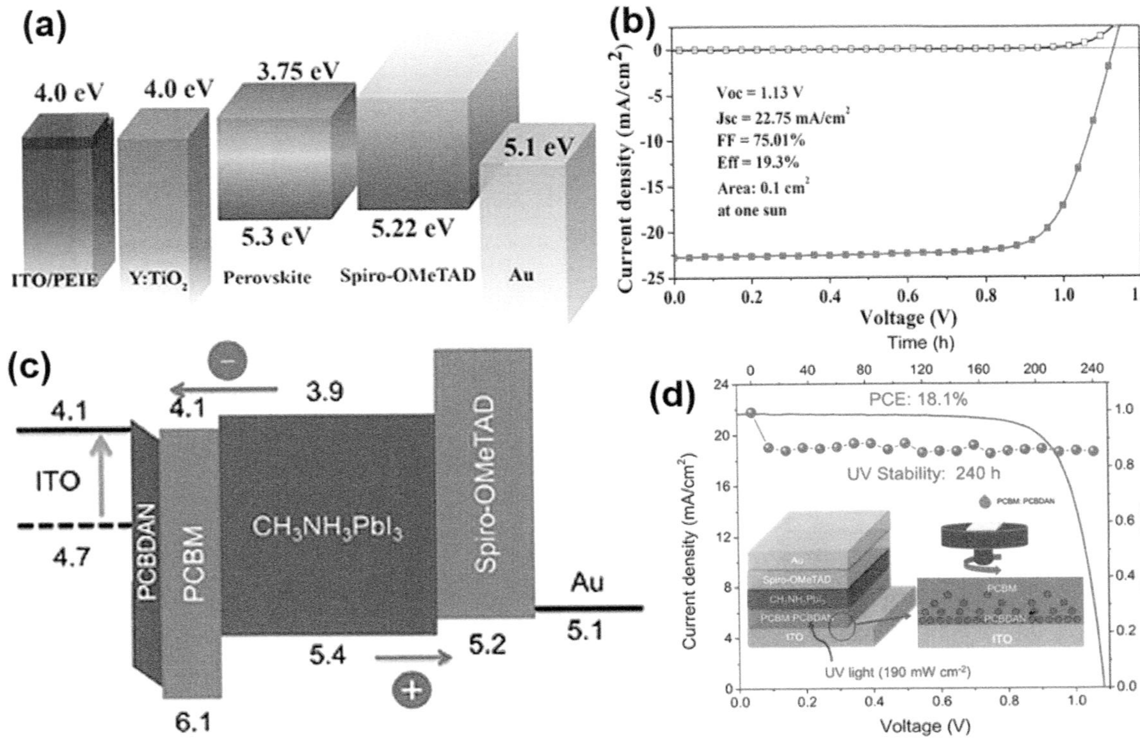

Figure 12.7 (a) Diagram of energy levels of device ITO/PEIE/Y-TiO_2/perovskite/Spiro-OMeTAD/Au. (b) *J-V* performance for interlayer PEIE-based champion device. Panels (a, b) reproduced with permission from [57], Copyright 2014, AAAS. (c) Energy-level diagram of the device with a self-organized PCBDAN interlayer. (d) The formation process of the PCBM:PCBDAN composites; device performance and UV stability. Panels (c, d) reproduced with permission from [58], Copyright 2017, WILEY-VCH.

PEIE-ITO can efficiently promote electron transport from ETL to the transparent conductive oxide (TCO). An improved V_{oc} of 1.13 V resulted in a high *PCE* of 19.3% (Fig. 12.7b). The optimized device showed considerable air stability that maintained 85% of the initial performance after 24 hours.

Xie et al. developed a fully organic ETL PCBM-PCBDAN composites-based PSCs, where PCBDAN is [6,6]-phenyl-C61-butyric acid 2-((2-(dimethylamino)-ethyl) (methyl)amino)ethyl ester [58]. They achieved a *PCE* of 18.1% with remarkable photo-stability over 240 h under continuous UV-light soaking (Fig. 12.7d). The PCBDAN through self-organization can tailor the work function of ITO, resulting in enhanced charge transportation and reduced interface recombination (Fig. 12.7c). Such self-organization of organic material with metal oxide-free PSCs may also be compatible in roll-to-roll and fully printing device processes on flexible substrates for future production of large-scale flexible PSCs.

12.3.3 Interface Engineering at ETL/AL Interface

In general, the ETL in a PSC consists of compact TiO_2 (c-TiO_2) and mesoporous TiO_2 (mp-TiO_2). However, the photocatalytic effect of TiO_2 is the main reason for the photo-instability of PSCs. To solve the TiO_2-induced instability, the insertion of an interfacial layer (IL) between the ETL and active layer (AL, i.e., perovskite) is useful. This interface engineering has considerable benefits: (1) improvement of photo-stability by the use of UV-stable material; (2) carriers transport barriers and reduction of interfacial defects. Based on prominent features including controllable thickness and excellent uniformity, small-molecule organic or polymer materials often serve as IL at ETL/AL interfaces to improve interface contact and promote the growth of perovskite crystal grains.

To modify the c-TiO_2 layer and facilitates charge extraction and transport, a triblock fullerene derivative (i.e., PCBB-2CN-2C8) can be used as an IL [Fig. 12.8(a,b)] [56]. The planar PSCs with the triblock fullerene derivative IL showed a better UV stability over 500 h than that of only c-TiO_2 (i.e., w/o interface layer) based devices (Fig. 12.8c). The other fullerene derivative IL such as the insertion of PCBDAN at the c-TiO_2/perovskite interface produced high-quality perovskite films and improved both electron extraction efficiency and photo-stability [59].

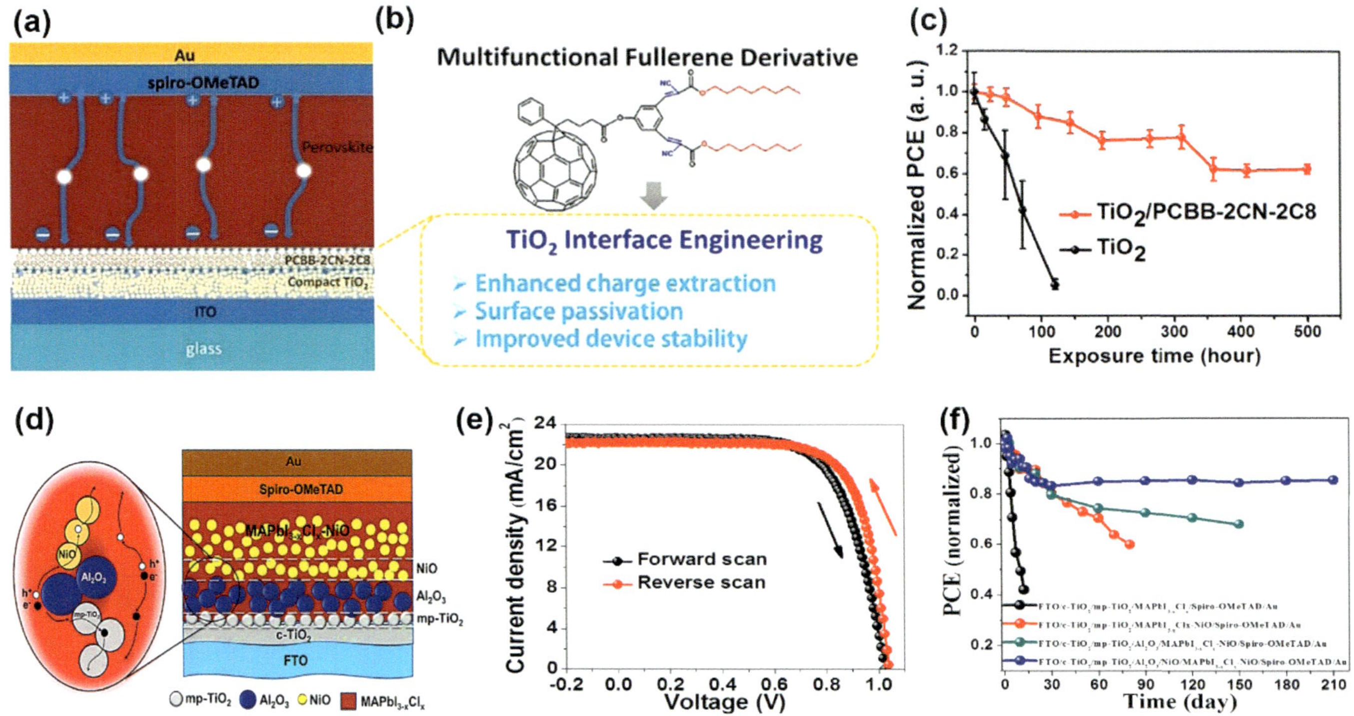

Figure 12.8 (a) and (b) Device configuration with multifunctional fullerene derivative interface engineering. (c) Air stability of with/without interface engineering-based devices. Panels (a–c) reproduced with permission from [59], Copyright 2015, American Chemical Society. (d) Device configuration by inserting Al_2O_3/NiO. (e) Low hysteretic behavior of *J–V* curve by interface engineering. (f) Air stability of PSCs by Al_2O_3/NiO interface engineering. Panels (d–f) reproduced with permission from [36], Copyright 2017, Elsevier.

Inorganic materials have also been introduced as IL in the ETL/AL interface due to their superior stability, low-cost, eco-friendly, and facile synthesized properties. For example, the insertion of CsBr between the c-TiO_2/perovskite interface reduced the chemical reactivity of TiO_2 and defects density under UV-light absorption, resulting in high-performance and optical stability planar PSCs [60]. Such a straightforward strategy is of great importance for solving photo-induced degradation at the TiO_2/perovskite interface. Other non-oxide-based inorganic materials (such as CuI [61], carbon/graphite [62], CdS [63], and KCl [64]) have also been used to enhance efficiency and long-term stability as well as to eliminate *J–V* hysteresis.

To avoid PSC degradation due to the mp-TiO_2 scaffold under UV light, more photo-stable inorganic materials can be introduced between c-TiO_2/perovskite layers. Examples are mesoporous SnO_2 film (mp-SnO_2) [65] and $SrAl_2O_4$: Eu^{2+}, Dy^{3+} (SAED) [66]. The mp-SnO_2 resulted in better photo-stability over 1200 h because of improved interfacial contact. Interestingly, the use of SAED showed substantial improvement in air stability over 2600 h and fast self-healing ability after illumination [66]. The excellent photo-stability and self-healing are attributed to characteristics of SAED material itself: (1) SAED exhibits a broad absorption in the UV region and emission in the green light region because of transition from 5d to 4f level of Eu^{2+}; (2) SAED has efficient photo down-conversion consequence from UV light into visible light; (3) SAED reduces performance recovery time due to trapped carriers which are induced by only visible light.

More interestingly, Al_2O_3 thin film can act as an insulator or protective layer through interface engineering at two different positions, such as ETL/Al_2O_3/perovskite and perovskite/Al_2O_3/HTL. Hahn et al. reported fully ambient-air-solution-processed PSCs with Al_2O_3/NiO interface engineering (Fig. 12.8d), which achieved a high *PCE* of 18.14% with low hysteresis *J–V* behavior (Fig. 12.8e) and excellent air stability over 210 days (Fig. 12.8f) [36]. They further fabricated fully-ambient-processed mesoscopic semitransparent PSCs by a morphology-controllable islands-structure-$MAPbI_{3-x}Cl_x$-NiO composite film with Al_2O_3/NiO interface engineering, resulting in better performance and robust air stability. This unconnected-islands-structure-$MAPbI_{3-x}Cl_x$-NiO composite can provide the desired transparency for the development of neutral-color semitransparent PSCs. The interfacial recombination can be reduced substantially with a thin Al_2O_3 scaffold interlayer. The Al_2O_3 IL forms a profitable barrier that separates two semiconductors between p-type NiO NPs and n-type mp-TiO_2, eliminating all probabilities of contacts between HTL and ETL as well as between perovskite and TCO. In addition, the thin interlayer of NiO has an important characteristic of electron blocking, which prevents electrons from moving to HTM and the top electrode. Undoubtedly, the interface engineering with metal oxides (such as SnO_2, Al_2O_3, $SrTiO_3$, and NiO) plays an important role in the performance and stability of PSCs.

12.3.4 Interface Engineering at AL/HTL Interface

Although it is quite difficult to insert an IL between AL and HTL by using solution-processed methods due to the incompatibility of perovskite materials with many polar solvents, there have been still several reports to overcome such barriers by careful selection of ideal materials and solvents. Zhang et al. developed a layer of C_{12}-silane with a self-assembled hydrophobic alkyl chain at the perovskite/HTL interface through dip-coating of the perovskite/TiO_2 film into C_{12}-silane isopropanol solution [67]. This C_{12}-silane layer with a self-assembled hydrophobic alkyl chain on the perovskite layer exhibits both passivation of electron recombination and high moisture resistance attributed to its insulating and hydrophobic properties.

Yang and co-workers reported highly efficient and stable planar PSCs by inserting polymers at the perovskite/HTL interface [Fig. 12.9(a–d)] [68]. The polymers not only can form a thin surface-passivation layer of perovskite but also commendably occupy the GBs of perovskite film. Such a result can effectively passivate recombination and hysteresis as well as improve the electronic properties of perovskite–polymer composites. The improved stability is ascribed to strong chemical interaction and cross-linking between polymers and perovskite [Fig. 12.9(e,f)] as well as the protective effect of polymer on the perovskite layer. Wang et al. also deposited a polymethyl methacrylate (PMMA) on the perovskite layer to fill pinholes or GBs of the perovskite surface, which not only eliminated shunt-current paths and reduced recombination but also protected the perovskite layer from oxygen and moisture [69].

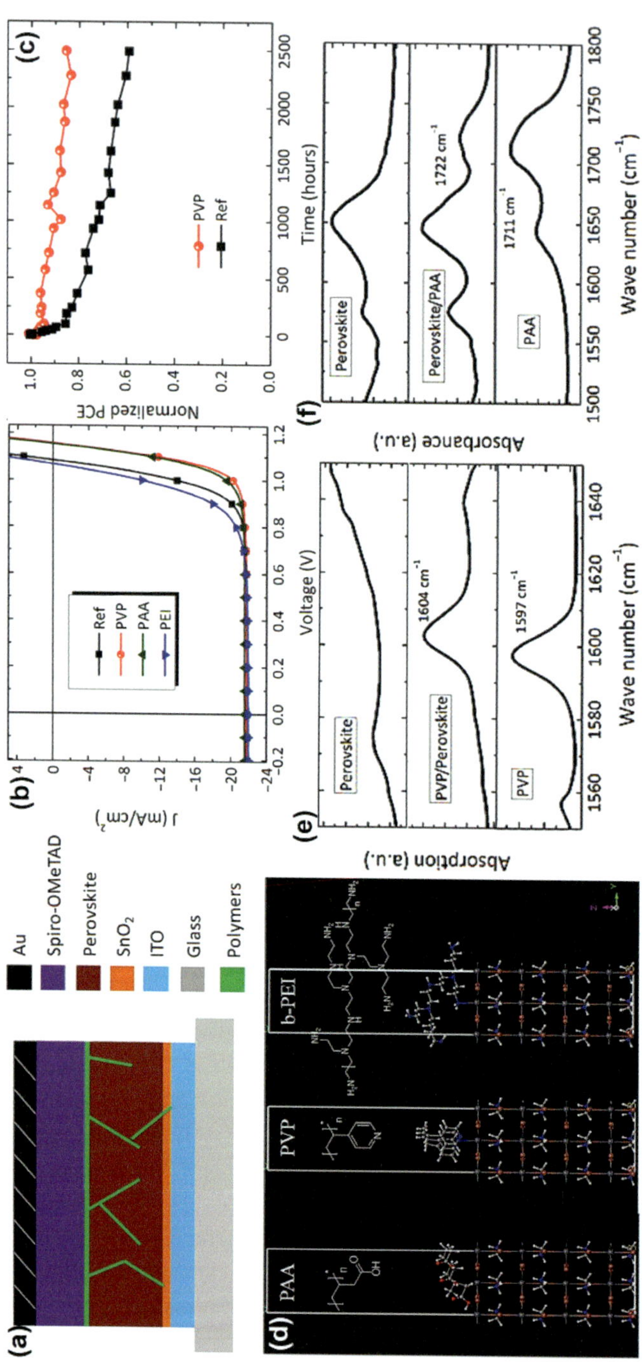

Figure 12.9 (a) Schematic diagram of the device architecture and (b) its performance with various polymers interface engineering. (c) Air stability of the solar cells with and without PVP treatment. (d) Images of polymer structures. (e) FTIR spectra of perovskite, PVP, and perovskite/PVP films. (f) FTIR spectra of perovskite, PAA, and perovskite/PAA films. Reprinted with permission from [68], Copyright 2017, AAAS.

Interestingly, Nazeeruddin et al. reported a highly efficient (*PCE* of 21.31%) and stable PSC with additional deposition of 3D structured $FAPbBr_{3-x}I_x$ layer on an active layer $(FAPbI_3)_{0.85}(MAPbBr_3)_{0.15}$ [70]. The $FAPbBr_{3-x}I_x$ layer has electron blocking properties with improved interface contact. Inspired by previous experience, they subsequently reported a strategy of 2D perovskite (PEA_2PbI_4) interface engineering (and/or surface passivation) to passivate 3D perovskite surface ($Cs_{0.1}FA_{0.74}MA_{0.13}PbI_{2.48}Br_{0.38}$), which yielded a *PCE* of 20.08% with improved photo-stability over 800 h under continuous illumination and 50 °C [71]. The insertion of small-molecule surfactants or polymers at the interface of perovskite/HTM is probably an effective strategy to passivate the surface of perovskites, to suppress ions migration and *J–V* hysteresis, to improve perovskite film morphology, and to protect perovskite crystalline grains from moisture and oxygen.

Snaith and co-workers reported operationally stable PSCs with the insertion of a solution-processed porous Al_2O_3 buffer layer between perovskite/Spiro-OMeTAD layers [72]. The porous Al_2O_3 layer effectively prevented moisture penetration and top metal electrode diffusion into the perovskite layer, resulting in improved FF and air stability. Afterward, Koushik et al. deposited an ultra-thin Al_2O_3 film using the atomic layer deposition (ALD) method at the perovskite/HTL interface and achieved high efficiency of 18% with enhanced air stability over 70 days [73]. The ultra-thin ALD Al_2O_3 film effectively protects the perovskite layer and avoids direct contact between Spiro-OMeTAD and TiO_2. Additionally, another interesting work was reported based on the distribution of ~7 nm inorganic perovskite ($CsPbBrCl_2$) quantum dots (QDs) which uniformly crossed a surface of $MAPbI_3$ perovskite film [74]. Such a strategy can effectively passivate defects at the perovskite/HTL interface and suppress interfacial recombination. The influences of quasi-Fermi level from the passivation strategy of $CsPbBrCl_2$ QDs effectively mitigate the energy disorders of $MAPbI_3$, narrow the band-tail electronic states, and reduce the mid-gap states of $MAPbI_3$. As a result, this strategy-based device exhibits an improved V_{oc} of 1.15 V and *PCE* of over 21% with negligible *J–V* hysteresis and excellent operational stability. Besides, the $CsPbBrCl_2$ QDs films also have a surface hydrophobic property.

12.3.5 Interface Engineering at HTL/Electrode Interface

Considering environmentally unstable small-molecule organic HTMs, and diffusion of electrode metal (such as silver, gold, and aluminum) and perovskite ions which are caused by light soaking and thermal stress, insertion of an IL at the HTL/electrode interface has been also proposed. For instance, Habisreutinger et al. [75] introduced an insulated polymer layer of poly(methyl methacrylate) (PMMA) on the P3HT-SWNTs composite layer and yielded highly stable PSCs. Arora et al. reported a highly efficient and stable PSC by inserting rGO at the CuSCN/Au interface (i.e., CuSCN/rGO/Au) [76]. Compared to the Spiro-OMeTAD/Au-based device, the CuSCN/rGO/Au-based cell showed excellent thermal stability over 1000 h under continuous illumination at 60 °C. Thus, the interlayer of stable rGO can substantially improve the interface contact and suppress ions migration, resulting in improved long-term photo-stability.

12.3.6 Interface Engineering at Multi-Interface Locations

The introduction of ILs at different locations (i.e., TCO/ETL/IL/Perovskite/HTL/IL) has also been explored. As discussed above, graphene and its derivatives effectively suppress ions migration and resist moisture penetration, which is beneficial in improving the stability. Agresti et al. introduced graphene and GO at mp-TiO_2/perovskite and Spiro-OMeTAD/Au interfaces, respectively, resulting in improvement of charge-carrier injection/collection and operational stability under continuous 1 sun illumination at 60°C [77]. Besides, the double ILs used in the solar cells can improve both efficiency and stability because of the improvement of energy-level alignment and charge transportation efficiency, modification of interface contacts, and reduction of interface degradation. For example, Hou et al. reported a low-temperature processed highly efficient stable, and cost-efficient planar PSC with insertion of C60-SAM/SnO_x/PC60BM and Ta-WOx on ITO and PDCBT, respectively [78]. In consequence, the champion device yielded a PCE of 21.2% with improved photo-stability over 1000 h. More interestingly, when PDCBT, P3HT, and Spiro-MeOTAD films were treated by I_2 vapor, the PDCBT showed superb stability but the other two films were destroyed after 2 min. Such a strategy by introducing interlayers at different interfaces substantially reduces interface barriers, mitigates ion migration, improves quasi-ohmic contacts, and protects perovskites from air moisture. In addition to organic materials, metal oxides (such as SnO_2 QDs and Cu_xO) as ILs with two different locations have also been reported to enhance device stability [79].

In this section, we conclude that interface stabilization is a feasible and effective method to improve the performance and stability of n-i-p-type PSCs. Interface stabilization has several advantages: (1) aligns band energy levels to facilitate carriers' extraction and transportation, (2) reduces trap density in perovskite film and at interfaces, (3) improves interfacial contacts from intimacy layers, (4) suppresses ions migration and recombination, and (5) protects perovskites from water and oxygen.

12.4 Composite-Based Charge-Transport Materials

PSCs have a structure in which a light absorber is sandwiched between an n-type ETL and a p-type HTL. A PSC has an n-i-p structure when ETL is located on the TCO electrode through which light enters, and a p-i-n structure when HTL is located on the TCO. Perovskite material has a large number of photo-generated electrons attributed to its high extinction coefficient, so the charge collection rate and overall efficiency are determined by the type of charge-transport layer (ETL or HTL) located on the side of light. The preparation of a perovskite thin film having high crystallinity also varies substantially depending on the surface characteristics of the charge-transport layer. Therefore, it is crucial to develop a charge selection material having high efficiency, high charge selectivity, and long-term stability for the production of high-performance PSC.

12.4.1 Composite-Based Electron-Transport Layer

Charge-transport layers are required to have specific properties. For example, they should have well-matched band energy levels with the perovskite layer to facilitate charge transfer and increase the built-in potential. Also, acceptable charge mobility is required to reduce charge recombination within the transport layer [80]. TiO_2 is an electron-transport material (ETM) widely employed in efficient PSCs for its suitable band alignment and high chemical stability. However, TiO_2 generally requires high-temperature processing for high crystallinity and charge-carrier mobility, which slows the production rate and is costly. To address this problem, a composite of conductive graphene and pre-synthesized TiO_2 NPs as low-temperature processed ETLs was introduced. The presence of graphene reduces the formation of energy barriers at interfaces because graphene provides a work function with an energy level between FTO and the conduction band of TiO_2. Moreover, the graphene-TiO_2 composites have a higher electrical conductivity than the low-temperature processed ETLs, improving the device fabrication process and performance by more than 15% [81–83].

The metal oxide/graphene composites have excellent electrical conductivity and carrier mobility, so the carrier transport is more efficient than that of the metal oxide layer alone [84]. According to simulation studies, the use of TiO_2-graphene composite as ETL can improve the performance of solar cell devices by 17.8% due to the increase in carrier mobility [85].

To develop a high-quality low-temperature processed ETL, a novel application of SnO_2 composites with graphene quantum dots (GQDs) showed effective performance in PSCs. In the GQD-based devices, the optimization of GQD size and concentration could improve the film morphology and conductivity as well as match Fermi energy level with the light absorber perovskite layer. GQD optimization could remarkably enhance electron transfer and suppress interfacial recombination, resulting in a significant improvement in photovoltaic performance. Compared to metal oxide ETL, metal oxide/GQDs composite ETL has high electron mobility, low-temperature processability, and high mechanical durability, which makes it suitable for flexible solar cells [86].

ZnO and SnO_2 are two of the most promising n-type semiconductors for ETL application in PV devices due to their high charge mobility and transparency. However, the chemical instability of ZnO and the inherent low conduction band of SnO_2 cause decomposition of active layer materials and low PCE in perovskite SCs. ZnO/SnO_2 composites could overcome the limitations of

ZnO and SnO_2, thus avoiding the photocatalytic activity and lowering the charge recombination. The ZnO/SnO_2 composites showed promising performance as ETL material in solid-state dye-sensitized solar cells (DSSCs) and PSCs with improved photo-conversion efficiency [87].

By combining the GQDs and the metal oxide double layer (SnO_2/ZnO) as ETL, charge-carrier mobility and recombination inhibition are further improved. A comparative study of PSCs with SnO_2/ZnO ETL and their combination with various concentrations of GQDs showed an enhancement in optoelectronic properties of a photovoltaic device. Device fabrication with the GQDs-modified ETL is an effective technology for designing charge selection materials for manufacturing PSCs close to theoretical performance [88, 89].

To solve the instability issues related to ETL and to fabricate highly stable and efficient PSCs, Hahn et al. developed $SrTiO_3$/Al_2O_3-graphene as an ETM [90]. Figure 12.10 shows schematic cell configuration (a), illustrative mechanism of charge separation and conduction through composites layer (b), cross-sectional FESEM image of the cell (c), and corresponding energy band diagram (d). The mesoporous Al_2O_3-graphene (mp-AG) composite layer is quite conductive because of the presence of highly conductive graphene sheets and thus provides both high electron concentration and electron mobility, leading to efficient charge extraction and conduction as well as enhancing V_{oc}.

The mp-AG acts as an ETL scaffold not only to improve the quality of perovskite film in terms of large grain size and homogeneous morphology but also to enhance charge transport due to the presence of conductive graphene. Furthermore, it eliminates the chance of contact between perovskite and the FTO electrode, which helps to impede shunt paths. Compared to the control cell with TiO_2 ETL, the $SrTiO_3$/mp-AG ETL-based PSCs exhibited overall a best performance (i.e., V_{oc} = 1.057 V, J_{sc} = 25.75 mA/cm^2, FF = 75.63%, and PCE = 20.58%). More importantly, the composites-based device without encapsulation showed not only remarkable thermal- and photo-stability but also long-term stability with retaining 97–99% of the initial values of photovoltaic parameters with sustaining ~93% of initial PCE over 300 days under ambient conditions.

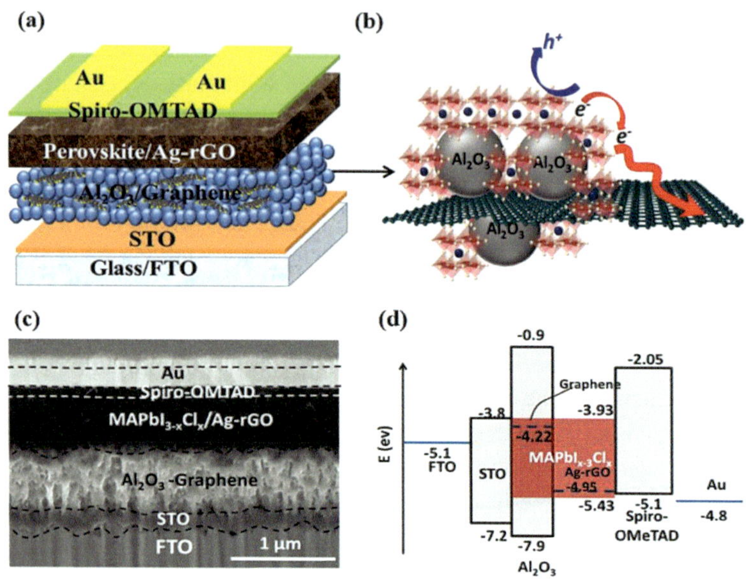

Figure 12.10 (a) Schematic configuration of FTO/STO/mp-AG/$MAPbI_{3-x}Cl_x$: Ag-rGO/Spiro-OMeTAD/Au, (b) Illustrative mechanism of charge separation and conduction through the mp-AG composite layer, (c) Cross-sectional FESEM image of the cell, and (d) Corresponding energy band diagram. Reprinted with permission from [90].

12.4.2 Composite-Based Hole-Transport Layer

Organic p-type materials such as poly(3,4-ethylenedioxythiophene): poly(styrene sulfonate) (PEDOT:PSS) has high conductivity and excellent solution processability, so PEDOT:PSS has been commonly used in HTL of a planar perovskite solar cell to transport holes from the perovskite active layer to the electrode. However, PEDOT:PSS is not an ideal HTL because of its inhomogeneous electrical property, poor electron blocking ability, high acidity, and hygroscopic nature. Application of PEDOT:PSS and perfluorinated ionomer (PFI) composite with a high work function level could improve the performance of PSCs by enhancing built-in potential [91]. PEDOT:PSS/GO composite, is considered to be a promising material that can compensate for the shortcomings of PEDOT:PSS as HTL in a planar PSC. A well-matched work function energy level between GO

(5.0 eV) and PEDOT:PSS (5.1 eV) results in efficient hole transport and collection to the electrode [92].

PEDOT:PSS-metal oxide composite film can be used for HTL in an inverted PSC. In particular, planar devices based on PEDOT:PSS-MoO_x composite film shows ~60% improvement in efficiency compared to PEDOT:PSS control device. Presence of metal oxides in PEDOT:PSS modifies surface free energy and roughness of hole-transport layer and affects the crystallization of perovskite films to form larger grains with fewer GBs. Moreover, with a small amount of incorporation of metal oxide, the work function of PEDOT:PSS could be tuned to match the energy band levels of the active layer, which can significantly boost device photovoltage [93].

Inorganic p-type nickel oxide (NiO_x) has been explored as an HTL material to replace PEDOT:PSS in planar PSCs. However, due to insufficient electrical conductivity, NiO_x-based devices show high series resistance and low fill factor (FF). To solve such a problem with NiO_x, carbon nanotubes (CNTs)-NiO_x composite HTL with superior optical transparency and electrical conductivity was developed to improve conductivity and hole-transport property. The CNTs-NiO_x composite HTL-based PSCs enhanced the device performance by ~15% [94]. Another problem with NiO_x hole selective material is that its energy band is not well-matched with perovskite absorbers, but this can be solved by using an additive in NiO_x HTL. For example, amino-functionalized GQDs added in NiO_x enhanced device performance owing to high conductivity, chemical stability, and tunable electronic properties [95].

References

1. Chen, B., Rudd, P. N., Yang, S., Yuan, Y., et al. (2019) *Chem. Soc. Rev.*, **48**, pp. 3842–3867.
2. Mahmoudi, T., Wang, Y., and Hahn, Y.-B. (2020) *Adv. Energy Mater.*, **10**, pp. 1903369.
3. Wang, Y., Yang, Y., Wu, S., Zhang, C., et al. (2020) *Nanoscale Adv.*, **2**, pp. 5883–5889.
4. Shao, Y., Fang, Y., Li, T., Wang, Q., et al. (2016) *Energy Environ. Sci.*, **9**, pp. 1752–1759.

5. Mahmoudi, T., Wang, Y., and Hahn, Y.-B. (2018) *ACS Energy Lett.*, **4**, pp. 235–241.
6. Hashmi, S. G., Tiihonen, A., Martineau, D., Ozkan, M., et al. (2017) *J. Mater. Chem. A*, **5**, pp. 4797–4802.
7. Tress, W., Yavari, M., Domanski, K., Yadav, P., et al. (2018) *Energy Environ. Sci.*, **11**, pp. 151–165.
8. Ni, Z., Bao, C., Liu, Y., Jiang, Q., et al. (2020) *Science*, **367**, pp. 1352–1358.
9. Li, F., Deng, X., Qi, F., Li, Z., et al. (2020) *J. Am. Chem. Soc.*, **142**, pp. 20134–20142.
10. Chen, Y., Yang, J., Wang, S., Wu, Y., et al. (2020) *iScience*, **23**, pp. 100762.
11. Park, C., Ko, H., Sin, D. H., Song, K. C., and Cho, K. (2017) *Adv. Funct. Mater.*, **27**, pp. 1703546.
12. Zhang, F., Shi, W., Luo, J., Pellet, N., et al. (2017) *Adv. Mater.*, **29**, pp. 1606806.
13. Xu, J G., Xue, R., Chen, W., Zhang, J., et al. (2018) *Adv. Energy Mater.*, **8**, pp. 1703054.
14. Liu, C., Huang, Z., Hu, X., Meng, X., et al. (2018) *ACS Appl. Mater. Interfaces*, **10**, pp. 1909–1916.
15. Wu, W.-Q., Wang, Q., Fang, Y., Shao, Y., et al. (2018) *Nat. Comm.*, **9**, pp. 1625.
16. Wu, C., Li, H., Yan, Y., Chi, B., et al. (2018) *Sol. RRL*, **2**, pp. 1800052.
17. Zhao, Y., Wei, J., Li, H., Yan, Y., et al. (2016) *Nat. Comm.*, **7**, pp. 10228.
18. Li, X., Dar, M. I., Yi, Y. C., Luo, J. S., et al. (2015). *Nat. Chem.*, **7**, pp. 703–711.
19. Zheng, X., Hou, Y., Bao, C., Yin, J., et al. (2020) *Nat. Energy*, **5**, pp. 131–140.
20. Wang, R., Xue, J., Meng, L., Lee, J.-W., et al. (2019) *Joule*, **3**, pp. 1–14.
21. Liu, Z., Cao, F., Wang, M., Wang, M., and Huang, S. (2020) *Angew. Chem.*, **132**, pp. 4190–4196.
22. Chen, H., Liu, T., Zhou, P., Li, S., et al. (2020) *Adv. Mater.*, **32**, pp. 1905661.
23. Wang, R., Xue, J., Wang, K.-L., Wang, Z.-K., et al. (2019) *Science*, **366**, pp. 1509–1513.
24. Bi, D., Gao, P., Scopelliti, R., Oveisi, E., et al. (2016) *Adv. Mater.*, **28**, pp. 2910–2915.
25. Smith, I. C., Hoke, E. T., Solis-Ibarra, D., McGehee, M. D., and Karunadasa, H. I. (2014) *Angew. Chem. Int. Ed.*, **126**, pp. 11414–11417.

26. Tsai, H., Nie, W., Blancon, J.-C., Stoumpos, C. C., et al. (2016) *Nature*, **536**, pp. 312–316.

27. Ren, H., Yu, S., Chao, L., Xia, Y., et al. (2020) *Nat. Photon.*, **14**, pp. 154–163.

28. Mei, A., Li, X., Liu, L., Ku, Z., et al. (2014) *Science*, **345**, pp. 295–298.

29. Hu, Y., Si, S., Mei, A., Rong, Y., et al. (2017). *Sol. RRL*, **1**, pp. 1600019.

30. Grancini, G., Roldán-Carmona, C., Zimmermann, I., Mosconi, E., et al. (2017) *Nat. Commun.*, **8**, pp. 15684.

31. Zhou, Y., Wang, F., Cao, Y., Wang, J.-P., et al. (2017) *Adv. Energy Mater.*, **7**, pp. 1701048.

32. Lin, Y., Bai, Y., Fang, Y., Chen, Z., et al. (2018) *J. Phys. Chem. Lett.*, **9**, pp. 654–658.

33. Fan, J., Ma, Y., Zhang, C., Liu, C. (2018), *Adv. Energy Mater.*, **8**, pp. 1703421.

34. Liu, P., Xian, Y., Yuan, W., Long, Y., et al. (2020), *Adv. Energy Mater.*, **10**, pp. 1903664.

35. Wang, Y., Rho, W.-Y., Yang, H.-Y., Mahmoudi, T., et al. (2016) *Nano Energy*, **27**, pp. 535–544.

36. Wang, Y., Mahmoudi, T., Rho, W.-Y., Yang, H.-Y., et al. (2017) *Nano Energy*, **40**, pp. 408–417.

37. Wang, Y., Mahmoudi, T., Yang, H.-Y., Bhat, K. S., et al. (2018). *Nano Energy*, **49**, pp. 59–66.

38. Nie, W., Tsai, H., Blancon, J.-C., Liu, F., et al. (2018) *Adv. Mater.*, **30**, pp. 1703879.

39. Li, L., Mi, J., Yong, Y., Mao, B., and Shi, W. (2018) *J. Mater. Chem. C*, **6**, pp. 8226–8233.

40. Yao, K., Li, F., He, Q., Wang, X., et al. (2017) *Nano Energy*, **40**, pp.155–162.

41. Wang, Y., Mahmoudi, T., and Hahn, Y.-B. (2020) *Adv. Energy Mater.*, **10**, pp. 2000967.

42. Yun, J., Jun, J., Yu, H., Lee, K., et al. (2017) *J. Mater. Chem. A*, **5**, pp. 21750–21756.

43. Lin, M.-W., Wang, K.-C., Wang, J.-H., Li, M.-H., et al. (2016) *Adv. Mater. Interfaces*, **3**, pp. 1600135.

44. Mahmoudi, T., Wang, Y., and Hahn, Y.-B. (2018) *Nano Energy*, **47**, pp. 51–65.

45. Chung, C.-C., Narra, S., Jokar, E., Wu, H.-P., and Diau, E. W.-G. (2017) *J. Mater. Chem. A*, **5**, pp. 13957–13965.

46. Yavari, M., Mazloum-Ardakani, M., Gholipour, S., Marinova, N., et al. (2018). *Adv. Energy Mater.*, **8**, pp. 1702719.

47. Mahmoudi, T., Wang, Y., and Hahn, Y.-B. (2019) *ACS Energy Lett.*, **4**, pp. 235–241.

48. Eperon, G. E., Burlakov, V. M., Goriely, A., and Snaith, H. J. (2014) *ACS Nano*, **8**, pp. 591–598.

49. Roldán-Carmona, C., Malinkiewicz, O., Betancur, R., Longo, G., et al. (2014) *Energy Environ. Sci.*, **7**, pp. 2968–2973.

50. Ye, S., Rao, H., Zhao, Z., Zhang, L., et al. (2017) *J. Am. Chem. Soc.*, **139**, pp. 7504–7512.

51. Zhang, H., Wang, H., Williams, S. T., Xiong, D., et al. (2017) *Adv. Mater.*, **29**, pp.1606608.

52. Jin, J., Li, H., Chen, C., Zhang, B., et al. (2017) *ACS Appl. Mater. Interfaces*, **9**, pp. 42875–42882.

53. Bi, C., Zheng, X., Chen, B., Wei, H., and Huang, J. (2017) *ACS Energy Lett.*, **2**, pp. 1400–1406.

54. Boopathi, K. M., Mohan, R., Huang, T.-Y., Budiawan, W., et al. (2016) *J. Mater. Chem.*, **4**, pp. 1591–1597.

55. Jahandar, M., Heo, J. H., Song, C. E., Kong, K.-J., et al. (2016) *Nano Energy*, **27**, pp. 330–339.

56. Zimmermann, I., Gratia, P., Martineau, D., Grancini, G., et al. (2019) *J. Mater. Chem. A*, **7**, pp. 8073–8077.

57. Wang, Y., Arumugam, G. M., Mahmoudi, T., Mai, Y., and Hahn, Y.-B. (2021) *Nano Energy*, **87**, pp. 106141.

58. Xie, J., Yu, X., Huang, J., Sun, X., et al. (2017) *Adv. Sci.*, **4**, pp. 1700018.

59. Li, Y., Zhao, Y., Chen, Q., Yang, Y., et al. (2015) *J. Am. Chem. Soc.*, **137**, pp. 15540–15547.

60. Zhang, Y., Wang, P., Yu, X., Xie, J., et al. (2016) *J. Mater. Chem. A*, **4**, pp. 18509–18515.

61. Li, W., Zhang, W., Reenen, S. V., Sutton, R. J., et al. (2016). *Energy Environ. Sci.*, **9**, pp. 490–498.

62. Byranvand, M. M., Kim, T., Song, S., Kang, G., et al. (2018) *Adv. Energy Mater.*, **8**, pp. 1702235.

63. Ma, Y., Deng, K., Gu, B., Cao, F., et al. (2016) *Adv. Mater. Interfaces*, **3**, pp. 1600729.

64. Liu, X., Zhang, Y., Shi, L., Liu, Z., et al. (2018) *Adv. Energy Mater.*, **8**, pp. 1800138.

65. Lee, Y., Paek, S., Cho, K. T., Oveisi, E., et al. (2017) *J. Mater. Chem. A*, **5**, pp. 12729–12734.

66. Chen, C., Li, H., Jin, J., Chen, X., et al. (2017) *Adv. Energy Mater.*, **7**, pp. 1700758.

67. Zhang, J., Hu, Z., Huang, L., Yue, G., et al. (2015) *Chem. Commun.*, **51**, pp. 7047–7050.

68. Zuo, L., Guo, H., deQuilettes D. W., Jariwala, S., et al. (2017) *Sci. Adv.*, **3**, pp. e1700106.

69. Wang, F., Shimazaki, A., Yang, F., Kanahashi, K., et al. (2017) *J. Phys. Chem. C*, **121**, pp. 1562–1568.

70. Cho, K. T., Paek, S., Grancini, G., Roldán-Carmona, C., et al. (2017) *Energy Environ. Sci.*, **10**, pp. 621–627.

71. Cho, K. T., Grancini, G., Lee, Y., Oveisi, E., et al. (2018) *Energy Environ. Sci.*, **11**, pp. 952–959.

72. Guarnera, S., Abate, A., Zhang, W., Foster, J. M., et al. (2015) *J. Phys. Chem. Lett.*, **6**, pp. 432–437.

73. Koushik, D., Verhees, W. J. H., Kuang, Y., Veenstra, S., et al. (2017). *Energy Environ. Sci.*, **10**, pp. 91–100.

74. Zheng, X., Troughton, J., Gasparini, N., Lin, Y., et al. (2019) *Joule*, **3**, pp. 1963–1976.

75. Habisreutinger, S. N., Leijtens, T., Eperon, G. E., Stranks, S. D., et al. (2014) *Nano Lett.*, **14**, pp. 5561–5568.

76. Arora, N., Dar, M. I., Hinderhofer, A., Pellet, N., et al. (2017) *Science*, **358**, pp. 768–771.

77. Agresti, A., Pescetelli, S., Taheri, B., Castillo, A. E. D. R., et al. (2016) *ChemSusChem*, **9**, pp. 2609–2619.

78. Hou, Y., Du, X., Scheiner, S., McMeekin D. P., et al. (2017) *Science*, **358**, pp. 1192–1197.

79. Guo, Y., Lei, H., Xiong, L., Li, B., and Fang, G. (2018) *J. Mater. Chem. A*, **6**, pp. 2157–2165.

80. Shin, S. S., Lee, S. J., and Seok, S. I. (2019) *Adv. Funct. Mater.*, **29**(47), pp. 1900455.

81. Tan, H., Jain, A., Voznyy, O., Lan, X., et al. (2017) *Science*, **355**, pp. 722.

82. Liu, M., Johnston, M. B., and Snaith, H. J. (2013) *Nature*, **501**, pp. 395

83. Zhou, H., Chen, Q., Li, G., Luo, S., et al. (2014) *Science*, **345**, pp. 542.

84. Wang, J. T.-W., Ball, J. M., Barea, E. M., Abate, A., et al. (2014) *Nano Lett.*, **14**, pp. 724.

85. Dadashbeik, M., Fathi, D., and Eskandari, M. (2020) *Sol. Energy*, **207**, pp. 917–924.

86. Zhou, Y., Yang, S., Yin, X., Han, J., et al. (2019) *J. Mater. Chem. A*, **7**(4), pp. 1878–1888.

87. Shalan, A. E., El-Shazly, A. N., Rashad, M. M., and Allam, N. K. (2019) *Nanoscale Adv.*, **1**(7), pp. 2654–2662.

88. Nagaraj, G., Mohammed, M. K. A., Shekargoftar, M., Sasikumar, P., et al. (2021) *Mater. Today Energy*, **22**, pp. 100853.

89. Shin, D. H., Kim, J. M., Shin, S. H., and Choi, S. H. (2019) *Dyes Pigm.*, **170**, pp. 107630.

90. Mahmoudi, T., Wang, Y., and Hahn, Y. B. (2020) *Adv. Energy Mater.*, **10**(2), pp. 1903369.

91. Lim, K.-G., Kim, H.-B., Jeong, J., Kim, H., et al. (2014) *Adv. Mater.*, **26**, pp. 6461.

92. Lee, D. Y., Na, S. I., and Kim, S. S. (2016) *Nanoscale*, **8**(3), pp. 1513–1522.

93. Duan, C., Liu, Z., Yuan, L., Zhu, H., et al. (2020) *Adv. Opt. Mater.*, **8**(17), pp. 2000216.

94. Ryu, J., Yoon, S., Park, J., Jeong, S. M., and Kang, D. W. (2020) *Appl. Surf. Sci.*, **516**, pp. 146116.

95. Wang, Z., Rong, X., Wang, L., Wang, W., et al. (2020) *ACS Appl. Mater. Interfaces*, **12**(7), pp. 8342–8350.

Problems

12.1 Why are composite materials appropriate for improving the performance and stability of PSCs?

12.2 What is the advantage of organic small-molecule-based perovskite composites?

12.2 Why do $MAPbI_3$–PEG composites exhibit self-healed properties?

12.4 What is the role of ammonium halides as additives in perovskite films?

12.5 Why LD-3D perovskite heterojunctions are useful for both efficiency and stability in PSCs?

12.6 What is the role of NiO_x as an additive in PSCs?

12.7 Why graphene and its derivatives can be used for highly efficient and stable PSCs?

12.8 Describe briefly the merits of interface engineering strategy in perovskites solar cells.

12.9 What are the advantages of inserting multi-interface layers in PSCs?

Chapter 13

Characterization of Solar Cell Materials and Devices

To develop materials and devices, it is necessary to analyze the materials' properties and devices' performance. Characterization of solar cell materials includes three categories—chemical, physical and structural analysis. For chemical and physical analysis of materials, spectroscopic techniques are useful to identify atomic and molecular structures, study the flow of energy in chemical reactions, and perform a kinetic energy analysis of electrons ejected from an atom or molecule.

For structural analysis, X-ray diffraction (XRD) and electron microscopy (EM) are useful tools to determine the crystallographic structure of a material and to investigate the microstructural characteristics of organic and inorganic materials, respectively. To examine the photovoltaic parameters of a solar cell, the basic characteristics are the short-circuit current (I_{sc}), the open-circuit voltage (V_{oc}), the fill factor (FF), and the solar energy conversion efficiency (η). In addition, space-charge-limited current (SCLC) measurements are commonly used to examine the carrier mobility and trap density of transport materials in a solar cell.

In this chapter, typical characterization techniques are introduced for solar cell materials and devices.

Next-Generation Solar Cells: Principles and Materials
Yoon-Bong Hahn, Tahmineh Mahmoudi, and Yousheng Wang

ISBN 978-981-4968-66-9 (Hardcover), 978-1-003-37238-7 (eBook)
www.jennystanford.com

13.1 Spectroscopic Techniques

Spectroscopy is the measurement of spectra produced by matter involving either absorption, emission, or scattering of radiation. Atomic and molecular spectra can provide detailed information about the structure and chemical properties of materials and devices. Spectroscopic techniques are one of the main sources to determine molecular geometries, such as bond lengths, bond angles, and torsion angles. Those techniques can also provide significant information about molecular symmetry, energy-level distribution, electron density, and electric and magnetic properties. Spectroscopy has been an essential tool in the development of models for atomic and molecular structure, which prompts scientists to refine existing models to accurately reproduce the experimentally observed spacing between energy states.

Nowadays, spectroscopic tools are routinely used for the quantitative and qualitative chemical analysis and the characterization of new molecules and materials, which have been widely applied for elaboration and testing of theoretical models, synthetic chemistry, the study of reaction mechanisms, or biochemistry and materials science. Atomic and molecular spectroscopies mainly relate to the absorption or emission of electromagnetic radiation and the changes taking place in systems as a consequence of the radiation. Diffraction methods, however, involve the wave nature of the radiation and rely on the interpretation of the interference patterns between waves upon their encounter with obstacles of dimensions roughly close to their wavelength. When these waves correspond to electromagnetic waves in the X-ray region of the spectrum or particles such as electrons or neutrons, their wavelengths are similar to the spacing between atoms in condensed phases such as liquids or crystals and they can be employed to obtain useful information about the atomic positions in these materials [1].

The strength of the radiation energy interacts with the molecules in different ways (Fig. 13.1):

- High-energy sources produce the breaking of bonds: X-rays, γ-rays
- Medium-energy sources induce the excitation of electrons: UV–Vis

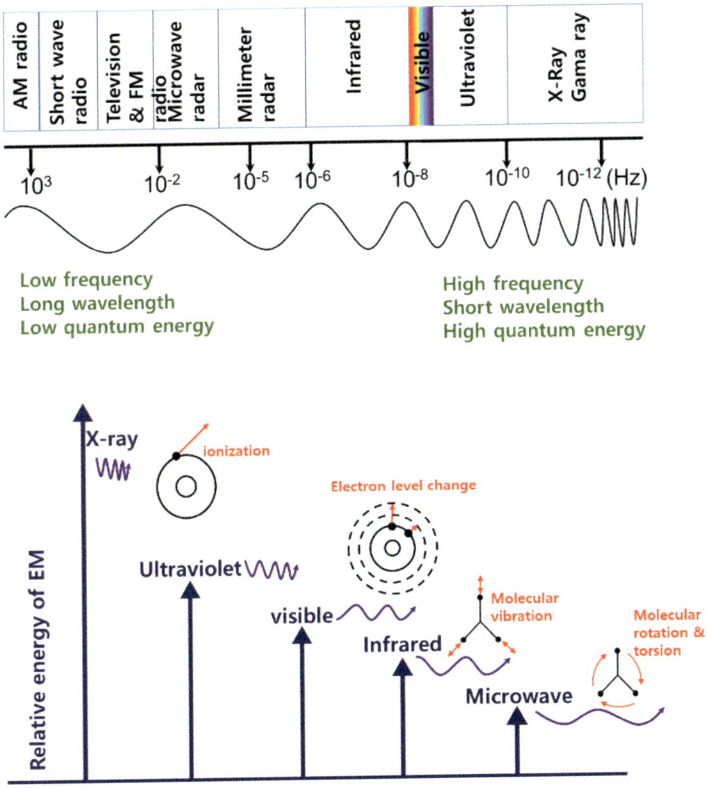

Figure 13.1 Interaction of radiation with matter.

- Low-energy sources produce vibrations in chemical bonds: Infrared
- Very low–energy sources produce rotation of the chemical bonds: microwave/radio waves.

13.2 Chemical Analysis

13.2.1 Fourier Transform Infrared (FTIR) Spectroscopy

As one of the most prevalent and widely used spectroscopic techniques, infrared (IR) spectroscopy can determine the structure and identity of basic compounds in the field of inorganic and organic chemistry. The presence of functional groups in chemical compounds gives different chemical properties. Each molecule has a unique spectrum, which is often referred to as its fingerprint. A molecule's absorption peak can be identified by comparing it with a database of spectra.

IR radiation has a wavelength range of 2~15 μm, which shows lower energy than visible or ultraviolet light. There are three principal IR regions: near-IR (12800~4000 cm^{-1}), mid-IR (4000~200 cm^{-1}), and far-IR (200~10 cm^{-1}). In general, the mid-IR region is used for the measurement of chemical molecule's covalent bonds and the FTIR technique uses the mid-IR region (4000~670 cm^{-1}).

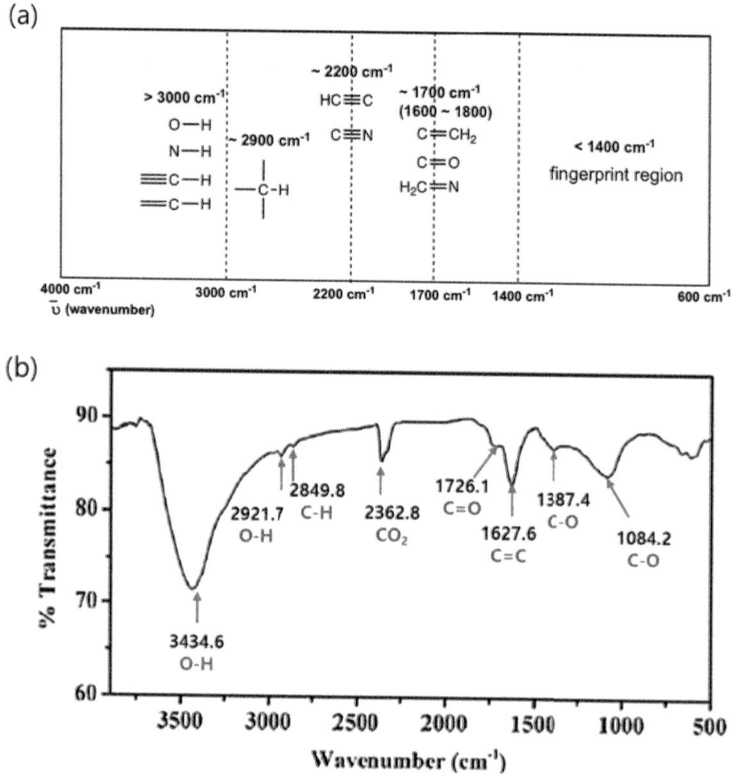

Figure 13.2 (a) Typical IR absorption bands. (b) FTIR spectrum of graphene oxide.

The covalent bond of a molecule selectively absorbs radiation of a particular wavelength and changes the vibration energy of the bond. IR radiation induces various types of vibration (stretching or bending) depending on the atoms in the bond. Each molecule shows different transmittance patterns, attributed to the different absorption frequencies for different bonds and functional groups. The spectrum can be represented by a graph with a wavenumber (cm^{-1}) in terms of transmittance [2]. (Wavenumber corresponds to the molecular bond vibrational energy, 1/wavelength.)

How to read an FTIR spectrum? Reading the spectrum is a matter of determining which groups and bonds correspond to which peaks. A simple reference table for the various chemical bonds can be found in Fig. 13.2a. For example, the FTIR spectrum of graphene oxide is shown in Fig. 13.2b. Different functionalities in the graphene oxide including aliphatic C-O stretch (1100–1300 cm^{-1}), alkoxy C-O stretch (1250–1310 cm^{-1}), anhydride C-O stretch (1100–1400 cm^{-1}), aromatic C=C bending (1500–1700 cm^{-1}), aldehyde C=O stretch (1690–1740 cm^{-1}), alkyl C-H stretch (2950–2850 cm^{-1}), carboxylic acid O-H stretch (2500–3000 cm^{-1}), and alcohol/phenol O-H stretch (3200–3550 cm^{-1}) are determined from the FTIR spectrum.

13.2.2 X-ray Photoelectron Spectroscopy (XPS)

XPS is a surface-sensitive spectroscopic technique based on the photoelectric effect. XPS can identify the elements present within or covering the surface of a material (i.e., elemental composition), chemical state, overall electronic structure, and density of electronic states of a material. When a material is irradiated by an X-ray beam, the energy from the X-ray is absorbed by the core electrons of each element. Since the energy required for each element that is emitted and subsequently detected is *unique*, "binding energy" can be used to identify the element's presence in the analyzed material [3].

How to analyze XPS spectra? In XPS analysis, the position of a peak on the x-axis indicates the elemental and chemical composition. The x-axis is traditionally displayed as "binding energy" in electron volt (eV). The y-axis shows the intensity of a particular element that can be found at the surface.

Figure 13.3a shows a survey spectrum of graphene oxide at a wide energy range. The spectrum shows carbon and oxygen peaks. A significant background beneath the peaks in the spectrum can be observed, which is caused by scattered electrons as scattering reduces the kinetic energy and intensity from the peaks. The background can be removed when quantifying the spectrum because the obtained spectrum can only perform qualitative analysis and degree of energy level by itself. Gaussian curve fitting can analyze the number of components in detail. It is also possible to determine chemical states of each element. For example, the C1s region of the XPS spectrum from graphene oxide is shown in Fig. 13.3b. The carbon atoms in graphene oxide have three chemical states that show three peaks in XPS spectrum.

13.2.3 Energy-Dispersive X-ray Spectroscopy

Energy-dispersive X-ray spectroscopy (EDS, also abbreviated XEDS or EDX) used in conjunction with scanning electron microscopy (SEM) is a chemical microanalysis technique, which enables the chemical characterization or elemental analysis of materials. It relies on the interaction of X-ray excitation with material. In a sample, atoms have ground state electrons bound to nucleus with discrete energy levels. The incident beam excites an electron in an inner shell to the outer shell and creates an electron–hole pair [4]. An electron from an outer, with higher energy, fills the hole, and the difference in energy between the higher-energy shell and the lower-energy shell releases in the form of an X-ray. The number and energy of the emitted X-rays from a sample are measured by an EDS spectrometer. As the emitted X-ray energies are characteristic of the atomic structure of the emitting element, EDS measures the elemental composition of the sample.

A typical EDS spectrum of $SrBaSnO_3$ is shown in Fig. 13.4a. The EDS analysis also provides elemental mapping based on compiling specific elemental composition data across an area of a sample. For every pixel in the digital image collected, a complete elemental spectrum is also collected. By processing these spectra, the relative intensity of spectral features associated with several elements of interest can be translated into computed colorization layers, which color-codes the electron photomicrograph signifying layers and sites of elemental compositional information in a sample (Fig. 13.4b).

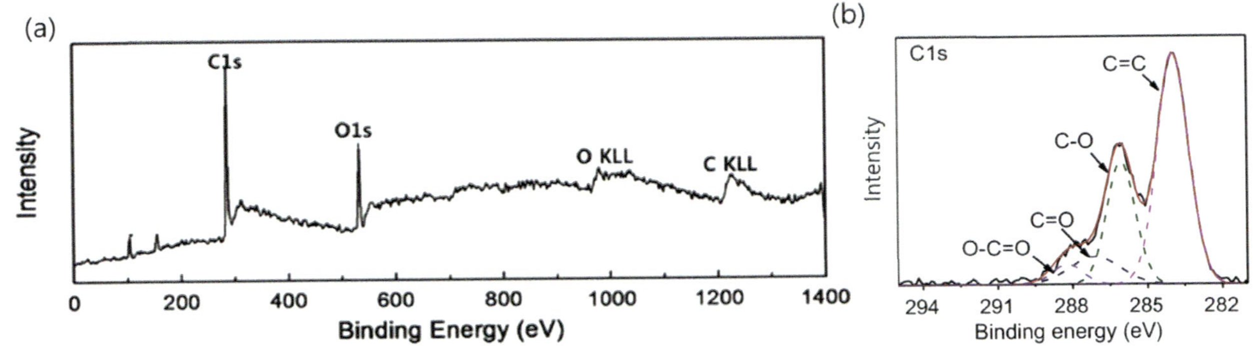

Figure 13.3 (a) Survey XPS spectrum of graphene oxide. (b) XPS core-level spectra of a C1s for graphene oxide.

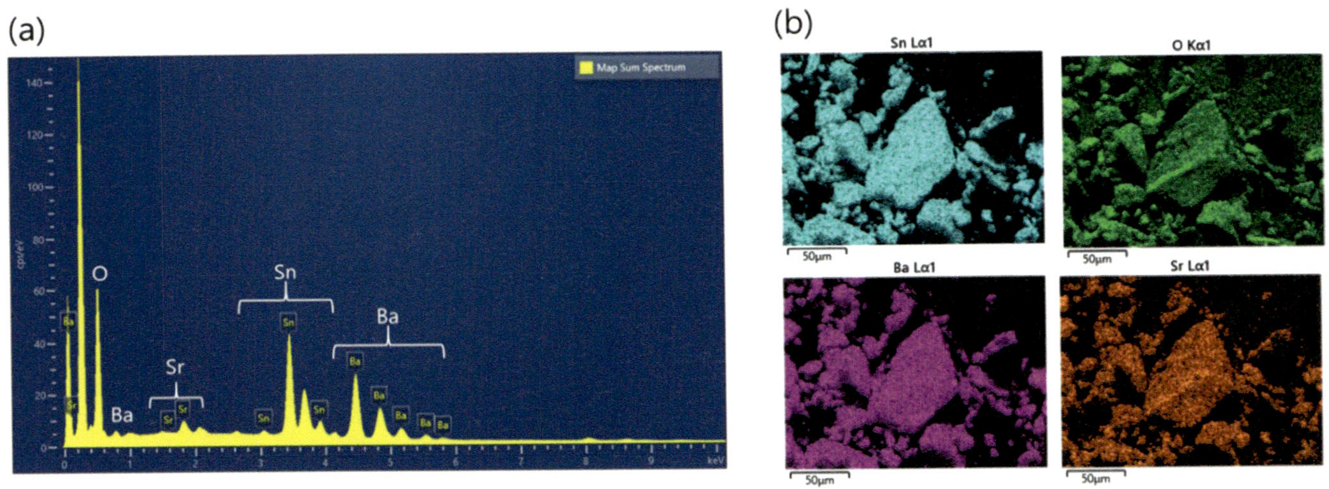

Figure 13.4 (a) EDS spectra and (b) EDS mapping of $SrBaSnO_3$ material.

13.3 Physical Analysis

13.3.1 Raman Spectroscopy

Raman spectroscopy is used to identify molecules and study chemical bonding and intramolecular bonds. Because vibrational frequencies are specific to a molecule's chemical bonds and symmetry (the fingerprint region of organic molecules is in the wavenumber range 500~1500 cm^{-1}), Raman spectroscopy provides a structural fingerprint to identify molecules.

Raman spectroscopy is a well-known technique for the analysis of graphene-related materials. There are three main bands of interest in the Raman spectra of graphene, the G-band (1580 cm^{-1}), the D-band (1350 cm^{-1}), and the 2D-band (2690 cm^{-1}). These bands are indicative of the quality of the sample and the number of layers of graphene. While the G-band and 2D-band will always be present, the D-band can only be seen for samples containing defects.

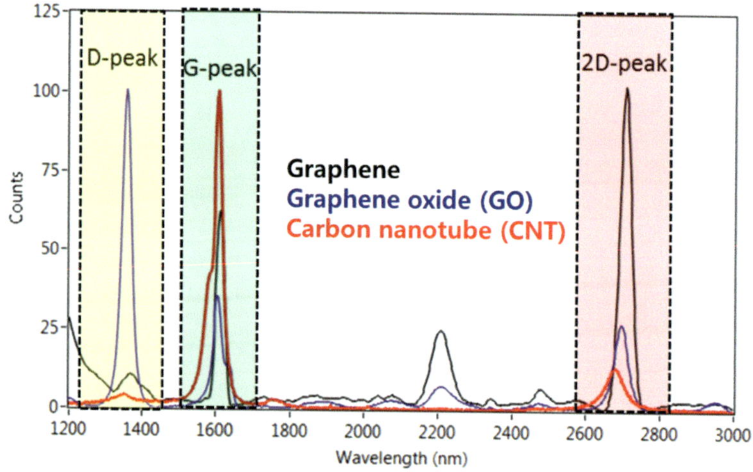

Figure 13.5 Raman spectroscopy of pristine graphene, graphene oxide, and carbon nanotube [7].

Figure 13.5 shows the difference in Raman spectrum in graphene, graphene oxide, and carbon nanotube. The D-band peak highlights defects in the graphene. The higher intensity of the D-band is attributed to the higher level of disorder within the graphene oxide.

The D-band is a 1-phonon lattice vibrational process and a double resonance band. The sharp G-band at ~ 1585 cm^{-1} represents the in-plane stretching vibrations of C=C bonded carbon atoms. The G-peak shifts to lower wavenumbers as the number of graphene layers increases. The 2D-band (also referred to as the G′-band) is also indicative of the number of layers of graphene sample. The 2D-band originates from a double resonance-enhanced two-phonon lateral vibrational process [8]. The 2D-band is the second-order overtone of the D-band, but it does not arise from a defect and therefore it is not indicative of defects.

13.3.2 Photoluminescence (PL) Spectroscopy

PL spectroscopy is a widely used technique to examine the optical and electronic properties of semiconductors and molecules. It is often referred to as fluorescence spectroscopy in chemistry. The relaxation processes can be studied by using time-resolved photoluminescence (TRPL) spectroscopy, which can measure the decay lifetime of the photoluminescence [9]. The mechanism of PL is by light emission from a matter after the absorption of photons. It is one of the forms of luminescence (light emission) which is initiated by photoexcitation. Following excitation, relaxation processes occur in which photons are re-radiated. In a typical PL measurement, a semiconductor is excited with a light source that provides photons with higher energy than the bandgap energy. After photo-absorption, electron–hole pair (exciton) can be formed in the conduction and valence bands. The excitations undergo energy and momentum relaxation toward the bandgap minimum, and then the electrons recombine with holes under the emission of photons.

The PL peak location determines the bandgap of semiconductor materials. The PL peak intensity represents the concertation of the electron–hole pair formed under illumination and the quenching speed of photo-excited exciton. Figure 13.6a shows PL spectra of ZnO and its doped materials (Al:ZnO and Mg:ZnO), showing different PL peak onsets. Such results indicate that the three materials have different bandgaps. The bandgap of ZnO (3.23 eV) changed to 3.44 eV and 3.35 eV by doing Mg^{2+} and Al^{3+}, respectively.

Time-resolved photoluminescence spectroscopy. Time-resolved photoluminescence (TRPL) can analyze the fast, electronic deactivation kinetics of photoluminescent material. The lifetime of electron–hole and its radiative recombination is usually in the range of pico- to nanoseconds (Fig. 13.6b). The TRPL decay can be influenced by the radiative and non-radiative recombination processes in the material environment as well as interactions with other molecules.

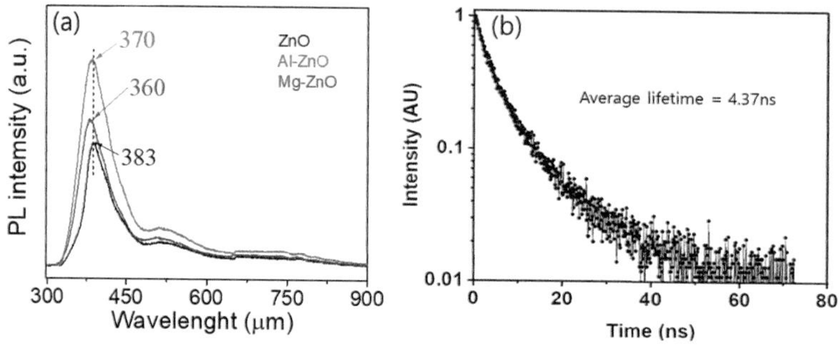

Figure 13.6 (a) PL spectra of ZnO, Al-doped ZnO, and Mg-doped ZnO. (b) TRPL decay spectra of the nanocrystal.

The decay kinetics is measured by time-related photon counting in which the material is repeatedly excited by a pulsed laser and the irradiated emission is detected at a specific wavelength. The obtained lifetime can therefore provide information about the local chemical environment or insights into reaction mechanisms [10]. The emission lifetime can change from nanoseconds to milliseconds in metal–organic complexes, inorganic crystal structures, and semiconductors. In this case, the luminescent species relaxes from its lowest excited triplet state by emitting a photon in a process called phosphorescence. The TRPL decay curve needs to be normalized and plotted on the semi-log scale graph (Fig. 13.6b). Besides, the decay lifetime also is calculated by the software of the measurement instrument.

13.3.3 UV–Vis Absorption Spectroscopy

UV–Vis spectroscopy measures the number of discrete wavelengths of UV–Vis light that are absorbed by or transmitted through the material in comparison to a reference sample. This property is influenced by material compositions. In semiconductors, this spectrum provides information about electron transition between the valence and conduction bands, and it can be used for the estimation of optical bandgap of a semiconductor.

UV–Vis measurements are straightforward and highly reproducible. To obtain a bandgap from a UV–Vis curve, the data must be interpreted. The bandgap can be estimated by drawing a tangent line to the curve. The bandgap of the material will be underestimated if a tangent line is fitted to a point within this tail, as seen in Fig. 13.7a [11].

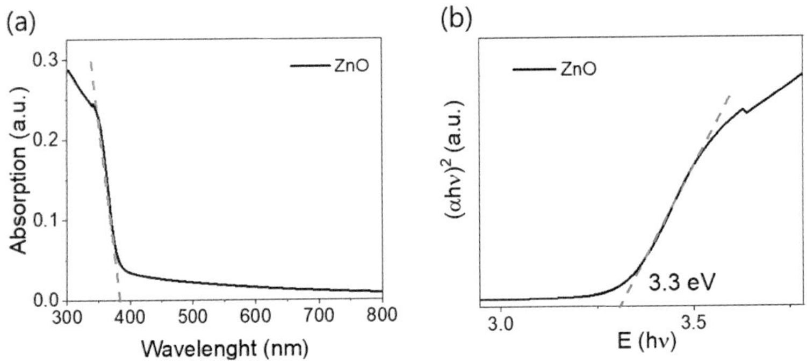

Figure 13.7 (a) UV–Vis absorption spectra of ZnO. (b) Calculation of optical bandgap from UV–Vis absorption spectra using the Tauc method.

Optical bandgap energy can be estimated using the relation between the incident photon energy ($h\nu$) and the absorption coefficient (α), given by the following relation (i.e., Tauc method):

$$(\alpha h\nu)^{1/n} = A(h\nu - E_g) \tag{13.1}$$

where h is Planck constant, ν is photon's frequency, A is constant, E_g is the bandgap energy, and n depends on the type of transition. The bandgap is estimated by plotting $(\alpha h\nu)^2$ vs. $h\nu$ and extrapolating the straight portion of the curve on the energy ($h\nu$) axis, as shown in Fig. 13.7b.

13.3.4 Ultraviolet Photoelectron Spectroscopy (UPS)

The principle of UPS is the same as XPS, which uses ultraviolet photons instead of high-energy X-rays. The energy of UV photons interacts with the valence band, providing information about the electronic structure of a semiconductor. UPS is an important technique to calculate the work function, valence band energy, and conduction band energy of a semiconductor material. Figure 13.8 shows a typical UPS analysis of ZnO nanocrystals.

The main objective of UPS measurements is to obtain iodization energy (E_i) and cutoff-edge energy (E_{cutoff}). Work function (Φ) is estimated by subtracting the difference of the secondary electron cut-off from the incident light energy (21.2 eV):

$$\Phi = 21.2 - E_{cutoff} \tag{13.2}$$

The position of the valence band (VB) relative to the Fermi level which is the midpoint of baseline in the spectrum can be obtained by taking the onset of the valence-band edge. The valence-band energy (E_{VB}) and conduction-band energy (E_{CB}) are calculated, respectively, by [12]:

$$E_{VB} = 21.2 - (E_{cutoff} - E_i) \tag{13.3}$$

$$E_{CB} = E_{VB} - E_g \tag{13.4}$$

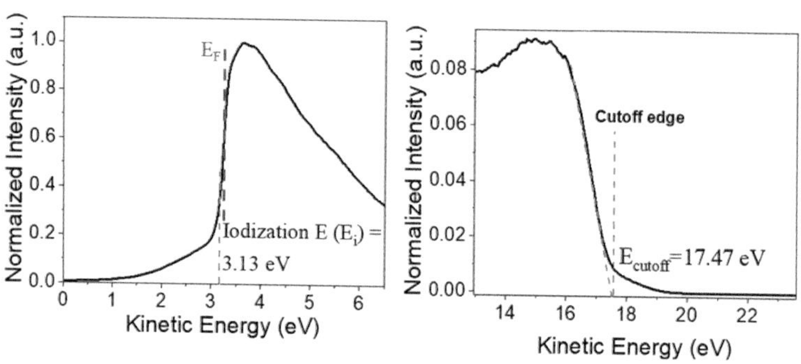

Figure 13.8 UPS spectra of ZnO thin films.

13.4 Structural Analysis

13.4.1 X-ray Diffraction Analysis

XRD is a technique for determining the crystallographic structure of a material. It works by irradiating a material with incident X-rays and then measuring the intensities and scattering angles of the X-rays that leave the material. The intensity of the scattered X-rays is plotted as a function of the scattering angle. The structure of the material is determined from the analysis of the location, angle, and intensities of scattered peaks. Beyond being able to measure the average positions of the atoms in the crystal, information on how the actual structure deviates from the ideal one, resulted from internal stress or defects, can be determined.

The diffraction of the X-rays, which is central to the XRD method, is a subset of the general X-ray scattering phenomena. XRD, generally used to mean wide-angle X-ray diffraction (WAXD), belongs to several methods that use elastically scattered X-ray waves. Other elastic scattering-based X-ray techniques include small-angle X-ray scattering (SAXS), where the X-rays are incident on the sample typically over the small angular range of 0.1–10°. SAXS measures structural correlations of the scale of several nanometers or larger (such as crystal superstructures), and X-ray reflectivity measures the thickness, roughness, and density of thin films. WAXD covers an angular range beyond 10°.

Bragg's law. Bragg's law refers to the simple equation: $n\lambda = 2d_{hkl} \sin \theta$, and explains why the cleavage faces of crystals appear to reflect X-ray beams at certain angles of incidence (θ). In Bragg's equation, d_{hkl} is the distance between atomic layers in a crystal, and the λ is the wavelength of the incident X-ray beam (Fig. 13.9).

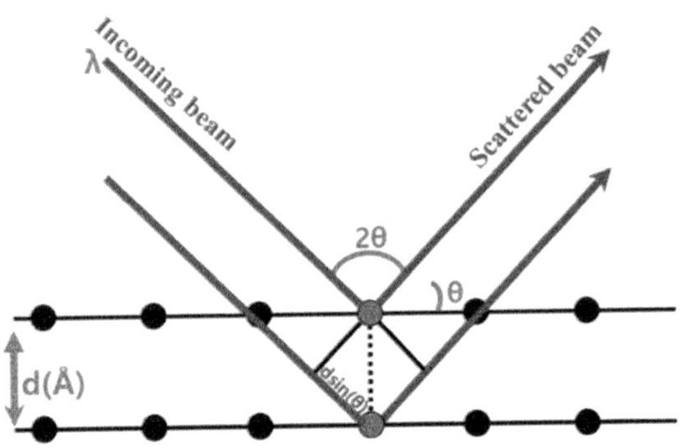

Figure 13.9 Schematic illustration of Bragg's law condition.

For directions, the ($h\ k\ l$) Miller indices represent the normalized difference in the respective x, y, and z coordinates (in a cartesian coordinate) of two points along the direction. The Miller indices ($h\ k\ l$) of a plane are simply the h k l values of the direction perpendicular to the plane [5]. For example, Fig. 13.10 shows the XRD pattern for the ZnO powder sample. The diffraction peaks at angles (2θ) of 31.78°, 33.9°, 36.8°, 47.63°, 56.67°, 62.94°, and 67.98°and correspond to the reflection from (100), (002), (101), (102), (110), (103), (112) crystal planes of the hexagonal wurtzite zinc oxide structure.

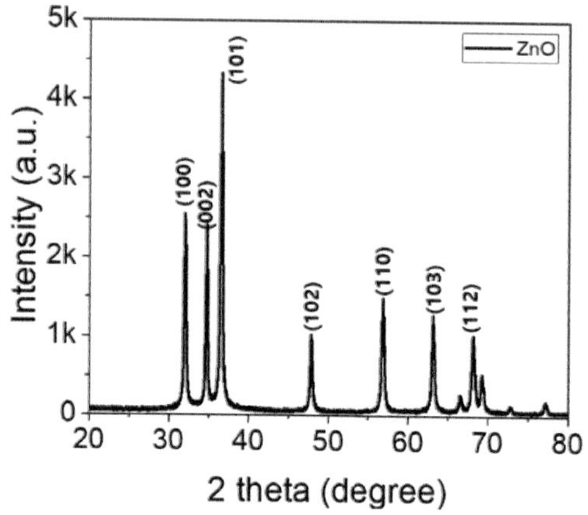

Figure 13.10 XRD pattern of ZnO nanoparticles.

13.4.2 Electron Microscopy

Electron microscopy (EM) uses a beam of accelerated electrons as a source of illumination. As the wavelength of an electron can be up to 100,000 times shorter than that of visible light photons, EM gives high-resolution images of materials and reveals the structure of smaller objects. There are two main types of electron microscopes, transmission electron microscope (TEM) and scanning electron microscope (SEM).

13.4.2.1 Transmission electron microscope (TEM)

TEM uses electrons to interact with a sample, providing information about its morphology, crystallography, and composition. The electron emitted from filament passes through the multiple electromagnetic lenses and makes contact with the screen where the electrons are converted into light and an image is obtained.

The high-resolution transmission electron microscopy (HRTEM) uses both the transmitted and the scattered beams to create an interference image. It is a phase-contrast image and can be as small as the unit cell of the crystal. In this case, the outgoing modulated electron waves at very low angles interfere with themselves during propagation through the objective lens. All electrons emerging from the specimen are combined at a point in the image plane. HRTEM has been extensively and successfully used for analyzing crystal structures and lattice imperfections in various kinds of material on an atomic resolution scale. It can be used for the characterization of point defects, stacking faults, dislocations, grain boundaries, and surface structures. For example, Fig. 13.11 shows TEM and HRTEM images of ZnO nanorods with atomic layer distancing of 0.45 nm.

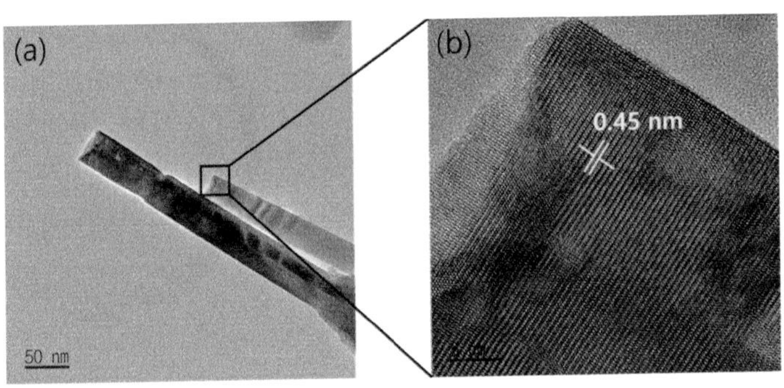

Figure 13.11 TEM (a) and HRTEM (b) of a ZnO nanorod.

TEM also provides crystallographic information on a selected area of the sample, which is called selected area (electron) diffraction (SAED). In a semiconductor, the SAED pattern is a specific case of electron diffraction which is used to determine crystal orientation, measure lattice constant, and examine its defects. SAED pattern of an amorphous material looks like diffuse rings, but for well crystalline materials it is bright spots and polynanocrystalline materials show small spots making up rings, each spot arising from Bragg reflection from an individual crystallite, as shown in Fig. 13.12.

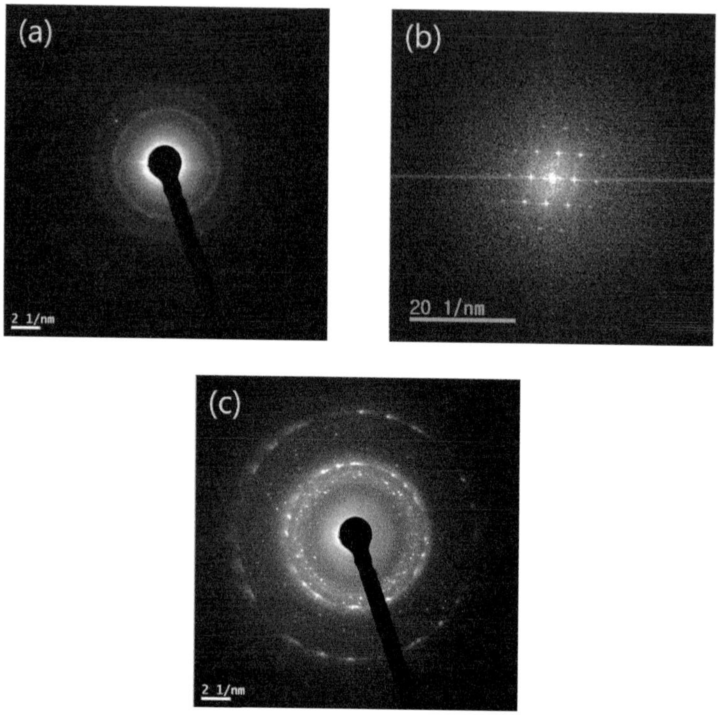

Figure 13.12 SAED patterns of (a) amorphous materials, (b) well crystalline materials, and (c) polynanocrystalline materials.

SAED pattern provides information about the crystalline phase of a sample (each phase has a characteristic SAED pattern). The step-by-step method to index the SAED pattern is [13]:

- First, the diameter of each ring, 2R, needs to be measured by image processing software. For example, TiO_2 sample: in the first ring, the 2R = 5.37 [1/nm].
- Obtain the value of the radius, R, with respect to central spot: R = 2.68 [1/nm], the unit 1/nm means the distance is in reciprocal lattice.
- Obtain the interplanar distance (d), in real space, as 1/R: in the case of TiO_2, d = 1/R = 1/2.68 =0.372 nm. Note that the unit is nm. 0.372 nm is the d value for one ring which corresponds to a certain set of hkl indices in the crystal structure. Based on the literature the d value of 0.372 nm corresponds to the (101) plan of anatase TiO_2.

- Obtain *d* values for other rings.
- Compare the *d* value with the diffraction data from the literature to determine the composition of the sample.

13.4.2.2 Scanning electron microscope (SEM)

An SEM uses a focused beam of electrons to provide high-resolution, three-dimensional images, and information on topography, morphology, and composition of materials. It is called SEM because an image is formed by scanning an electron beam focusing on the surface of the sample.

The main difference between SEM and TEM is that SEM creates an image by detecting reflected or electrons from the surface, while TEM uses transmitted electrons (those are passing through the sample) to create an image. As a result, TEM provides information on the inner structure of the sample, such as crystal structure, morphology, and stress state information, while SEM provides information on the sample's surface and its composition.

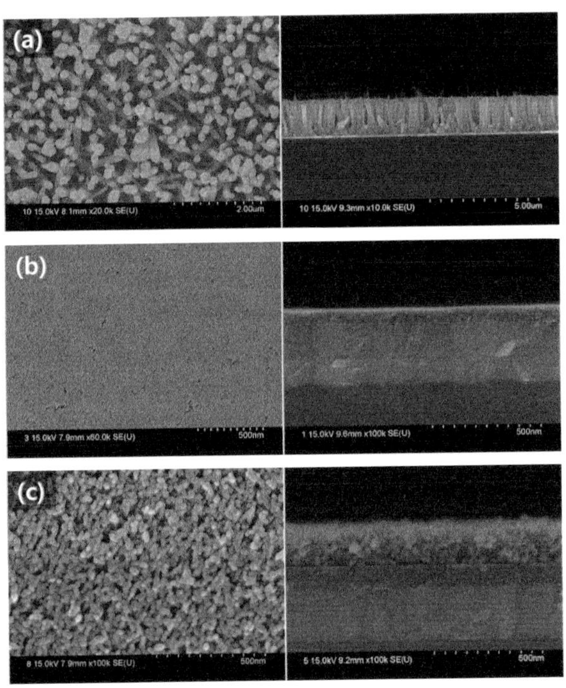

Figure 13.13 FESEM images of (a) as-grown ZnO nanorods, (b) metal oxide thin film, and (c) mesoporous nanoparticle film.

To obtain high-quality images of samples in SEM analysis, a thin layer of gold/palladium coating is used to improve image quality to enable nanometer-scale measurements. Figure 13.13 shows a typical example of SEM images of the surface and cross-sectional morphology of as-grown nanorods and metal oxide thin film and mesoporous nanoparticle films. SEM imaging techniques provide detailed information about nanostructure morphology, size distribution and thickness, surface attachment, and cross-sectional properties of nanostructures.

13.4.3 Atomic Force Microscopy (AFM)

Atomic force microscopy (AFM) or scanning probe microscopy (SFM) is a high-resolution form of scanning probe microscopy. Cantilever-style instruments scan sample surfaces by using sharp tips at the end. While scanning over the sample surface, van der Waals or other forces (i.e., electrostatic and hydrophobic or hydrophilic forces) occur between the probe and the sample, resulting in a deflection of the probe. AFM is particularly versatile in its applications since it can be used at ambient temperatures under different environments.

Different types of samples such as semiconductors, polymers, nanoparticles, biotechnology, and cells can be analyzed by AFM. Generally, AFM is used to examine the topography of a sample. It also provides information about the ability of the material to be dispersed as well as the homogeneity of the particles within the dispersion. Additionally, it can provide information on particle properties such as particle size, surface area, electrical properties, and chemical composition. There are specific detection tips that can be used to determine the material's principal mechanical, magnetic, and electrical properties.

Unlike TEM and SEM, AFM images contain three-dimensional spatial information. For AFM, the vertical dimension (z-axis) is typically used to determine the diameter of spherical particles or roughness of thin films due to probe-sample convolution. In addition, *z*-axis data from AFM image typically has lower noise levels and higher resolution than lateral dimensions. In the cases of SEM and TEM, dimensions are measured directly from the *x*–*y* distances in the images, thus lateral magnification is an important factor in optimizing resolution

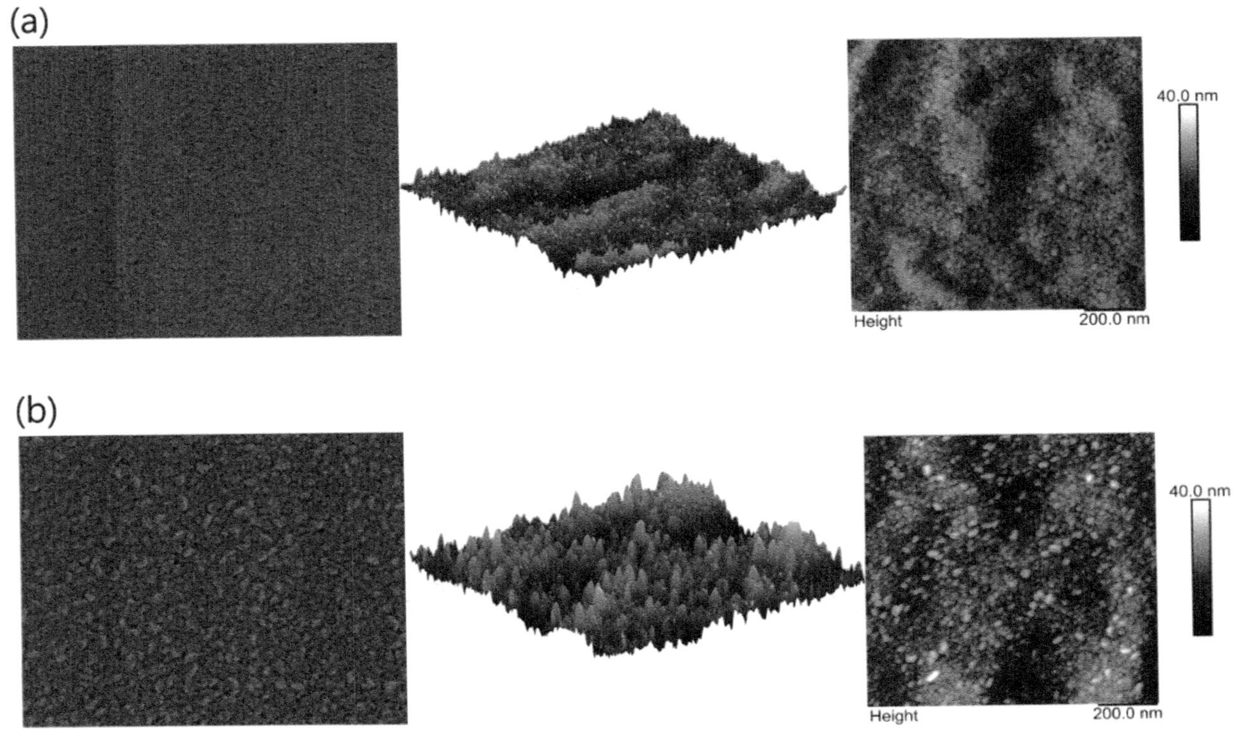

Figure 13.14 Comparative images of ZnO thin film with (a) smooth surface and (b) rough surface, analyzed by SEM and AFM. (left) SEM image, (middle) AFM 3D-image, and (right) AFM topography image.

in the data obtained.

SEM is suitable for imaging a large-area sample. On the other hand, AFM can offer very strong contrast on smooth samples. Figure 13.14 shows comparative images of ZnO thin film with a smooth surface (Fig. 13.14a) and rough surface (Fig. 13.14b) analyzed by SEM and AFM, respectively.

Both techniques can provide different types of information. SEM images generally depict the shape and dimensions of a sample. Moreover, coupled EDS detectors can produce contrast based on atomic numbers and measure the types of elements and their concentrations. In AFM, the most common mode is the topographic mode, which gives fully three-dimensional data. However, to obtain such information from SEM image, a cross section must be produced or the sample has to be tilted to take a cross-sectional image. AFM can also provide a mechanical characterization of the sample, hardness, viscoelastic properties, and friction measurements. Advanced techniques on the AFM can be further used to measure other properties like magnetic fields, electrical properties, or the force of molecular interactions.

13.5 Characterization of Photovoltaic Parameters

13.5.1 Current–Voltage Analysis

The current–voltage (I–V) plot of a photovoltaic cell is used to calculate photovoltaic parameters. The main electrical characteristics of a solar cell or module are determined by the relationship between the current and voltage produced from the I–V curve. The intensity of the solar radiation and photo-generated electrons control the current density. Thus, an I–V curve is a graphical representation of the operation of a cell and provides information on the relationship between the current and voltage at operating conditions. I–V curves provide the information required to configure a solar cell system so that it can operate as close to its maximum power point (MPP) as possible (Fig. 13.15).

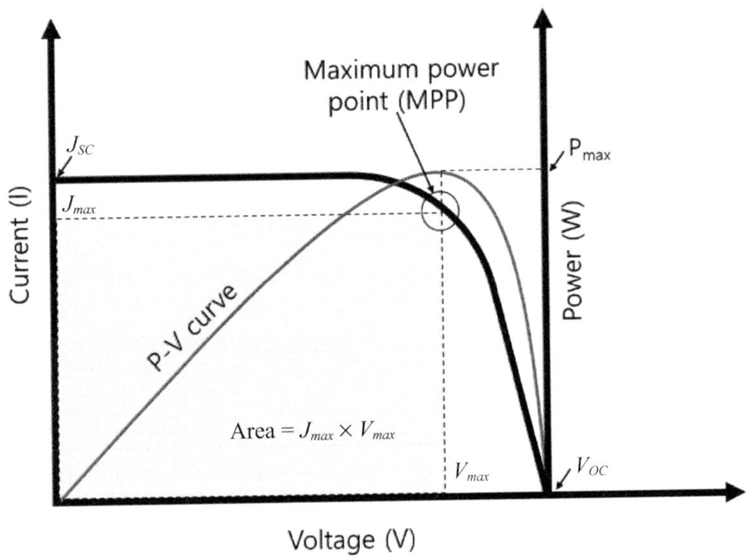

Figure 13.15 Current–voltage (*J–V*) characteristics of a typical solar cell operating under normal conditions.

Photovoltaic parameters are determined from the *J–V* curve:

- V_{oc} (open-circuit voltage): This is the maximum voltage that the device provides when the terminals are not connected to any load (i.e., an open-circuit condition). This value is much higher than V_{mp} which relates to the operation of the PV array which is fixed by the load.
- J_{SC} (short-circuit current): The maximum current provided by a device when the output connectors are shorted together (i.e., a short-circuit condition). This value is much higher than J_{max} which relates to the normal operating circuit current.
- MPP (maximum power point): This relates to the point where the power supplied by the device is at its maximum value, where MPP = $J_{max} \times V_{max}$. The maximum power point of a photovoltaic array is measured in Watt (W).
- FF (fill factor): The fill factor, a measure of quality of the solar cell, is calculated by comparing the maximum power to the

theoretical power that would be output at both the open circuit voltage and short circuit current together. FF can be calculated by Eq. (3.1) or Eq. (4.1). The closer A is to unity (1), the more power is generated. FF is typically between 0.7 and 0.8.

- PCE (power conversion efficiency): Solar cell efficiency is defined as the ratio of energy output from the solar cell to input energy from the sun. PCE is calculated by Eq. (4.2).

13.5.2 Incident Photon-to-Current Conversion Efficiency Analysis

Incident photon-to-current conversion efficiency (IPCE) measurement is used to correlate the efficiency of SC as a function of wavelength with the short-circuit current measurements of a cell under one-sun illumination. IPCE analysis measures quantum efficiency (QE). The concept of QE is explained in Chapter 2. IPCE measures the effectiveness of a solar cell device to convert incident

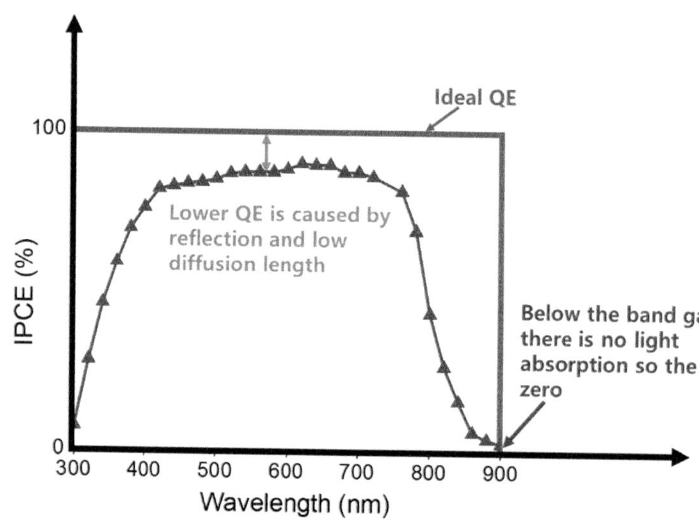

Figure 13.16 The QE of a perovskite solar cell.

photons into electrons. IPCE indicates the integrated QE over the whole solar electromagnetic spectrum, which shows the amount of current that the cell will produce when exposed to sunlight. If all photons of a certain wavelength are absorbed and the resulting carriers are collected, then the QE at that particular wavelength is unity. The QE for photons with energy below the bandgap is zero. A QE curve for an ideal solar cell is illustrated in Fig. 13.16.

13.5.3 Impedance Spectroscopy

In electrochemical and solid-state devices, impedance spectroscopy (IS) is used to profile the electronic structure and monitor electrical and electrochemical processes. As the frequency varies, the electrochemical impedance spectroscopy (EIS) measurement records the response of the sample to a small-signal, sinusoidal stimulus. Based on the characteristic dynamics being measured, the device dissipates and/or stores the electrical energy of the excitation signal. Different dynamics, or relaxation processes, can be distinguished from each other by their unique relaxation times (or time constants), i.e., how long it takes for the device to return to its initial equilibrium after it has been excited. Equivalent circuit models (ECMs) are used to extract and quantify information from spectral data, and their values offer insights into the underlying physical nature of these processes. EIS can be used as a complement to conventional electrical characterization and has shown to be very effective in providing insight into different performance-relevant processes and parameters in energy conversion and storage devices [14].

EIS consists of a so-called Nyquist plot representing negative of the imaginary part (Z'') versus the real part (Z') of the complex impedance of individual electrodes or electrochemical cells. Common features of Nyquist plots obtained from perovskite SCs have a low-frequency (LF) and high-frequency (HF) components that are separated by several orders of magnitude. Figure 13.17a illustrates a typical example of Nyquist plots for perovskite SCs in HF and LF regions. There is no consensus in the literature concerning the nature of the high and low-frequency loops, though recent studies suggest that the HF signature is associated with electrical transport and the LF signature with electrochemical dynamics [14, 15].

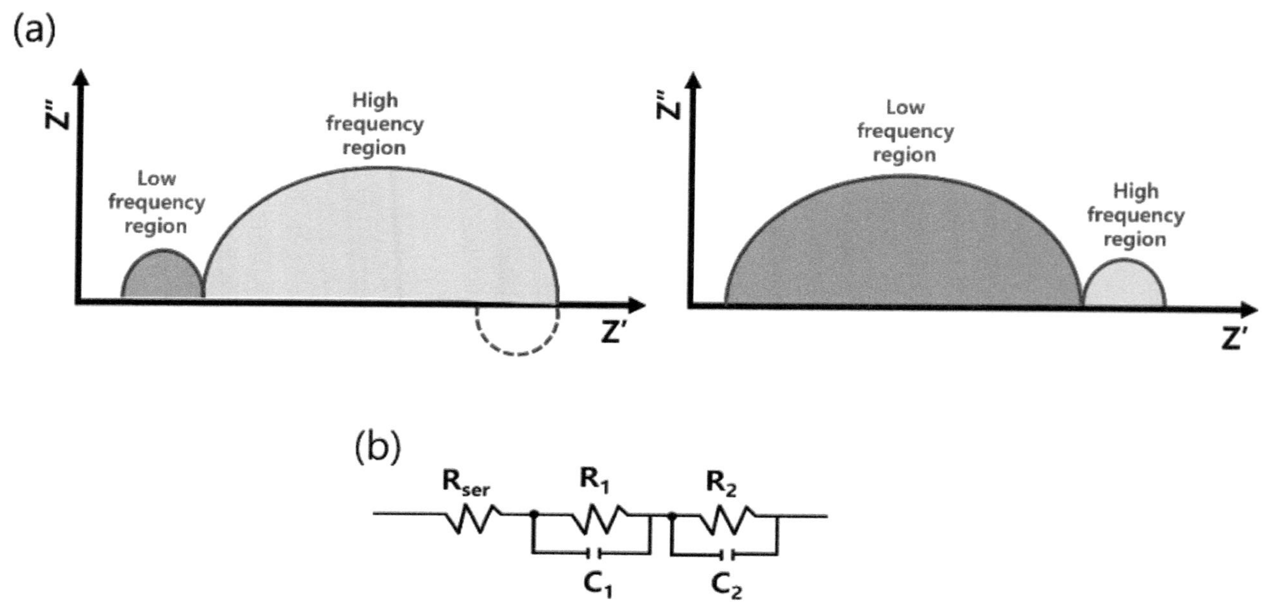

Figure 13.17 (a) Example of Nyquist plots for a perovskite solar cell with high frequency contribution that is smaller or larger than low frequency contribution. (b) Two-component Voigt equivalent circuit.

Various fitting parameters can be extracted from the impedance spectra including resistance, capacitance, and their product, time constant, with respect to the operation model of the device. Several equivalent circuits have been proposed for the impedance spectrum of a PSC. The most commonly used is the series circuit of RC element, known as Voigt circuit. The Voigt equivalent circuit is shown in Fig. 13.17b. In this circuit, R_{ser} is the series resistance, R_1 and R_2 are the first and second impedance components' resistances, and C_1 and C_2 are the respective capacitances.

The parameter's value can be detected from a graph or given by fitting software while analysis. The following steps are proposed to analyze Nyquist plots obtained from a solar cell:

- Identify the number of arcs of an impedance spectrum.
- Estimate the resistance values from the plot, as intercepts with the $Z' = 0$ axis. Estimate the values of the time constants from the peak positions in the frequency plot.
- Select the equivalent circuit for fitting the data.
- Verify that the fit is corrected by comparing the model obtained from the fit to the resulted data. And extract the parameters from the data table.

13.5.4 Space-Charge-Limited Current

To examine the intrinsic transport properties such as carrier mobility and trap density of a semiconductor, space-charge-limited current (SCLC) measurements are commonly used. The measurement involves the dark current–voltage characteristic of a single-carrier device, i.e., a device in which contact on both sides of a semiconductor is oriented toward the conduction band so that only electrons or holes are injected. With the SCLC method, the carrier mobility and trap density can be estimated separately. The device configuration for SCLC measurements has a sandwich structure similar to that of a solar cell with vertical charge transport, in contrast to a field-effect transistor with lateral charge transport [16]. SCLC measurements can be performed ideally on a single-carrier device with symmetric ohmic contacts on both sides of a semiconductor, as shown in Fig. 13.18a.

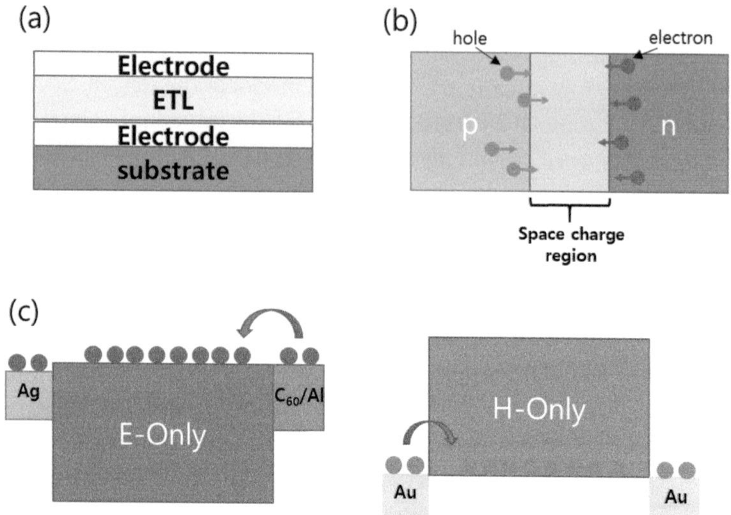

Figure 13.18 (a) Structure of SCLC device with symmetric ohmic contacts. (b) Space charge accumulation in the depletion region of a p-n junction. (c) Schematic of electron-only and hole-only devices.

When a solid has an ohmic contact with a metal (especially in a dielectric), the current flows in proportion to the square of the voltage due to the repulsion of the space charge. This is often referred to as SCLC. Space charge is an accumulation of charges in a specific region and is often found in the depletion region of a p-n junction and near the cathode of an electron tube (Fig. 13.18b).

In general, diffusion and recombination occur because the carrier concentrations in the n-type semiconductor and the p-type semiconductor are different, and as a result, atoms fixed in the lattice form space charges. However, in a vacuum state, it takes some time for electrons to flow from the cathode under the electric field in a p-n junction.

Child's Law (or the Child–Langmuir law or three-halves-power law) gives the maximum SCLC in a planar diode of infinite radius (i.e., one-dimensional beam) as a function of the length and potential difference between anode and cathode.

$$J = \frac{4\varepsilon_0}{9}\left(\frac{2e}{m}\right)^{\frac{1}{2}} \frac{V^{\frac{3}{2}}}{d^2} \tag{13.5}$$

where J is current density, V is applied potential, d is film thickness, ε_0 is the permittivity of free-space (8.85×10^{-12} F/m), m is electron mass and e is an elementary positive charge. The Child's law applies only to vacuum diodes and generally does not apply to semiconductors/ insulators in a single-carrier device.

SCLCs are studied by the electron-only device (EOD) or hole-only device (HOD), in which only electrons or holes are injected into the semiconductor toward electrodes (Fig. 13.18c). In these devices, the maximum current is limited by the buildup of space charge and the density of space charges varies according to the permittivity of the semiconductor. In this case, the current will be determined by the conductivity, i.e., the product of the charge density and the charge mobility.

A typical relationship of current density-voltage (J–V) in a carrier-transport layer (i.e., ETL or HTL) has been known to be composed of four regions: Ohmic current, trap-limited space charge limited current (T-SCLC), trap-filled limited (TFL) current, and space-charge-limited current (SCLC) regions (Fig. 13.19) [17]. The space-charge-limited (or trap-free) SCLC region follows the Mott-Gurney law, and it is observed at sufficiently high bias voltage such that the trap states are completely filled and no longer affect the current conduction.

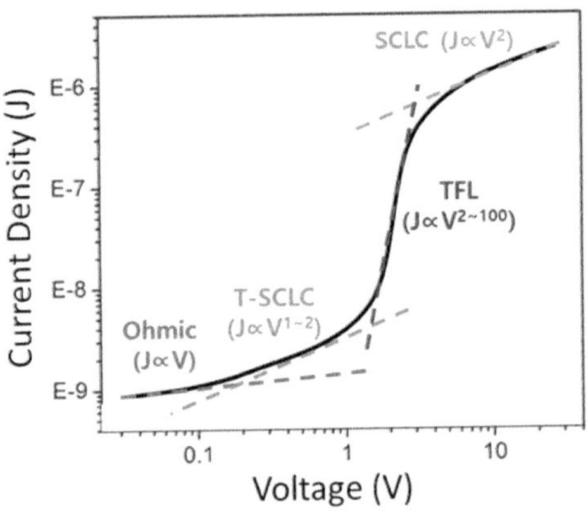

Figure 13.19 Typical current density-voltage (J–V) relationship in a semiconducting layer. Four different regions are characterized by different exponents of J–V relation.

In the trap-free SCLC regime, the current is dominated by charge carriers injected from the contacts and the *J*–*V* characteristics become quadratic (i.e., $J \sim V^2$). The current depends only on mobility and does not depend on the charge-carrier density; therefore, the mobility can be estimated from simple *J*–*V* measurements. To extract the mobility from SCLC measurements, it is important to ensure that it is in the SCLC regime. Therefore, the scaling exponent in a log–log plot has to be analyzed with a value of 2.

For a semiconductor sandwiched between two selective ohmic contacts, the current density (*J*) flowing through the device is given by Mott–Gurney law [18]:

$$J = \frac{9}{8}\varepsilon\mu\frac{V^2}{L^3} \tag{13.6}$$

where *V* is the applied voltage, ε is the dielectric constant of the active layer, *L* is the thickness of the active layer, and μ is the mobility. Charge mobility is calculated by the Mott–Gurney equation.

Moreover, from the *J*–*V* plot in a log–log scale, the trap density can be estimated from the trap-filled-limit voltage (V_{tfl}) [19, 20]:

$$V_{\text{tfl}} = \frac{qn_{\text{t}}L^2}{2\varepsilon} \tag{13.7}$$

where q is the elementary charge and n_{t} is the trap density. This formula can easily be derived at any point of the *J*–*V* curve. The trap-filled-limit voltage (V_{tfl}) generally refers to the crossing point between the low-voltage regime and the trap-filled-limited regime.

References

1. Pavia, D. L., Lampman, G. M., Kriz, G. S., and Vyvyan, J. A. (2014) *Introduction to Spectroscopy*. Cengage Learning.
2. Dutta, A. (2017) Fourier transform infrared spectroscopy. In: Thomas, S., Thomas, R., Zachariah, A. K., and Mishra R. K. (eds.), *Spectroscopic Methods for Nanomaterials Characterization.* Elsevier, pp. 73–93.
3. Fadley, C. S. (2010) *J. Electron. Spectrosc. Relat. Phenom.*, **178**, pp. 2–32.
4. Hodoroaba, V.-D. (2020) Energy-dispersive X-ray spectroscopy (EDS). In: Hodoroaba, V.-D., Unge, W. E. S., and Shard, A. G. (eds.) *Characterization of Nanoparticles.* Elsevier, pp. 397–417.

5. Elton, L. R. B. and Jackson, D. F. (1966) *Am. J. Phys.*, **34**(11), pp. 1036–1038.
6. Mulvaney, S. P. and Keating, C. D. (2000) *Anal. Chem.*, **72**(12), pp. 145–158.
7. CRAIC Technologies (2021) The Science of Raman Spectrometer Design. https://www.microspectra.com/technical-support/283-raman-spectrometer-design
8. Venezuela, P., Lazzeri, M., and Mauri, F. (2011) *Phys. Rev. B Condens. Matter*, **84**(3), pp. 035433.
9. Gilliland, G. D. (1997) *Mater. Sci. Eng. R Rep.*, **18**(3–6), pp. 99–399.
10. Han, N. S., Shim, H. S., Seo, J. H., Kim, S. Y., et al. (2010), *J. Appl. Phys.*, **107**(8), p. 084306.
11. Pathak, C. S., Mishra, D. D., Agarwala, V., and Mandal, M. K. (2012) *Ceram. Int.*, **38**(7) pp. 5497–5500.
12. Greczynski, G. and Hultman, L. (2020) *Prog. Mater. Sci.*, **107**, p. 100591.
13. Yang, W., Shen, H., Min, H., and Ge, J. (2020) *J. Mater. Sci.*, **55**(2), pp. 701–712.
14. von Hauff, E. and Klotz, D. (2022) *J. Mater. Chem. C*, **10**(2), pp. 742–761.
15. Todinova, A., Contreras-Bernal, L., Salado, M., Ahmad, S., et al. (2017) *ChemElectroChem*, **4**(11), pp. 2891–901.
16. Kokil, A., Yang, K., and Kumar, J. (2012) *J. Polym. Sci. B: Polym. Phys.*, **50**, pp. 1130–1144.
17. Zhang, P., Ang, Y. S., Allen, L., Garner, A. L., et al. (2021) *J. Appl. Phys.*, **129**, 100902.
18. Corre, L., Duijnstee, V. M., Tambouli, E. A., Ball, O. E., et al. (2021) *ACS Energy Lett.*, **6**, pp.1087–1094.
19. Lampert, M. A. and Mark, P. (1970) *Current Injection in Solids*. Academic Press: New York.
20. Jain, A., Kumar, P., Jain, S. C., Kumar, V., et al. (2007) *J. Appl. Phys.*, **102**(9), 094505.

Problems

13.1 What is spectroscopic analysis and what is the main reason for using spectroscopic techniques for characterization?

13.2 In material characterization via spectroscopy, the strength of the radiation energy interacts with the molecules in different ways. What are the energy sources and their roles?

13.3 For chemical analysis FT-IR is most applicable spectroscopic technique, explain how FT-IR helps analyze chemical compounds.

13.4 What is the purpose of X-ray photoelectron spectroscopy (XPS) in material detection?

13.5 Raman spectroscopy is well-known technique to analyze graphene materials, explain what are the main three peaks in Raman spectroscopy of graphene and what do the peaks represent?

13.6 Briefly explain the basics of photoluminescence spectroscopy and what does the PL peak determine?

13.7 Explain, what is the purpose of time-resolved photoluminescence (TRPL) spectroscopy?

13.8 What is the application of ultraviolet photoelectron spectroscopy (UPS) in semiconductor analysis?

13.9 Explain how to calculate work function and HOMO-LUMO levels of a semiconductor using UPS data?

13.10 In solar cell analysis, what is the purpose of using incident photon-to-current conversion efficiency (IPCE) measurements?

13.11 Explain what SCLC is and how to measure the mobility of solar cells?

13.12 In SCLC analysis of EOD of $FAPbI_3$ perovskite device, with 300 nm thickness, the V_{TFL} is measured 2.02 V. Calculate the trap density and mobility. Relative dielectric constant for $FAPbI_3$ ($\varepsilon = 62.23$).

Index

For Product Safety Concerns and Information please contact our EU representative GPSR@taylorandfrancis.com Taylor & Francis Verlag GmbH, Kaufingerstraße 24, 80331 München, Germany